Don Scroggin

DON G. SCROGGIN
THOMPSON CHEM. LAB.
WILLIAMS COLLEGE
WILLIAMSTOWN, MASSACHUSETTS 01267

David O. Johnston, Ph.D.

Department of Chemistry
David Lipscomb College
Nashville, Tennessee

John T. Netterville, Ph.D.

Chairman, Dept. of Chemistry
David Lipscomb College
Nashville, Tennessee

James L. Wood, Ph.D.

Department of Chemistry
David Lipscomb College
Nashville, Tennessee

Mark M. Jones, Ph.D.

Chairman, Dept. of Chemistry
Vanderbilt University
Nashville, Tennessee

Chemistry and the Environment

W. B. SAUNDERS COMPANY Philadelphia London Toronto
1973

W. B. Saunders Company: West Washington Square
                             Philadelphia, Pa.   19105

                             12 Dyott Street
                             London, WC1A   1DB

                             833 Oxford Street
                             Toronto 18, Ontario

Chemistry and the Environment   ISBN   0-7216-5185-2

Print No.:   9   8   7   6   5   4   3   2   1

PREFACE

Ours is a most fascinating window of time in the history of man's understanding of himself and his environment. It is a fascinating time because we have so many interesting things to use and can understand a great many of them. Chemistry and chemical technology have helped to provide us with a way of life never known before, and chemical theory has matured sufficiently to explain many of the phenomena which only a century ago lay in the realm of mystery. With relatively little knowledge of atoms and molecules, considerable understanding of many relevant chemical phenomena can be ours. This brief text brings to focus a blend of some of the interrelationships among atoms, molecules, man, and his chemical products and problems in the areas of air and water pollution, consumer products, chemical action of the human body, drugs and medicines, and chemical energy.

This book is intended for one-quarter or one-semester courses in chemistry for liberal arts students. A sufficient variety of topics is included to cater to the interests of almost any class. Once the first seven chapters are completed, the order of material can be varied without loss of continuity of presentation. It might be preferable, however, to take Chapter 9 before taking the other chapters on chemicals and life (10 through 12) and to take Chapter 19 (air pollution) before Chapter 22 (chemistry of the automobile).

The approach in this book is to add a measure of chemical understanding to the things we see and use. This is more fun and intellectually more satisfying than simply learning facts without reason. For example, it is a very simple thing to learn that butylated hydroxytoluene (BHT) is put in potato chips and other fatty foods to prevent rancidity. But, why? The application of a few chemical principles can explain in a satisfying way what BHT does and how it works. In order for the study from this text to be a very worthwhile experience, you will have to let your curiosity and natural tendency to ask "why" flow freely.

This book would never have left the scribbled page without the help of many competent and talented people. We are deeply indebted to the staff of W. B. Saunders Co., including Grant Lashbrook, Herb Powell, Mary Agre, Tom O'Connor, and Joan Garbutt, who guided and directed this effort to whatever place of respectability it may have. We are especially indebted to Professors Bill Masterton, Gene Rochow, and Norman Juster for invaluable criticism of the manuscript in its various stages. Of course, responsibility for the contents of the text rests on the authors of the text alone and not on our esteemed colleagues.

There was no physical way for the manuscript to be read without the efficient typing and manuscipt assembling of Mamie Johnston, Beverly Loring, Charlotte Patillo, and Jane Arnold Williams. Thanks to talented Susan Johnston for preparing preliminary sketches of most of the new illustrations. We greatly appreciate the contribution of those students at David Lipscomb College who were taught this material in its preliminary form and who, by their comments and puzzled looks, helped to evolve the manuscript into a more teachable form.

Finally, it is to our wives that we dedicate this effort and gratefully acknowledge their support and understanding during the preparation of this manuscript.

DAVID O. JOHNSTON
JOHN T. NETTERVILLE
JAMES L. WOOD
MARK M. JONES

CONTENTS

PART 1   THE BASIS OF UNDERSTANDING MATTER

PART 2   A STAR PERFORMER—CARBON

PART 3   LIFE AND CHEMISTRY

PART 4   CHEMISTRY FOR BETTER LIVING— CONSUMER PRODUCTS

PART 5   CHEMISTRY FOR WORSE LIVING

PART 6   CHEMISTRY FOR BETTER OR WORSE

PART 7   CHEMISTRY OUT BEYOND

Prologue

SCIENCE AND TECHNOLOGY AS A NEW PHILOSOPHY

A PHILOSOPHICAL AND HISTORICAL BACKGROUND

The various factors which led to the systematic development of science and its associated technologies in the Western World are both numerous and complex. One of the key points seems to be the development of printing techniques which greatly increased the availability of the accumulated knowledge of mankind. Another seems to be the refusal to accept *"authority"* as the ultimate judge of truth; this indeed still seems to be one of the major differences between the sciences and other areas of human knowledge. No one can state successfully, as proof of the validity of a piece of *scientific* information, that so-and-so said this and therefore it must be true. This skeptical attitude is evident in all nations and at all times; however, it was only in Western Europe that such a frame of mind came to be cultivated systematically and ultimately formed the basis of certain intellectual organizations, such as the Royal Society, whose primary purpose was to search for and propagate a kind of impersonal knowledge.

The most important aspects of this type of knowledge are the ways in which it changes. With the passage of time, it becomes more extensive, more accurate, more broadly disseminated, *and*, most important of all, more concisely summarized in terms of generalizations or their equivalent: mathematical equations giving the quantitative relationships connecting a set of properties.

Because the physical sciences deal with our environment, this growth of

1

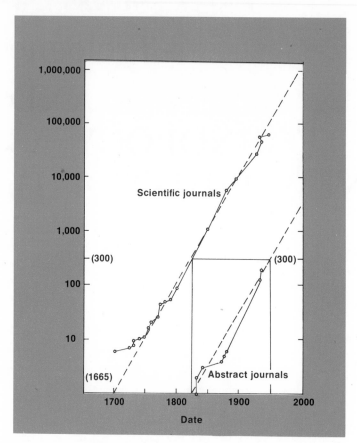

FIGURE P-1   Rate of increase in number of scientific journals since 1665. (Reproduced with permission of Yale University Press.)

knowledge leads to an ever increasing understanding of the physical world and the ways it can be manipulated to obtain desired ends. This manipulation of our environment can attain a high degree of sophistication. Presently, modern technology has two types of roots. The one derives from the practical knowledge of craftsmen and frequently has been transmitted orally or by apprenticeships. In contrast, information in the sciences has been transmitted by publications of various sorts and formal education. The dual roots tend to fuse in practice and reinforce each other. The older method of oral tradition had its ups and downs, as is especially obvious in certain areas where knowledge of techniques has died out (e.g., the early Middle Ages in Europe as contrasted with the Roman Empire). Transmission by publication makes knowledge available to all who can read and who have the required education for understanding.

Over the last 200 years, accumulated knowledge has been put to use on a very extensive scale in this country and in other areas of the world which have had the means to follow the example set by England, which underwent this "industrial revolution" first. The result has been the development of a society largely dependent upon and supported by a technology which is itself undergoing constant change. The first consequence of this technology has been to increase the *rate* at which things can be produced. This in turn has continually changed the occupational patterns of millions of human beings, and has brought forcefully to mind the persistence of *change* in our pattern of life.

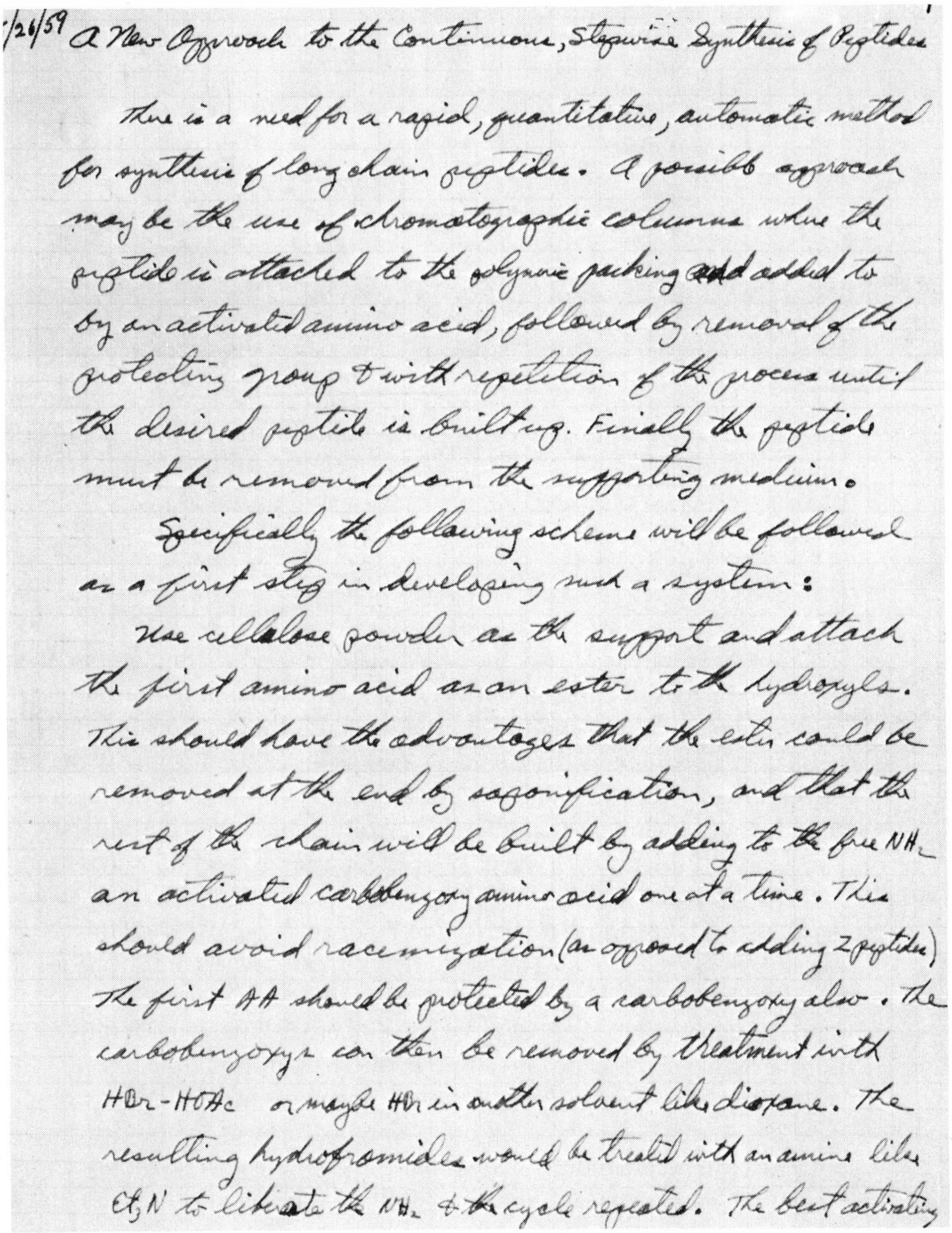

[Handwritten notes by Dr. Bruce Merrifield, dated 7/26/59:]

> 7/26/59 A New Approach to the Continuous, Stepwise Synthesis of Peptides
>
> There is a need for a rapid, quantitative, automatic method for synthesis of long chain peptides. A possible approach may be the use of chromatographic columns where the peptide is attached to the polymeric packing and added to by an activated amino acid, followed by removal of the protecting group & with repetition of the process until the desired peptide is built up. Finally the peptide must be removed from the supporting medium.
>
> Specifically the following scheme will be followed as a first step in developing such a system:
>
> Use cellulose powder as the support and attach the first amino acid as an ester to the hydroxyls. This should have the advantages that the ester could be removed at the end by saponification, and that the rest of the chain will be built by adding to the free NH_2 an activated carbobenzoxy amino acid one at a time. This should avoid racemization (as opposed to adding 2 peptides). The first AA should be protected by a carbobenzoxy also. The carbobenzoxy can then be removed by treatment with HBr–HOAc or maybe HBr in another solvent like dioxane. The resulting hydrobromides would be treated with an amine like Et_3N to liberate the NH_2 & the cycle repeated. The best activating

FIGURE P-2   A page from the notes of Dr. Bruce Merrifield (Professor of Biochemistry at Rockefeller University), the designer of a protein-making machine. Progress in science is made through disciplined human thought, taking advantage of previous knowledge and useful theories. (From Chemical and Engineering News, August 2, 1971.)

These changes have influenced profoundly the way in which people think about their material wants and the ways in which they can be satisfied. For example, there seems to be little argument with the statement, "If the number of human beings on the earth could be stabilized, a much higher standard of living could prevail over most of the earth." A statement such as this would have been greeted with widespread derision 500 years ago. Today, people *expect*

FIGURE P–3   Name some discoveries that led to this change in transportation. How many of them are chemical in nature? In this text we will discuss some changes in transportation.

the continuation of technological innovation to solve many of their own and the world's problems. This attitude has given rise to the phrase, "the revolution of rising expectations," to describe such a state of mind.

SCIENCE AND TECHNOLOGY—A DIFFERENCE

We have already used the words science and technology in their proper context, but the distinction between the two words may not be clear. Science and technology almost defy definition because they are distinct fields but have similar goals, differ in the layman's and scientist's concepts of the words, and cover broad and varied activities. Precise definitions of the concepts are so detailed and technical that there is a danger they will be understood only by specialists in the areas involved. For our purposes, working definitions of these concepts will be sufficient. A satisfactory working definition of science is one such as the following: *science is the body of knowledge obtained by methods based upon observation.* This definition carries several implications. First, the practice of science is a human activity. Human beings do the observing and gain the knowledge. Second, there is an inherent limitation in science. Anything outside or beyond our powers of observation is, in principle, outside the bounds of science. Third, there is authority in science. The authority is observation.

The term "scientific method" refers more to a mental attitude than to a procedure. The attitude is one which requires the strictest intellectual honesty in the collection of data and, hopefully, in the arrangement of these data in a pattern that reveals underlying principles. The data normally must be collected under conditions which can be reproduced anywhere in the world, so that new data can be obtained to confirm or refute the old theories (geologists and astronomers are excused from these rigid requirements). The results obtained are thus independent of differences occurring in language, culture, religion, or economic status among various observers and represent a unique type of truth.

One dictionary definition of technology is "applied science." This definition is too limited, as we have already seen. More and more, technology is building on the knowledge of science, but it is progressing also from the arts and crafts. A working definition that is satisfactory for our purposes is: *technology embraces the means employed by people to provide material objects for human sustenance*

4

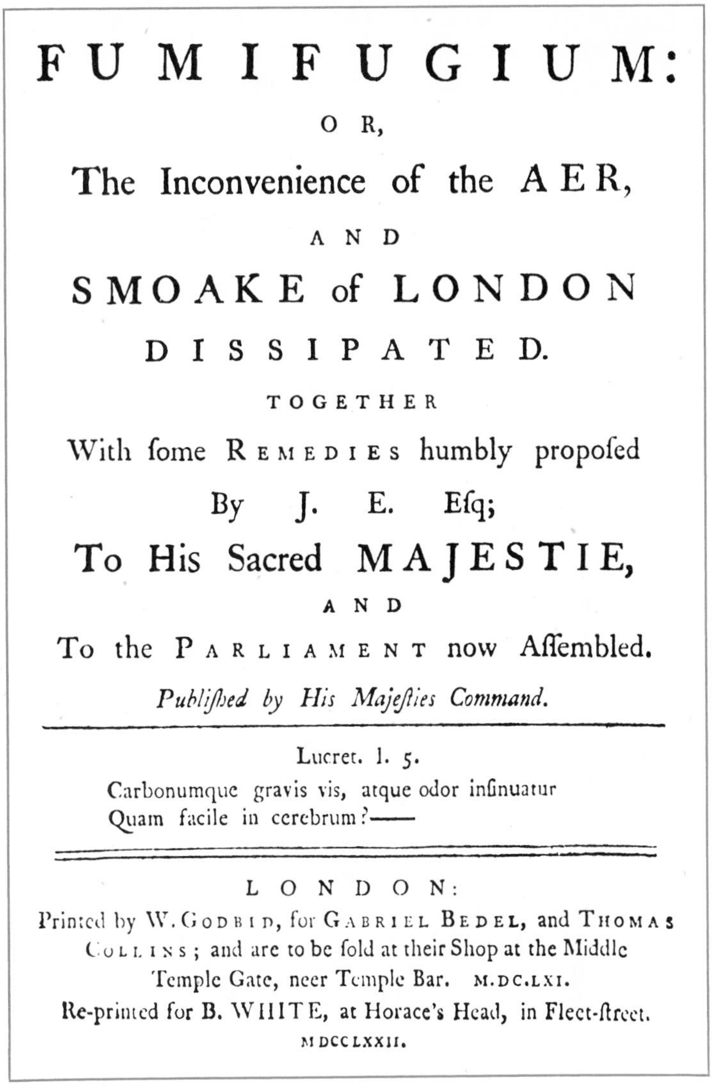

FUMIFUGIUM:

OR,

The Inconvenience of the AER,

AND

SMOAKE of LONDON

DISSIPATED.

TOGETHER

With some REMEDIES humbly propoſed

By J. E. Eſq;

To His Sacred MAJESTIE,

AND

To the PARLIAMENT now Aſſembled.

Publiſhed by His Majeſties Command.

Lucret. l. 5.

Carbonumque gravis vis, atque odor inſinuatur
Quam facile in cerebrum?——

LONDON:

Printed by W. GODBID, for GABRIEL BEDEL, and THOMAS COLLINS; and are to be ſold at their Shop at the Middle Temple Gate, neer Temple Bar. M.DC.LXI.
Re-printed for B. WHITE, at Horace's Head, in Fleet-ſtreet. M DCC LXXII.

FIGURE P-4   Title page from J. Evelyn, F. R. S., *The Smoake of London*. The Latin quotation is from the Roman poet Lucretius (97–53 B.C.). It may be translated, "How easily the heavy potency of carbons and odors sneaks into the brain!" (Courtesy A. E. Gunther and the University Press, Oxford.)

and comfort. One implication of this definition is that technology, like science, is a human activity. People invent new methods and devices, bringing about technology, and people use the products of technology.

It is important to recognize that science and technology are not one and the same. Science is knowledge, and technology is *doing* something with one's physical and material environment. The basic motive for "bringing about technology" is the desire to obtain more or better material things. The "spirit of science" is the spirit of rational inquiry through observation. In scientific work, knowledge is the important product. In technology, new or improved material products are the goal. Thus, widespread and rapid publication of definitive results is sought in science, whereas secrecy may be sought in technology until patent claims are established or until the product is ready for commercial exploitation.

Science and technology reinforce each other by way of complex, two-way interactions. Each one can build upon itself or upon a cross-linkage tying one

5

to the other. Technology is dependent upon science for knowledge of the properties of materials and of sources of energy, and for predicting the behavior of natural forces. Science is dependent upon technology for its tools and instruments, for the preparation of materials, and for the storage and dissemination of information.

Now, equipped with working concepts of science and technology, we shall look at some of their contributions and effects on society. In Chapter one, some of these concepts will be applied to chemistry as a segment of science.

TECHNOLOGY: ITS TRIUMPHS AND PROBLEMS

Almost as soon as the industrial revolution began in England, the public realized that technological progress brought with it a series of problems. The first to be noticed was the necessity for progress to be accompanied by changing patterns of employment.

It is obvious that if a machine makes as much thread as 100 men can make, the men are released to do other work. The 100 men, however, do not look on this as an advantage, especially if they are settled in their place of employment with their families. The new opportunities that result from such a machine are rarely of benefit directly to the men displaced. The wealth of their country is increased since there are now 100 men able to do other work. However, the initial reaction of the men in 19th century England was to riot and break up the machinery.

The increased use of fuels of all sorts, especially the introduction of coal and coke into metallurgical plants, led to widespread problems with air pollution which were recognized and discussed over 100 years ago.

The most important point of these observations is the realization that technological progress is always obtained at some cost, and the cost may not be obvious at the outset.

A very important technological development was recognized as necessary in 1890 by Sir William Crooks, who addressed the British Association for the Advancement of Science on the problem of the fixed nitrogen supply (that is, nitrogen in a chemical form plants are capable of using in growth). At that time, scientists recognized that nitrogen compounds were necessary in fertilizers and that the future food supply of mankind would be determined by the amount of nitrogen compounds which could be made available for this purpose. The source of these nitrogen supplies was then limited to rapidly depleting supplies of guano in Peru and to sodium nitrate in Chile. It was realized that when these were exhausted, widespread famine would result *unless* an alternative supply could be developed. This problem was recognized first by English scientists as a potentially acute one because by the 1890's England had become very dependent upon *imported food supplies*.

Widespread interest in this problem led to research on a number of chemical reactions by which nitrogen can be obtained from the relatively inexhaustible supply present in the air. Air is 21 per cent oxygen (O_2) and 79 per cent nitrogen (N_2). The nitrogen in the air is present as the rather unreactive molecule N_2,

FIGURE P-5 An early German ammonia plant. (From "The Realm of Chemistry," Econ-Verlag Gmb H. Düsseldorf, 1965.)

and in this form it can be used as a source of other nitrogen compounds only by a few kinds of bacteria. Some is also transformed into NO by lightning, and when this is washed into the soil by rain it can be utilized by plants. Needless to say, the amount of nitrogen transformed into chemical compounds useful to plants by these processes is quite limited.

Several chemical reactions were developed to form useful compounds from atmospheric nitrogen, but the best known and most widely used one has an ironic history. While England was interested in nitrogen for fertilizers, Germany was interested in nitrogen for explosives. The German General Staff realized that the British Navy could blockade German ports and cut them off from the sources of nitrogen compounds in South America. As a consequence, when a German chemist named Fritz Haber showed the potential of an industrial process in which nitrogen reacts with hydrogen in the presence of a suitable catalyst to form ammonia, the German General Staff was quite interested and furnished support through the German chemical industry for the study of the reaction and the development of industrial plants based on it. The first such plant was in operation by 1911, and by 1914 such plants were being built very rapidly by Germany.

When the First World War broke out in August, 1914, many people thought that a shortage of explosives based on nitrogen compounds would force the war to end within a year. Unfortunately, by this time the nitrogen fixing industry in Germany was capable of supplying the needed compounds in large amounts. This process thus prolonged the war considerably and resulted in an enormous increase in mortality. Subsequently, the ammonia process has been used on a huge scale to prepare fertilizers and now is largely responsible for the fact that the earth can support a population of four billion. Ammonia production by this process exceeds 30,000 tons per day in the United States alone.

Problems seem to arise from the development of many technological processes. The control of nuclear energy brings with it the ability to make nuclear explosives. The development of rapid and convenient means of transportation such as the automobile and the airplane also brings forth new weapons of war and air pollution problems. Man, however, must learn to control his technology in such a manner as to maximize its benefits and minimize its disadvantages.

FIGURE P-6   Man usually has the capability of making very large quantities of materials that he finds useful. Sometimes mass production precedes a complete understanding of the consequences. In this industrial process, ribbons of soap are produced and are subsequently converted into flakes or bars. (Courtesy of the Proctor and Gamble Company.)

These problems arise with *all* technological developments, even the most primitive. The discovery of the techniques necessary to the manufacture of iron led first to the development of new weapons (swords) by their discoverers, the Hittites, who then proceeded to conquer their neighbors and lead the first successful invasion of Egypt (ca. 1550 B.C.).

TECHNOLOGY AND THE HUMAN ENVIRONMENT

The growth in large scale technology has a very large number of effects, both direct and indirect, on the human environment. The examination of a few of these shows just how complex these consequences can be.

An obvious case is the development of atomic energy. When the incredibly large amounts of energy which could be released by nuclear reactions were first recognized, the development of such energy sources capable of providing energy for mankind on a scale previously thought to be impossible was placed on a top priority basis. After some nuclear reactors had been built and actually placed in operation, it was evident that their operation was accompanied by some serious potential risks to their human users. The first was the danger of some potential disaster such as the explosion of a boiler, with the consequent possible dispersal of radioactive material. The second danger was in the generation of radioactive products as the nuclear reaction proceeded. The uranium used in such reactors was transformed into a wide variety of fission products which made the operation of the pile more difficult as time went on. The re-purification of the uranium could be accomplished chemically, but what was to be done with the radioactive wastes generated in the process? Obviously they could not be dumped into a sewer since many of them have relatively long half-lives. Present practice calls for solid radioactive wastes to be placed in vaults and buried at carefully chosen sites. Liquid wastes are placed in underground storage tanks. The problem which faces us here is obtaining the benefits of the technology with the minimum disruption of our own environment.

FIGURE P-7   Land burial trench at the Oak Ridge National Laboratory reservation. Each day's accumulation of waste containers is buried by 3 or more feet of earth. (From *Radioactive Wastes*. Courtesy of the U.S. Atomic Energy Commission, Washington, D.C.)

The same type of problem arises whenever we introduce a specific chemical compound into our environment to accomplish one single thing. The compound is often capable of a variety of actions and can lead to consequences undreamed of by those who introduce it. Many such examples can be found in the fields of drugs and insecticides. Thalidomide was designed to be an effective tranquilizer, and for this purpose it is an unqualified success. Unfortunately, when taken by pregnant women, it often results in abnormal development in the children they bear.

Man must always combat insects in his struggle for a food supply, and the

TABLE P-1   ESTIMATED ANNUAL CROP LOSS DUE TO PESTS AND DISEASE (PER CENT OF TOTAL U. S. CROP DESTROYED)°

Crop	Insects	Weeds	Disease
Corn	12	10	12
Rice	4	17	7
Wheat	6	12	14
Potatoes	14	3	19
Cotton	19	8	12

° From Scientific Aspects of Pest Control, National Academy of Science Publication No. 1402.

FIGURE P-8   Is DDT a real hazard to birds of prey and to man? Does DDT produce more harmful effects than helpful? Data are still being collected to answer these weighty questions.

most effective aid he now receives in this struggle is from insecticides. Unfortunately, many insecticides are rather unspecific poisons.

Arsenic compounds were used on a very large scale as insecticides, but a realization of their great toxicity for man has led to their replacement by other compounds equally effective but far less toxic to man. However, many of the compounds which are most effective in protecting crops for man's use are also very toxic to man and capable of causing death when improperly handled or ingested. Even those insecticides which are not obviously harmful to man often have side effects which make them unattractive. DDT, for example, is very effective for the control of a wide variety of insects and, as far as can be ascertained, is nontoxic for men. Because of its widespread use and the fact that it is accumulated in fish and animal fats, it is found to concentrate enormously in birds of prey (eagles, pelicans, and ospreys), whose diet is principally fish, and may disrupt their reproductive cycles. A side effect such as this has led to extensive agitation to replace DDT with other compounds which are equally effective as insecticides yet free from this particular side effect. It is possible, however, that the compounds used to replace DDT will have other side effects which will not become obvious until after they have been used for some time. It would seem that the use of any insecticide will have a considerable effect upon the bird population of an area, if only through the changes it introduces into the food supply of the birds.

FIGURE P-9   (From "The Washington Star." Cartoon by Gib Crockett.) Strangely enough DDT is much safer for *humans* than practically any other insecticide.

Chemistry, however, is not the only science which can furnish technological processes for the control of insects. There is a wide variety of biological processes which can be used for the same purposes. One is to furnish assistance to the biological species which destroy insects, such as other parasitical insects. Another is the introduction of large numbers of sterile insects. When these mate with normal insects, no offspring are produced; the number of a type of undesirable insect can be reduced considerably by this process, at least temporarily. These examples are given to emphasize the fact that there are usually several very different kinds of processes which can be used to solve any given practical problem. In the future these will be assessed, at least in part, on their ability to leave the environment undamaged as well as on their monetary cost.

TECHNOLOGICAL DEVELOPMENT AND ITS ENVIRONMENTAL CONSEQUENCES

A very good example of a highly desirable technological development which has consequences that are obviously not so desirable is seen in the development of fertilizers. Most men agree that an abundant food supply is good. Most would

11

FIGURE P-10   Human thought versus chemicals. Will man ever completely control chemicals—or will they control him? (Courtesy of Varian, Palo Alto, California.)

admit, too, that every crop harvested from a field removes essential nutrients from that field: nitrogen, phosphorus, potassium, and so forth. By replacing these lost elements with fertilizer, we can restore or enhance the amount of food we obtain from the field. The increased yields obtained with increased application of fertilizers have been well established and commonly held to be desirable. It is also well established that increased fertilization of a field *increases* the concentration of these essential nutrients in the rain water which runs off such land.

The change in the mineral concentration in the runoff water is capable of causing drastic changes in the rivers and lakes into which it drains. By increasing the amount of minerals available, we have greatly stimulated the growth of surface algae in rivers and lakes, and as these grow they choke out other forms of life. The growth of such slimes makes the water much less capable of supporting its normal population of fish, and ultimately these die off. The algae also make the water more difficult to purify for drinking purposes. After a time the increased food supply is paid for by a general disruption of the biology of the waterways which drain an area. This disruption has occurred in many areas in the United States where the movement of water through lakes and rivers is slow.

Other examples of this same kind of process can be seen in the development

FIGURE P-11   This pond was fertilized with chemicals to produce an excessive growth of algae. Part *A* shows the water covered with the plant growth; Part *B* shows the decay that follows when the concentrated form of life cannot be sustained. (Courtesy of Dr. D. L. Brockway, Federal Water Quality Administration.)

of energy sources, the use of antibiotics, weed killers, and, in fact, any kind of process which releases chemical compounds into the environment *in sufficient quantities* to cause an appreciable percentage *increase* in its composition in the environment. This holds for the hydrocarbons released from automobile engines and for the carbon dioxide and sulfur dioxide released by electric power plants. The change which these *increased* concentrations of various compounds cause cannot always be estimated on the basis of experiments covering only a short period of time since some of the effects may be very long range ones which build up slowly. It is the *level* and not usually the *nature* of the products that makes pollutants. The level is related to population.

What are we striving for in solutions to environmental problems? Do we want immediate solutions at all costs? There are already several cases where hasty solutions accomplished nothing and even stood in the way of making rational, sensible, fruitful decisions. One example will illustrate this fact. Phosphates in detergents were singled out early as one of our top ten pollution villains. State and local governments all over the country rushed to make laws that would ban their sale, without much thought of the consequences. It now appears that substitutes can be even more dangerous than phosphates themselves. Production of one such substitute, nitrilotriacetate (NTA), had to be suspended at the request of the U.S. Surgeon General because of unresolved questions concerning its long-term effects on health and the environment. Other substitutes contain caustic materials that can cause blindness. The irony is that far more phosphorus is pouring into the nation's waters from human wastes, the runoff of agricultural fertilizers, and from natural soil erosion than from detergents. The surgeon general has now advised state and local governments to reconsider the banning of phosphates, and the Environmental Protection Agency has announced that it would recommend spending $500 million to improve sewage-treatment plants to prevent phosphates from all sources from flowing into affected lakes.

The problem of phosphates has gone full circle. We are back roughly where

13

we were a couple of years ago. Other similar situations involve DDT and nuclear power. What should we do? Ban all potentially hazardous chemicals? Reduce their use to some "acceptable" level? Wash no clothes and have no detergents in our waterways? Build no power plants and have no pollution from utilities? Have no cars on the street and have no automobile pollution? Hardly, unless we wish to sacrifice the present "quality" of our lives to achieve a different "quality." Perhaps we need to take reasonable time to assess the consequences of our actions—a better balance of values, a weighing of priorities, a wiser measuring of the social and economic costs against benefits. Let us fight pollution but let us be as knowledgeable as we can be about what we do. In this course, our goal will be to equip you with principles and facts that can help you understand and respond rationally to problems of our environment.

IS TECHNOLOGY ESCAPING CONTROL?

It is quite easy to imagine a situation in which a technological process can introduce into our environment drastic and irreversible changes which set a chain of events into action before we can stop them. This is especially easy to visualize in the case of changes which might be provided by the continuous increase in the carbon dioxide content of our atmosphere. This is caused by the increased use of "fossil" fuels, such as coal and petroleum, to furnish power for energy generation and transportation. What will be the long range effects, if any, from this steadily increasing amount of carbon dioxide in our atmosphere? At the present time, no one really knows.

There are other situations where the consequences of technology seem more obviously to be moving beyond the control of man. The ability to build nuclear weapons is now spreading quite rapidly and soon may be well within the power of all but the smallest nations. How are these to be controlled? What mechanisms can be developed to prevent mankind from destroying itself in an atomic holo-

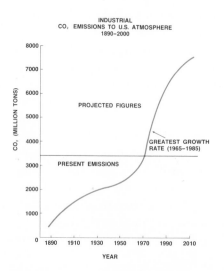

FIGURE P-12 CO_2 industrial emissions in millions of tons per year.

caust triggered by some insignificant local dispute? This is obviously a very urgent problem.

The application of technology to the problems of warfare has already produced some frightening developments, including chemical and biological warfare agents whose discovery has closely followed the extension of human knowledge in these sciences. The same kind of knowledge that allows more effective drugs and insecticides to be synthesized facilitates the synthesis of more effective agents for gas warfare. The understanding of cellular behavior that allows us to produce new varieties of high yielding grains also can be used to develop new techniques of biological warfare.

Obviously, a distinction must be made between scientific knowledge which can be used for good or evil purposes and the types of technology that are developed specifically for destructive purposes. The kind of emotional thinking which puts all technological developments under some kind of moral ban can only lead back to a new Dark Age of pestilence, disease, and famine. Most of the materials of the world can be used for a variety of purposes which are ethical to varying degrees. In our upward struggle we must always use our intelligence as well as our emotions if we are to succeed. We must be able to select the good developments from the indifferent ones and from the bad ones. This has *always* been a problem for mankind and probably always will be. In such a situation, ignorance can be catastrophic; only by a study of such processes and an evaluation of their probable consequences can rational selections be made. In this case it makes more sense to ask if we are being given control over things beyond our understanding than to ask "Is Technology Escaping Control?" Technology is always *initiated* by men or women and is under some sort of actual or potential control by them.

There is still another manner in which fears arise over the developing course of technology and our ability to control it. The vast majority of mankind has no knowledge of the scientific principles upon which technology is based and accordingly accepts the opinions of others on technological matters. This dependence upon experts, in turn, arouses fears of these experts and the damage they can do either by error or by evil intent. This fear, in some people, is an unreasoning, blind, driving force which leads them to condemn all technology. The only antidote to such fear is an understanding of technology based on study. Mankind's perennial enemy has been ignorance and the prejudice it generates. Human progress has always been the result of activities of that small percentage of people who accept neither the ignorance nor the popular prejudices of their fellow human beings.

In all fairness, it must be noted that the benefits of technology far outweigh its disadvantages, and also that its transformation of the human condition is still in its infancy. There is, even now, an enormous number of practical humanistic problems facing mankind which can be solved only by new scientific discoveries and the technological advances which they will make possible. It is the responsibility of all of us to learn about the sciences and to develop an understanding of the possible ways in which this knowledge may be developed to the advantage of mankind. In these learning processes, we will discover an avenue to responsible participation in decision making, an understanding of technological advances, and a thorough enjoyment of our investigations.

15

QUESTIONS

1. Select a law or generalization from a book on chemistry and trace down the supporting evidence. Do the same from a book on economics or sociology.

2. Cite three technological advances, give their direct benefits, and list a problem arising from each.

3. List 10 technological advances since 1940 which affect your life.

4. Name two specific changes that were made in the fall of 1970 to decrease the amount of air pollution caused by automobiles.

5. Try to think of a reason not to dump liquid chemical wastes in abandoned mine shafts.

6. Distinguish between science and technology.

7. Discuss three of the most important practical problems facing mankind, in terms of how science and technology have helped to bring them about and how they are being solved.

8. Look up the number of deaths in the U.S. due to malaria for the past century. Did a "break" occur? When? Why?

9. Is there any human evidence of technological advances resulting in the decline of civilization? Consider the article in the *Journal of Occupational Medicine: 7*:53–60 (1965).

10. Over coffee one morning, a friend states, "The results of chemical technology have all been harmful to mankind!" He goes on to list smog, water pollution, DDT, chemical and biological warfare agents, and so forth. Could you balance the argument by listing some advances in this area which have been for the general good of man?

SUGGESTIONS FOR FURTHER READING

Two references deserve special mention. The column "Eco-Chem" in the *Journal of Chemical Education* is a source of particularly interesting and appropriate questions.

One of Professor Isaac Asimov's many enjoyable books is excellent for its warmth and for information that it provides about scientists. Its title is *Asimov's Biographical Encyclopedia of Science and Technology*, Doubleday and Co., Inc., Garden City, N.Y.

Commoner, B. (Ed.), "Science and Survival," Viking Press, New York, 1966.

Kranzberg, M., and C. W. Pursell, Jr. (Eds.), "Technology in Western Civilization," Oxford University Press, New York, 1967. (In 2 volumes; a comprehensive collection of well-illustrated essays.)

Novich, S., "A New Pollution Problem," *Environment*, Vol. 11, No. 4, p. 2 (1969).

U.S. Atomic Energy Commission, "Radioactive Wastes," 1969.

Vavoulis, A., and A. Colver (Eds.), "Science and Society—Collected Essays," Holden-Day, Inc., San Francisco, 1966.

Wagner, Richard H., "Environment and Man," W. W. Norton, Inc., New York, 1971.

Woodwell, G. M., C. F. Wooster, Jr., and D. A. Isaacson, "DDT Residues in an East Coast Esturary: A Case of Biological Concentration of a Persistent Insecticide," *Science*, Vol. 156, p. 821 (1967).

Part 1
The Bases of Understanding Matter

1
CHEMISTRY— A HUMAN ACTIVITY

Every human being is intimately involved in chemistry. Some spend a working lifetime learning about chemicals and making chemical products. These people are generally classified as "chemists." All of us use chemicals in food, medicines, cosmetics, clothing, cleansers, and the many other products which we buy. Because our bodies, the materials we use, and our total environment are chemical in nature, we are partners with chemistry.

The word, "chemistry," comes from an Arabic word, Khem, which describes the rich, black soil of the Nile Delta where some of the earliest chemists, the ancient alchemists, worked. But the etymology of the word doesn't suffice for modern times. Chemists do far more than work with dirt. They are interested in all the matter of Nature. Specifically, chemists are interested in characterizing the various kinds of matter and in changing material into different kinds of matter. They also search for theoretical explanations for observed chemical phenomena.

Observation is the absolute authority in chemistry, as it is in all the physical sciences. The philosophical assumption that underlies all of the natural sciences

17

is that the facts are there, and that they can be revealed through observation. The chemist, as a human being, is limited in his powers of observation by his natural senses of sight, sound, smell, touch, and taste. Of course, he may use instruments and devices of many different types to aid his senses, yet ultimately it is only through his senses that he has any capacity for observation.

One way of attaining truth through observation is to study a phenomenon in a variety of ways, as well as to make repeated observations. A chemist is usually not content with a single observation. Rather, he checks his experiment and he reports his results in such a way that other persons may repeat his observations and, hopefully, verify his results.

THE METHODS OF CHEMISTRY

There is no standard procedure or method whereby we can uncover the secrets of nature. However, all investigations of the properties and changes in matter have a few features in common. For example, observation is necessary and interpretation of the observations is generally included in every scientific investigation. Every investigation assumes that there is a physical cause for the observed phenomenon. Quite often some experiments are suggested by the results of other experiments. Tentative explanations of results or tentative predictions for new experiments, or *hypotheses*, are sometimes offered by the scientist who carries out the experiments. Sometimes experiments lead to dead ends and that line of pursuit is duly noted and dropped.

The complexities of observing a phenomenon in nature can be simplified through experimentation. An experiment is simply a way to observe nature under **controlled** conditions. For example, experiments can be designed to test *separately* the effect of water, temperature, or the action of a catalyst on the rate of a reaction. We could design experiments to test the effect of an intact husk on the popping ability of popcorn, for instance, or we could test the effect of moisture content on the ability of the corn to pop. Other factors could be tested one at a time to ascertain their effects. A necessary part of the experiment is the maintenance of constant factors, except for the ones whose influences are being studied. In every experiment, we collect data that lead to conclusions about each effect on the popping ability of the corn. Through our experiments, we might even arrive at an explanation of what causes popcorn to pop, if our methods have been ingenious enough to make the cause obvious. Antoine Lavoisier stated it this way:

> In performing experiments, it is a necessary principle which ought never to be deviated from, that they be simplified as much as possible, and that every circumstance capable of rendering their results complicated be carefully removed.

Our observations of matter and the changes it undergoes are generally enriched in two ways. They can be *summarized* into broad generalizations, known as scientific *laws*, or they can be *explained* by using prevalent *theories*. A scientific law is a statement used to describe observable regularities found in Nature. By way of contrast, the laws of a human government are prescriptive in that they prescribe how men should behave, but they are not necessarily

Antoine-Laurent
Lavoisier

Antoine Laurent Lavoisier (1743–1794) is often referred to as the "Father of Chemistry," primarily because he promoted and practiced accurate measurement—particularly weighing in chemistry—and because he was able, one by one, to break down the antique chemical notions of eighteenth-century chemists. The guillotining of Lavoisier on May 8, 1794 is believed by many to be the most deplorable single casualty of the French Revolution.

descriptive of how men do behave. The Law of Conservation of Matter and the Law of Definite Proportions are two important laws of chemistry which will be discussed later in this text. They, like all scientific laws, are concise, summarizing statements of many related facts.

A scientific theory is based on *interpretations* of a variety of observations and is convenient for explaining phenomena. The basic essence of a chemical theory is often not directly observable. The atomic theory, the kinetic molecular theory, and the theory of electrical conductivity are all important theories of chemistry. These, and other acceptable theories, have evolved from a large number of interpretations based on many observations into a "picture" or model of what appears to happen. Usually the model or picture is a visualization or an idea developed in a realm that is not directly accessible to our senses. A model is an imperfect representation (for instance, a road map) and, of course, it is not the real thing. Nevertheless, the model helps us to visualize and increase our understanding of chemical phenomena. A rather sophisticated model of the atom has been developed without direct observations of the atom itself. Subsequent observations have caused us to modify our concept of the atom to make it conform more closely to reality. A theory of chemistry relies heavily on imagination and creativity, while a law organizes and correlates experimental observations. A theory (model) is subject to continual modification or rejection, based upon newly acquired evidence (observations). We shall frequently call

19

upon scientific theories in this text to help us understand our modern world of chemistry.

Quite often, analogies are drawn between scientific interpretations and things in our experience. "The atom is like a miniature billiard ball or a tiny solar system," were two analogies used often in the past. Analogies are never perfect, but they are helpful in communicating scientific ideas, especially to the beginning student. We shall make frequent use of analogies in this text.

Chemical knowledge grows as more and varied observations are made, theories are revised, and laws are broadened. This has led some people to allege that scientists or chemists cannot be trusted because they continually change their minds about what is true. This allegation does not do justice to the meaning of "truth" in science. Consider as one example the scientific belief of a few years ago that "matter can be neither created nor destroyed" and that "energy can be neither created nor destroyed." As a result of studies carried out in the fields of nuclear chemistry, both of these expressions have been replaced by the concept that "the sum total of matter and energy in the universe is constant." This change was necessary when matter was observed to be converted into an equivalent amount of energy. This development should not be viewed as a case of a scientific "truth" being found false, but rather as a less refined understanding being replaced by more refined knowledge.

What happens when apparently contradictory data are obtained? Common sense and experience call for the worker to reexamine the evidence, seek additional evidence, or suspend judgment for a while. Consider as an illustration some determinations of the density of nitrogen performed in the laboratory of Lord Rayleigh in 1894. Lord Rayleigh found that the density of nitrogen prepared from various nitrogen-containing compounds differed slightly from that prepared by extraction from the atmosphere. Here, apparently, was a contradiction. Rayleigh became so frustrated that he wrote to the journal *Nature* and asked for suggestions on how to resolve the contradiction. Upon reexamining the evidence and seeking additional evidence of several kinds, it was found that the nitrogen from the air was not pure but was contaminated by some unreactive and previously unknown gases. As a result, the inert gases of the atmosphere were discovered—a most noteworthy advance in scientific knowledge.

FRITZ HABER—A CHEMIST AND A REMARKABLE PERSONALITY

A Brief Biography

Of the great German chemist, Fritz Haber (Hah'ber) (1868–1934), one of his associates wrote, "Every golden grain in the fields displays to him the gratitude of the soil he so richly endowed." He made contributions to virtually every field of chemistry, and he received a Nobel Prize for his synthesis of ammonia. During his time he was acclaimed to be the greatest authority in the world on the relation between scientific research and industry. He helped his country significantly in a time of need. He saw great opportunities to use the processes of nature to better mankind.

20

FIGURE 1-1   Fritz Haber and some cities important in his life.

The life and works of Haber communicate something of the nature of chemistry—and of the people who work in this field. In a few rare cases, chemical discoveries are based on the works of a single gifted individual, a genius. More often, however, they are the results of people being in the right place at the right time and having the right interests. In recent years, discoveries have come from the cooperation of teams working under effective direction and from work based on that of predecessors in the field. Fritz Haber illustrates well some of the spirit, attitudes, and methods that have produced modern chemistry. He was often successful (and we like to talk about successes) although he also had his failures. A glimpse at Haber's life is one way to understand some fundamental aspects of human endeavor and the growth of chemistry.

Haber's Nobel prize came in 1918 for his work in devising the proper conditions for preparing ammonia from nitrogen and hydrogen. Ammonia was as important a chemical in Haber's day as it is today. (In recent years, ammonia has become the second most abundantly produced chemical in the United States, second only to sulfuric acid.) Ammonia is still used, as it was in Haber's time, in the production of fertilizers and explosives.

Before looking at Haber's contribution to chemistry, let us consider him as a human being; one who showed emotion, competitiveness, prejudice, and interest in political affairs. **Haber was not the "typical" chemist, for there is no such being.** In the areas of social matters, political opinion, and religion, for example, chemists, like the general population, vary from extreme liberalism to extreme conservatism. Several episodes in Haber's life prominently show his human weaknesses, and together, they demonstrate that scientists can excel in their fields, although not necessarily in other areas of life.

In many respects Haber is a contemporary, and we can identify with him. He was a victim of racial prejudice because he was a Jew. At least two professorships were denied him because of antisemitism. This was also the cause for his exile from his beloved Germany by Hitler in 1933. Like many of our young people today, Haber educated himself out of a job. When he finished his doctorate in 1891, he could not find work. Schools took only the very best students to train for professorships (Haber barely passed his final oral examination), and most of the more than 5000 chemical factories in Germany did not need doctors of philosophy. He left the country to find employment.

Fritz Haber wanted science to have a practical nature. He once said, "It is not enough to seek and to know; we must also apply." One of the major concerns today is that not enough scientific endeavors are being directed toward the practical problems of pollution, food supply, disease, transportation, and population control. It has been said that Haber promoted the practical application of scientific discoveries to industrial use more than any other man in his time.

Haber, like some youths now, served a year of compulsory military training and began drinking heavily. (He broke this habit after he drunkenly walked through a plate glass window.) He liked to express his opinions bluntly and he did not seem to know how to use flattery to achieve his ends. Haber tired of following directions and wanted to be creative. This was fully evident in his first academic job. In the fall of 1892, Haber became an assistant in Ludwig

TABLE 1–1 SOME IMPORTANT EVENTS IN THE LIFE OF FRITZ HABER.

Dec. 9, 1868 born in Breslau; mother died	1907 meeting of Bunsen Society at Hamburg; confrontation with Nernst
1886 entered University of Berlin (studied under Hofman and Helmholtz)	1911 became professor of physical chemistry at University of Berlin and director of the Dahlem Kaiser Wilhelm Institute for Physical Chemistry
1887 enrolled at Heidelberg University (studied under Bunsen)	1913 first commercial production of ammonia by Haber process
1889 military training	April 22, 1915 directed gas attack at Ypres
Oct. 16, 1889 entered Charlottenburg *Technische Hochschule*	1915 made a captain in the German army
1891 awarded the degree of Doctor of Philosophy—applied for work with Wilhelm Ostwald at Institute for Physical Chemistry—was rejected—took position at a Budapest distillery	1916 supervised gas shelling at Verdun Clara Haber commits suicide became chief of Chemical Warfare Service
1894 became an instructor at Karlsruhe Engineering College and did research with Hans Bunte	1917 married Charlotta Nathan
1896 accepted into regular faculty at Karlsruhe; published first book	1918 daughter, Eva, born
1901 married Clara Immerwahr	November, 1919 received Nobel prize
1902 toured United States chemical industries	1920 began experiments to recover gold from the oceans
June, 1902 son, Hermann, born	1921 son, Ludwig, born
1904 began study of ammonia synthesis	1927 Charlotta and Fritz were divorced
1906 named full professor at Karlsruhe and director of the Electrochemistry Institute	1933 Nernst and Haber buried the hatchet; exiled from Germany
	Jan. 29, 1934 died of a heart attack in Lugano, Switzerland

Knorr's laboratory at the University of Jena. His main duties were assigned by Knorr. He began to hate the tasks, the endless routine of mixing reagents and purifying products, and the monotony of mixing, heating, distilling, and crystallizing. He soon left Jena for the Karlsruhe *Technische Hochschule*, where he slowly gained prominence. He eventually was appointed director of the Kaiser Wilhelm Institute for Physical Chemistry and Electrochemistry at Dahlem, a position he held until he was exiled by Hitler. Some of the important events in his life are listed in Table 1–1.

A classic example of Haber's competitive nature was his battle with Walter Hermann Nernst (1864–1944). Nernst was one of the founders of electrochemistry and had an unmatched genius for mathematical reasoning. Nernst and Haber had conducted experiments on ammonia synthesis separately and had obtained different results. Under similar conditions, Haber had obtained product yields about 50 per cent larger than those obtained by Nernst.

A showdown between Nernst and Haber occurred at the Hamburg meeting of the Bunsen Society in the summer of 1907. Both Nernst and Haber were there, armed with data and explanations. Unfortunately for Haber, Nernst had performed high-pressure experiments that Haber had not tried for a practical reason—the apparatus was too cumbersome and difficult to construct. Nernst had the last word at the meeting. He suggested also that Haber perform high-pressure experiments.

Haber took Nernst's statements as a personal affront. Immediately upon his return to Karlsruhe, he focused his attention on the ammonia synthesis. It was his intention to have the last word. Working with higher pressures, Haber and LeRossignol confirmed Haber's earlier values. For a while it seemed that Haber had dethroned the king of physical chemistry, Walter Nernst. The conflict between the two men persisted through much of their careers. Haber, in private conversation, bitterly criticized Nernst. Nernst told his classes that the Haber process was really his. Occasionally, both men failed to appear at scientific gatherings when they learned that the other had been invited. In the mellowness and wisdom of old age, Haber and Nernst finally came to an understanding and had a warm conversation in the spring of 1933. The antagonism and rivalry which had existed for so long seemed to be ended.

Haber was too much the hard-working thinker, too much the objective scientist, and too easily annoyed to be an appreciative or appreciated husband. He sensed his disinclination for romance. In 1930 he sadly told the wife of a good friend that women to him were like butterflies—beautiful to look at, but if he tried to touch them only the colorful dust remained in his hand. His first marriage ended with the suicide of his wife and his second in divorce after ten years of marriage.

On the other hand, Haber showed great concern for others. Once during the war, when he heard that his secretary's mother, an elderly lady, was ill, he stopped all his activities, secured food and wine at a time when both were scarce, and delivered them to her personally. He was particularly considerate of his assistants who worked for him. It was the way he treated students and assistants that helped make the Dahlem Institute an incubator for young scientists. As chief and teacher, he encouraged and challenged his students; as their second father and good friend, he was kind and understanding. He had the ability to develop the potentialities of his students. He treated workers as men and equals; he tried, as much as he could, to prevent anyone from becoming a laboratory scapegoat. He was careful in scientific papers or talks to give credit to all those who contributed to the research.

Haber was a dedicated scholar. While he barely passed his oral examination for his degree of Doctor of Philosophy, he did develop the traits peculiar to scholars in any field. He had tremendous energy for work. He virtually lost himself in his duties and projects. On many nights he was busy until 2:00 or 3:00 A.M. Books were always piled about his night table. He was never satisfied with superficial knowledge. If someone made one error in a talk, gave one false interpretation, repeated one sentence which was not clear, Haber called attention to the flaw.

Once, an enterprising student was the target of Haber's humor. Haber asked a candidate for a Doctor of Philosophy degree to state a method for preparing iodine. This element is usually manufactured by chemically treating sea plants, notably kelp. After some hesitation, the student answered, "It is obtained from a tree."

Unruffled, Haber continued, "Iodine tree? Describe one to me." The candidate expanded his guess, telling about the height of the tree, its shape, and the type of leaves.

"Where do these trees grow?" asked Haber, not changing his expression.

The student, knowing his answers were imaginative, was cheered by a line of questioning which seemed to indicate he might be correct. "In India and Brazil," he replied.

"Where else?" Haber asked immediately.

"The Dutch East Indies," claimed the candidate in full confidence.

"And when do these trees become mature?" continued Haber.

The student was no longer hesitating. "In 15 years," he said.

"And when do the iodine blossoms appear?" asked Haber.

"In the fall," was the answer.

Then the embarrassed student found a fatherly arm about him, saw a smile of recognition, and heard, "Well, my friend, I will see you again when the iodine blossoms appear once more."

Haber was most serious about his patriotism to Germany. His work in developing techniques for using chlorine gas in chemical warfare grew out of a sincere belief in his country. The stigma of having promoted chemical warfare followed Haber as long as he lived.

Haber had moments of deep despair and periods of withdrawal. The death of his mother at his birth left an indelible scar, as did the deaths of two others, his first wife and Emil Fischer, a good friend and great chemist. Fischer, who feared death from cancer, had asked Haber what would be a good chemical that he could use for suicide. Haber answered nonchalantly, "cyanide." Fischer used the poison and Haber felt responsible for his death. Later in life, he deeply mourned the deaths resulting from poison gas and the manufacture of explosives through ammonia synthesis. True, his accomplishments were also used for maintaining and safeguarding life, as in the invention of a coal gas detector and the production of fertilizers, but he was ever sensitive to the less humane aspects of his work.

Haber's work is illustrative of all chemical research procedures in that it consists of a series of experiments, observations, and interpretations which lead in turn to new experiments, and finally to an explanation of what is happening in theory. Haber's technique is not typical of every chemical investigation because methods vary from investigator to investigator. Haber's technique is vividly described in a portion of his Nobel address. In his words,

I therefore began tentatively to determine the approximate position of the ammonia equilibrium in the vicinity of $1000°C$. It now transpired that earlier trials had only proved negative by accident; it was easy, in the vicinity of $1000°C$ and using iron as a catalyst, to obtain the same ammonia content from both approaches.

It was further shown that the same results could be obtained with nickel as with iron, and it was found that calcium and in particular manganese were catalysts which would bring about a combination of the elements even at lower temperatures. . . By having a circulation system which alternately brought the gas at high temperature in contact with the metal and then washed out the ammonia at normal temperature, the conversion of a given mass of gas to ammonia could proceed stage by stage.

The most important point realized at that time was that . . . if one wished to obtain practical results with a catalyst at normal pressure, then the temperature must not be allowed to rise much beyond $300°$.

At that point it seemed to me, in 1905, useless to pursue the problem further. A combination of the elements had certainly been achieved, and the requirements for large-scale synthesis had been outlined; but these requirements appeared so unfavorable that they deterred one from a deeper study of the problem. . .

The synthesis of ammonia which had been demonstrated at normal pressure could be carried out at high pressure on a laboratory scale without any great difficulties. It needed only a slight modification of the pressure oven . . . But I did not think it worth the trouble; at that time I supported the widely-held opinion that a technical realization of a gas reaction at the beginning of red heat under high pressure was impossible. Here the matter rested for the next three years.

Further work followed, devoted to determining the equilibrium at normal pressure and at 30 atmospheres over an extended range of temperatures . . . During the course of these investigations, together with my young friend and co-worker Robert le Rossignol, whose work I would like to mention here with particular sincerity and gratitude, I took up once again, in 1908, the problem of ammonia synthesis abandoned three years earlier.

To begin with, it was clear that a change to the use of maximum pressure would be advantageous. It would improve the point of equilibrium and probably the rate of reaction as well. The compressor which we then possessed allowed gas to be compressed to 200 atmospheres, and thus determined our working pressure which could not easily be exceeded for any very large series of experiments. In the neighborhood of this pressure, the catalysts, with which we had become familiar in the course of our equilibrium determinations, very easily provided a rapid combination of nitrogen and hydrogen at about 700°C; this applied notably to manganese, followed by iron.

To achieve impressive results, however, we needed to discover catalysts which would induce rapid conversion at between 500° and 600°C. We hit upon the idea of searching the sixth, seventh and eighth groups in the Periodic System, whose principal metals chromium, manganese, iron and nickel possessed very definite catalytic properties, for metals which acted even more favourable; these we found in uranium and osmium. . .

Finally we needed an improvement in the circulation system which could act as a model for technical realization . . . The construction and operation (carried out in collaboration with Robert le Rossignol) of a small-scale plant which suited these requirements, together with the performance of the new catalysts mentioned, was indeed sufficient to persuade the *Badische Anilin- und Soda-fabrik* . . . to undertake high-pressure synthesis from the elements.

It may be that this solution is not the final one. Nitrogen bacteria teach us that Nature, with her sophisticated forms of the chemistry of living matter, still understands and utilizes methods which we do not as yet know how to imitate. Let it suffice that in the meantime improved nitrogen fertilization of the soil brings new nutritive riches to mankind and that the chemical industry comes to the aid of the farmer who, in the good earth, changes stones into bread.

FIGURE 1-2   Fritz Haber's pilot plant for producing ammonia. (From *Chemical and Engineering News*, Feb. 24, 1964, p. 108.)

There is one thing about Haber's investigation that is very typical. He did not discover the proper conditions for the ammonia synthesis accidentally. Very few, if any, scientific discoveries are accidental. There are at least three significant features of Haber's investigation that are almost invariably found in conjunction with "accidental" discoveries in science: (a) he was busily doing something at the time; (b) he was alert to notice something unusual; and (c) he was diligent in investigating what he had observed. "Accidental" discoveries are seldom made by idle persons. There may be an element of coincidence and luck in much scientific work, but advances are generally attributable to persons who are diligent, alert, and persistent.

MEASUREMENT

Haber's elucidation of the conditions needed for the synthesis of ammonia relied heavily on measuring the temperature, pressure, and amounts of chemicals used. Experiments frequently involve one or more quantitative measurements which can be repeated, verified and, hopefully, interpreted.

Measurement differs from counting in a fundamental but not entirely obvious way. Consider the determination of the number of lines of text on this page. An exact answer is possible simply by counting the lines. Consider next the measurement of the length of this page. It is impossible to get an exact answer, since repeated measurements with a meterstick will inevitably come out differently, even though the meterstick is read as carefully and exactly as possible. For example, the length of the page may be 23.4 cm., which does not mean that it is *exactly* 23.4 centimeters but rather that it is closer to 23.4 cm. than it is to either 23.3 or 23.5 cm. A second measurement will not necessarily be 23.4 cm., but may be 23.6 cm., and a third measurement may be 23.3 cm. In general, measurements, except where discrete objects are being counted, *are never exact* because they involve comparisons with standards, and such comparisons are inevitably subject to error. The error may be minimized by taking an average of a large number of measurements of the same phenomenon, and by using the most sensitive measuring device possible.

The heart and soul of reliable scientific facts, and, consequently, the basis for laws and theories, is the careful measurement of weights, volumes, times, lengths, and other quantities. Perhaps more than any other person, Antoine Lavoisier, in the late 1700's, put chemistry on a quantitative basis by very accurately measuring the weights of materials used in chemical reactions. Because of the significance of his work, Lavoisier, a French chemist, has been called the "Father of Modern Chemistry." The data that results from accurately recording measurements led directly to the formulation of several laws concerning chemical change.

The system of measurement used worldwide by scientists is the International System of Units (SI). This system is an extension of the metric system that began in France in 1790 and which is now the official system of weights and measures in almost all countries. Proposals to convert the United States' system from that of the English to SI are presently under study by Congress. By 1975, the United States will be the only country in the world which does not use SI, and our products will be much more difficult to sell elsewhere for that reason.

SI has one great advantage over older systems: it is simple. Its simplicity lies in its decimal counting, which requires that relatively few terms be remembered. Units are so defined that a larger unit is ten, or some power of ten, times larger than a smaller unit. A meter (a unit of length equal to 39.4 inches), is 100 centimeters, and a centimeter is 10 millimeters. A conversion from one unit of length to another in the SI system merely involves shifting the decimal point. Compare this with the complexity of converting miles to inches, wherein one would probably employ the factors 12 (12 inches = 1 foot) and 5280 (1 mile = 5280 feet). Instead of having a number of different root words for length, as in the English system (league, mile, rod, yard, foot, inch, mil), the metric system has just one, the meter. With suitable prefixes, the meter can express a unit useful to the watchmaker (millimeter = 0.001 meter) or the long distance runner (kilometer = 1000 meters).

In the SI system, only a few basic units are required; a unit of length is a meter; a unit of volume is a liter; a unit of mass (weight) is a gram; and a temperature unit is the Celsius degree. English equivalents of the first three are:

$$1 \text{ meter} = 39.4 \text{ inches}$$
$$1 \text{ liter} = 1.06 \text{ quarts}$$
$$1 \text{ gram} = 0.0352 \text{ ounce.}$$

Temperature scales are defined in terms of the behavior of samples of matter. For example, the familiar Fahrenheit scale defines the temperature at which water freezes as 32°F and the boiling point as 212°F. Thus, there are 180 (212 − 32) Fahrenheit degrees between the freezing and the boiling points of water. Most scientists use the Celsius scale, which defines the freezing point of water as 0°C and the boiling point as 100°C. Consequently, there are 100 Celsius degrees in this same temperature range.

For the interested reader and those who need additional aspects of SI to support laboratory work, more information is presented in Appendices A and B.

QUESTIONS

1. What traits of Haber's personality do you feel made him a successful chemist?

2. Distinguish between theory and law as used in chemistry.

   (a) Which has a better chance of being true? Why?
   (b) Which summarizes? Which explains?

3. How many times do you think a given experiment should give a result before a scientific fact is established? How many failures would you require before rejecting the "fact"?

4. Examine a newspaper or magazine article for scientific writing. Does the article present facts (observables) or draw conclusions for you? Does it base its validity on the reputation of a man or a profession? Does it play on emotion?

5. How does an experiment differ from what ordinarily occurs in nature?

6. Which of the following would be classified as observations and which as interpretations?

    (a) "That is a pretty dress."
    (b) Basketballs are round.
    (c) "That is a pretty, blue dress."
    (d) This book is published by W. B. Saunders Co.
    (e) Ice melts at 0°C.
    (f) Onions cause the eyes to water.
    (g) An apple made him sick.

7. In his Nobel address, Fritz Haber said, "In technical questions, where the scales oscillate between success and failure, the borderline between the two extremes is usually defined by modest differences in the consumption of energy and materials, and variations in these values which lie within one decimal power will determine the result." How might this apply to the wise use of our limited resources, such as coal and petroleum?

8. Why did others before Haber fail in their attempts to prepare ammonia by combining nitrogen and hydrogen?

9. It is apparent that an object to be measured does have a real and definite length. Why then do we say that a person who seeks to measure that length can never know exactly what it is?

10. What do you think of the argument that even the "observed facts" are interpretations of signals in the brain? Is there no basic difference between facts and theories?

11. It has been suggested that the sciences are ordered from the most basic field to the most complex and that the order is: mathematics, physics, chemistry, biology. How would you modify this list to include other natural and social sciences?

12. Which is the better definition of a scientist? "The scientist is a careful observer of nature who uses scientific attitudes and methods in discovering new things about nature," or "The scientist is one who happens to enjoy the study of science." Give reasons for your selection.

13. Debate the topics:

    (a) A free society must have an educated electorate that understands what the sciences are, how they progress, and how they may be applied to change man's life.
    (b) The sciences are far too complex for most people; hence, the scientific community alone should be given the responsibility of charting our scientific future.

14. How do scientific laws differ in principle from laws of government?

15. Does the process of observation lead to infallible truth?

16. There is a very interesting contrast in humanitarian responsiveness between Otto Hahn and Fritz Haber. Hahn had a very important part in the development of nuclear power and the atomic bomb. Like Haber, he was a German notable and a Nobel prize winner. While Haber made his contribution during World War I, Hahn's contribution came during World War II. Hahn's moral position would be praised by today's generation. You will find pertinent information on Hahn's position in the Asimov reference at the end of the Prologue and in *Chemistry*, Vol. 39, No. 12, p. 5 (1966). What was the main difference between Haber and Hahn in how they allowed the government to use their discoveries?

17. Are you willing for experts in scientific matters, such as Fritz Haber, to make decisions for you on moral and social issues related to the use of scientific knowledge?

NOTE: The following problems should be attempted after a more thorough study of the material presented in Appendices A and B.

18. Determine:

    (a) the mass in grams of a portion of water weighing 8 ounces.
    (b) the mass in kilograms of a 1 pound loaf of bread.
    (c) the length in centimeters of a pencil 9 inches long.
    (d) the volume in gallons of 10 liters of cider.
    (e) the speed in kilometers per hour of a car going 45 miles per hour.

19. If you prefer 70°F for your living quarters, what would be your preference on the Celsius scale?

20. While traveling in Europe, you have antifreeze put into your car radiator. The attendant tells you that your car will now be safe at temperatures down to $-10°C$. When you return to the United States, you hear on the radio that a low temperature of 20°F is expected for the coming evening. Do you need to worry about your car? Why?

SUGGESTIONS FOR FURTHER READING

References

Asimov, Isaac, "Realms of Measure," Fawcett Premier Paperback (1960).
Fischer, R. B., "Science, Man and Society," W. B. Saunders Co., Philadelphia, 1971. (Paperback)
Goran, M., "The Story of Fritz Haber," University of Oklahoma Press, Norman, 1967.
Nobel Foundation, "Nobel Lectures—Chemistry—1901—1921," Elsevier Publishing Co., New York, 1966.

Articles

"Chemists and Music: Deep Involvement," *Chemical and Engineering News,* Vol. 49, No. 52, p. 11 (1971).
Garrett, A. B., "The Discovery Process and the Creative Mind," *Journal of Chemical Education,* Vol. 41, p. 479 (1964).
Hildebrand, J. H., "It Ain't Necessarily So," *Chemistry,* Vol. 40, No. 9, p. 19 (1967).

2

ATOMS UNIQUE—
THE ELEMENTS

Atomic theory is one of the greatest achievements in the history of human thought. The evolution of a consistent concept of the atom is particularly amazing because, until 1970, no one had actually seen even the gross features of the atom. Instruments extending the senses (such as spectroscopes, cameras, weighing balances, and light microscopes) did not permit a glimpse of these invisible particles. In 1970, Dr. Albert V. Crewe, at the University of Chicago, used a specially designed electron microscope to "see" what amounted to rather crude shadows of single heavy atoms. The white dots in Figure 2–1 required enlargement five million times to be seen. These "shadows" are consistent with what was first postulated perhaps 2300 years ago and what we now have considerable, consistent evidence for—matter is granular or discontinuous, and atoms are single, identifiable, and very small bits of matter that collectively compose macroscopic (or seeable) pieces of matter.

It is a basic premise of chemistry that the *submicroscopic structure* of matter dictates its macroscopic properties. By submicroscopic structure we refer to atoms and molecules, which are too small to be directly observed with our senses or optical microscopes. The arrangement of atoms with respect to each other determines molecular or crystalline structure. The structure of atoms will be summarized in this chapter; molecular structure in the next; both will be used throughout this text to explain the nature of chemical substances.

Let's look at how knowledge of submicroscopic structure explains a familiar phenomenon. Perhaps you have had occasion to drop some bicarbonate of soda (baking soda) into a little vinegar and noticed the profuse fizzing. This same fizzing is seen when baking soda is dropped into orange juice, or into a mixture of water and vitamin C, or into hydrochloric acid. Vinegar, orange juice, vitamin C, and hydrochloric acid all have something in common which causes the soda to fizz. Each substance has a common structural feature. Each has at least one hydrogen atom loosely held to the rest of the molecule. These loosely-held hydrogen atoms trigger production of the fizz. If we find another substance with a loosely-held hydrogen atom, we could rather confidently expect the soda to fizz in contact with that substance. There are certain compounds in buttermilk

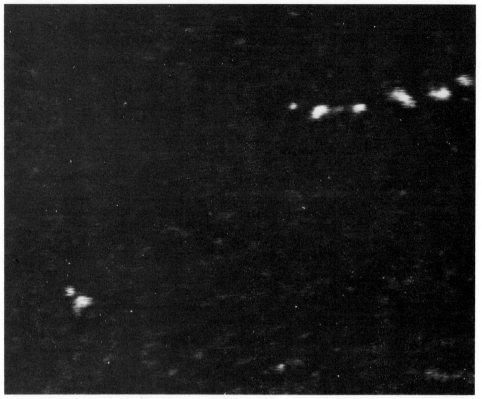

FIGURE 2-1   Chains of thorium atoms separated by an organic molecule. These chains were placed on a thin carbon film a tenth of a millionth of an inch thick. In each chain the smallest white dots represent single thorium atoms. The larger white dots are probably aggregates of a few thorium atoms very close together. (Courtesy of Professor Albert V. Crewe, Department of Physics, Enrico Fermi Institute, University of Chicago.)

and tomato juice that have loosely-held hydrogens. What do you expect will happen when baking soda is placed in a solution of these materials? Incidentally, the gas evolved is not hydrogen, it is carbon dioxide. The details of what happens are described later in the text.

GROSS FEATURES OF THE ATOM

The diameter of an average atom is about four hundred-millionths of an inch; that is, 0.00000004 or 4×10^{-8} inch. Atoms are so small it takes about one million atoms placed in a line to stretch across the diameter of the head of a straight pin. There are about one sextillion (1×10^{21}) atoms in a single drop of water, and more than a quintillion (1×10^{18}) atoms compose the period at the end of this sentence. (A quintillion people would populate almost a billion planets like earth with its 4 billion people!)

Atoms are discrete spherical particles. Just as oranges are distinct, separate units in a basket of oranges, atoms are separate pieces of matter in the macroscopic piece of matter that we can see. It is difficult for us to realize the discontinuous or granular nature of matter, although we live in a world replete with similar examples. Perhaps you have stood on a high hill or have been in an airplane and looked down on a "solid" patch of green which you know to be really a group of individual trees. Lucretius, the brilliant Roman poet of the first century B.C., once wrote about standing on a hill overlooking a plain covered by a vast array of soldiers. From his viewpoint, they looked like a solid, blended mass. There are other examples listed in Figure 2–2 where close observation is required to reveal true detail. In a similar fashion, when we look at bright, shiny metal, we "see" (that is, conclude, or interpret) small discrete particles called atoms.

The idea that atoms are spherical is suggested by the way they are packed in crystals of metals. Metals are nearly incompressible, which means that their atoms are about as close together as possible. When round objects, such as oranges, are packed close together, each orange has 12 oranges touching it. There are two, and only two, arrangements for the 12 closest neighbors, as shown in Figure 2–3. All experimental evidence indicates that incompressible metals crystallize in either one of these two ways. This implies that atoms, like oranges, are round.

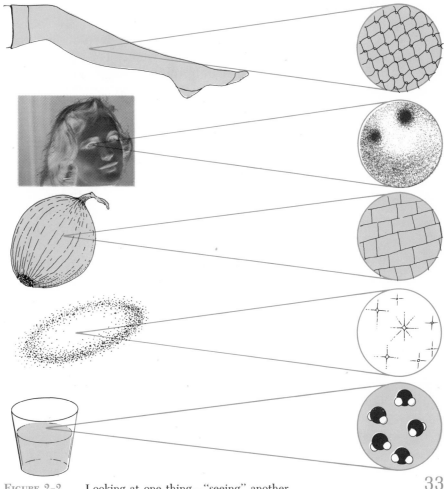

FIGURE 2-2   Looking at one thing—"seeing" another.

33

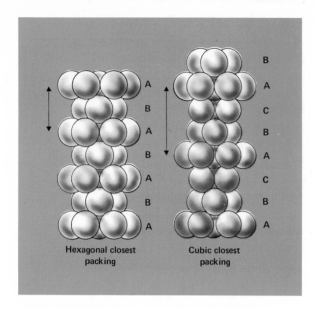

Hexagonal closest packing

Cubic closest packing

FIGURE 2–3   Close-packing of spheres. Each sphere is touching a maximum of 12 other spheres. In cubic close-packing, every other layer is exactly alike. In hexagonal close-packing, every third layer is exactly alike. Silver, aluminum, gold, copper, and lead atoms are cubic close-packed. Cadmium, magnesium, and zinc atoms are hexagonal close-packed.

MODERN ATOMIC THEORY IN CONTEXT

Concepts of the atom have evolved piece by piece in jigsaw-puzzle fashion. Each piece adds to the image of the whole. The "pieces" in the image came from precise experiments, careful observations, and consistent interpretations. Since even now most atoms are too small to detect by direct observation, the search for the atom has been a challenge to some of the best scientific minds. More than 30 Nobel prizes have been awarded since 1930 for specific contributions to atomic theory. Although this chapter is not intended to be a complete history of atomic concepts, a few of the key experiments supporting the atomic model will be described. The fascinating history of atomic theory is told in several of the works (mostly paperbacks) listed at the end of this chapter.

Figure 2–4 summarizes five major achievements in the development of atomic theory. The Greeks did not conceive of experimentation. They saw nature as beyond the control of man. Some Greeks surmised that matter is comprised of discrete pieces and others concluded that matter is continuous and has no discrete units. Leucippus and Democritus championed the atomistic idea; Plato and Aristotle promoted the concept of the continuous nature of matter. Apparently Plato and his student Aristotle made more forceful arguments because their opinion prevailed, and atomic theory would not interest scientists again for more than 2100 years. Thus, a triumph of argumentation and, possibly, of personality, defeated a concept of at least equal validity, impeding mankind's understanding of nature. A stubborn and unreasoning attitude which does not admit to new facts and which relies on persuasion, rather than evidence, has done as much damage to scientific progress as have political demogogues who use the same methods.

As the practice of experimentation grew, more and more details of the atom became clear. For example, John Dalton (1766–1844) had at hand the Law of Conservation of Matter (Chapter 4) and the Law of Definite Composition

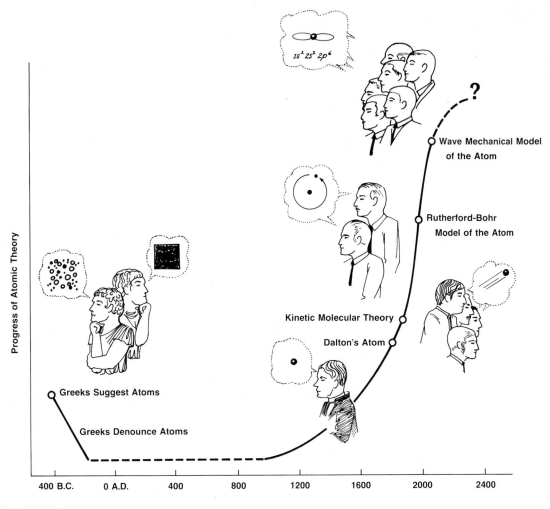

Progress of Atomic Theory (vertical axis)

Wave Mechanical Model
of the Atom

Rutherford-Bohr
Model of the Atom

Kinetic Molecular Theory

Dalton's Atom

Greeks Suggest Atoms

Greeks Denounce Atoms

400 B.C.　0 A.D.　400　800　1200　1600　2000　2400

Time (years)

FIGURE 2–4　Highlights of the progress of atomic theory.

John Dalton (1766–1844), a self-taught English school teacher, moved
to Manchester in 1793 and devoted the rest of his life to scientific
investigations. His presentation of atomic theory in the early part
of the 19th Century served as the basis from which modern chemical
theories have grown.

35

Niels Bohr (1885–1962) received his doctor's degree in the same year that Rutherford discovered the atomic nucleus, 1911. After studying with Thomson and Rutherford in England, Bohr formulated his model of the atom. Bohr returned to the University of Copenhagen and, as Professor of Theoretical Physics, directed a program that produced a number of brilliant theoretical physicists.

(Chapter 3) which could be logically explained by his concept of the hard, round spheres called atoms. The kinetic molecular theory assimilated many varied observations on gases into the concept of moving, colliding atoms (or combinations of atoms called molecules). Still other experiments, beginning around 1900, revealed details of the atom. Niels Bohr (1885–1962) envisioned the atom as a miniature solar system with a nucleus at the center and small bodies called electrons orbiting around it. Erwin Schrödinger later expanded Bohr's atomic concept and made it consistent with essentially all the known experimental observations of his day. Schrödinger's concept of the atom, referred to as the wave mechanical model, is largely mathematical and is the most difficult for the student to visualize.

The atom, its structure, and the relationships among atoms provide the foundation for our understanding of consumer products, medicine, heredity, pollution problems, and all the chemistry we depend on and are affected by each day. We shall use atomic theory time and again in this text to explain different subjects. We will now deal with some of the details of the theory.

SUBATOMIC PARTICLES

For our purposes in chemistry, we can consider atoms to be constructed from "elementary" particles: electrons, protons, and neutrons. The protons and neutrons are in the nucleus (at the center of the atom), and the electrons are distributed in the space around the nucleus.

Protons are submicroscopic particles with a mass of about 1 atomic mass unit (amu). The atomic mass unit is approximately 1.7×10^{-24} gram (Appendix A). Each proton exerts an effect that repels other protons, attracts electrons, and has no apparent effect on neutrons. This mysterious effect results from electrical charge. Two charges that nullify, or neutralize, each other are called "positive" and "negative," respectively. A proton possesses the smallest unit of positive charge which can just neutralize or cancel the effect of the smallest unit of negative charge, the charge of an electron.

Benjamin Franklin originated the terms positive and negative to describe the opposing but canceling effects of the two kinds of electrical charges. Franklin was not aware of electrons and protons; neither particle was discovered until

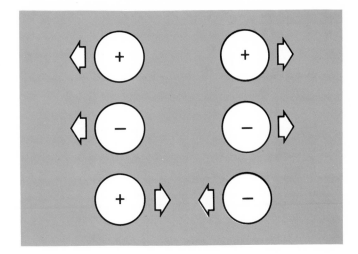

FIGURE 2-5   Like charges repel
and unlike charges attract.

more than 50 years after his death. Static electricity, however, and the phenomenon of electric charge had been observed and understood by the Egyptians
before the Christian era. It was to the gross properties of electric charge (the
effect caused by two nonconductors, such as shoe leather and a nylon rug, being
rubbed together) that Franklin applied his terms. Figure 2–5 illustrates the
actions charged particles can have on each other. As the figure shows, like charges
repel, and unlike charges attract.

The location of protons in an atom was ingeniously interpreted from an
experiment conducted by a group headed by Sir Ernest Rutherford in Manchester, England in 1911. The experiment is often repeated in school laboratories,
using the apparatus shown schematically in Figure 2–6. Alpha particles are

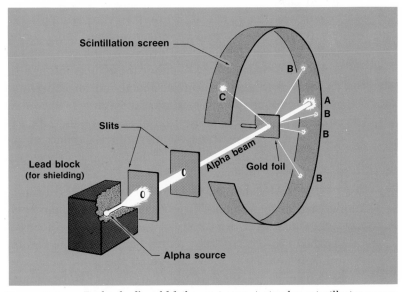

FIGURE 2-6   Rutherford's gold foil experiment. A circular, scintillation screen
is shown for simplicity; actually, a movable screen was employed. Most of the
alpha particles pass straight through the foil to strike the screen at point A.
Some alpha particles are deflected to points B, and some are even "bounced"
backwards to points such as C.

37

Lord Rutherford (Ernest Rutherford, 1871–1937) was Professor of Physics at Manchester when he and his students discovered the scattering of α particles by matter. Such scattering led to the postulation of the nuclear atom. For this work he received the Nobel prize in 1913. In 1919, Rutherford discovered and characterized nuclear transformations.

positively charged helium atoms that have a charge of $^+2$ and a mass of 4 amu. They are spontaneously given off at great speeds by a number of different radioactive elements, such as radium and uranium. In this experiment, the alpha particles are allowed to speed toward a very thin piece of gold foil. Their course can be followed by using a scintillation screen, similar to a television screen, which glows when the alpha particles hit it.

When the alpha particles were directed toward the gold foil, most of the particles went straight through it; their course was unaltered. However, the paths of a few alpha particles were changed drastically. Some even bounced back toward the source. This is comparable to shooting at a "solid" wall, only to have it act like a large-mesh wire fence. Most of the bullets go through the fence; only occasionally will one hit a strand of wire and change its direction of flight. This is what happened in Rutherford's experiment. Why?

By 1912, after extensive measurements of deflection angles and the number of alpha particles deflected, Rutherford and two of his students, Ernest Marsden and Hans Geiger, had fit together these facts into a nuclear model of the atom. The way in which this model accounted for experimental observations is suggested in Figure 2–7. The atom, according to this model, is mostly comprised of empty space; this allows the majority of the alpha particles to pass through unhindered. The relatively small, positively-charged nucleus near the center of the atom causes the deflection of the few positively charged alpha particles which pass near the nucleus.

With Marsden and Geiger's measurements, Rutherford was able to calculate the size of the nucleus. The radius of the nucleus was calculated to be about $\frac{1}{10,000}$ the size of the radius of the atom. This means that if the nucleus were expanded to the size of a tennis ball, the "edge" of the atom would be about one half mile away. The charge on the gold nucleus was roughly estimated to be 100 times greater than the charge on a hydrogen nucleus. (Later experiments brought about a more exact value of a charge of 79 on a gold nucleus.)

The Rutherford "gold-foil" experiment and its interpretation is a pivotal point in the development of the atomic theory. Not only was the nucleus "discovered," but it became obvious that the mass of the nucleus was about twice

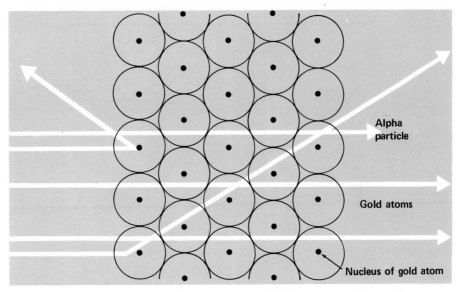

FIGURE 2-7   Rutherford's interpretation of how alpha particles interact with atoms in a thin gold foil. Actually, the gold foil was about 1000 atoms thick. For illustration purposes, points are used to represent the gold nuclei and the alpha particles are drawn larger than scale.

as great as the mass of the number of protons that were required to account for the positive charge on the nucleus. This prompted the search for another nuclear particle, the neutron. The neutron has no charge and has approximately the same mass as a proton, or about 1 amu. The discovery of the neutron is described in Chapter 6.

ELEMENTS AND ATOMIC NUMBERS

The number of protons in the nucleus of an atom characterizes a particular element. There are 105 different elements known today, and each of these elements has a different number of protons in its nucleus. Lead, for example, has 82 protons and gold has 79 protons. The number of protons in the nucleus of an element is called its *atomic number*. A list of atomic numbers for all the elements is given in the table inside the back cover.

Of the 105 elements known, 88 have been isolated in nature; 17 others have been made by the transmutation methods described in Chapter 6. A few of the naturally-occurring elements are listed in Table 2–1.

In chemical jargon, each element is represented by a symbol. A chemical symbol has one or two letters; the first letter is a capital and the second (if present) is lower case. A list of the symbols of the elements is also given in the table inside the back cover. Some symbols are derived from Latin words for the elements, such as Fe, from ferrum, Cu, from cuprum, Sn, from stannum, and K, from kalium.

In addition to the symbol representing the element itself, the symbol can represent the *amount* of the element as either an individual atom or as a mole

TABLE 2-1   SOME COMMON ELEMENTS°

Metals

(A metal is a good conductor of electricity, can have a shiny or lustrous surface, and in the solid form can be generally deformed without breaking.)

NAME	SYMBOL	PROPERTIES OF PURE ELEMENT AT NORMAL TEMPERATURE AND PRESSURE
Iron Latin, ferrum	Fe	strong, malleable, corrodes
Copper Latin, cuprum	Cu	soft, reddish-colored, ductile
Sodium Latin, natrium	Na	light silvery metal, very reactive, soft, low melting point
Silver Latin, argentum	Ag	shiny white metal, relatively unreactive, best conductor of electricity
Gold Latin, aurum	Au	heavy yellow metal, not very reactive, ductile
Chromium	Cr	resistant to corrosion, hard bluish gray

Nonmetals

(A nonmetal is often a poor conductor of electricity, normally lacks a shiny surface, and is brittle in crystal-solid form.)

Hydrogen	H	colorless, odorless (H_2), occurs as a very light gas, burns in air.
Oxygen	O	colorless, odorless gas (O_2), reactive, constituent of air
Sulfur	S	yellow solid (S_8), low m.p., burns in air
Nitrogen	N	colorless, odorless gas (N_2), rather unreactive
Chlorine	Cl	greenish yellow gas (Cl_2), very sharp choking odor, poisonous
Iodine	I	dark purple solid (I_2), sublimes easily

° Chemists usually use the symbol rather than the name of the element. In addition to denoting the element, the chemical symbol has a very specialized meaning which is described later in this chapter. A complete list of the elements with the symbols can be found inside the back cover of this book.

References explaining the historical development of chemical symbols are given in the suggestions for further reading at the end of Chapter 2.

of atoms. A mole is a specific number of things, just as a "dozen" is 12 items. A mole is a very large number; in fact, it is 602,000,000,000,000,000,000,000 (602 sextillion). While this would be a lot of elephants, a mole of atoms is usually a convenient number of atoms for laboratory work. One important way in which the chemist uses the concept of a mole will be described in Chapter 4.

ELECTRON CHARACTERISTICS

The mass of an electron is only 1/1840th of the mass of a proton, or about 0.00055 amu. The charge and mass are the same whether the electron comes from a battery, lightning, an electric power line, beta rays (from radioactive substances), or a cathode-ray tube.

FIGURE 2-8   Las Vegas, Nevada. A Neon Extravaganza, thanks to Cathode Ray Tube Technology. (Thanks to the Las Vegas News Bureau.)

Elements (and consequently atoms) are normally neutral. Since one electron will neutralize the electrical effect of one proton, it will take 79 electrons to neutralize the charges on 79 protons of a gold atom. The atomic number, then, represents not only the number of protons in an atom but also the number of electrons in a neutral atom.

Experiments with cathode-ray tubes, the forerunners of neon and fluorescent lights, led to the discovery of the electron. As early as 1748, scientists were involved in the study of electrical discharges in partially evacuated glass tubes. When approximately 1000 volts are applied to the metal electrodes, the gas in the tube glows very brightly (Figure 2-9). (The reason for this phenomenon will be given later in this chapter.) Through studies between 1870 and 1910 of cathode rays, electrons were discovered and characterized. A few cathode-ray tubes and their uses are described in Figure 2-9.

Regardless of the kind of metal used for electrodes, or the kind of gas used in the cathode-ray tubes, the particles moving from the cathode of the tube always had the same ratio of charge to mass. The ratio of charge of the cathode ray particles divided by the mass of the particle happened to be the one physical property of the particles that could be measured. More than 20 different metals were used with the same results. The simplest way to interpret this observation is to assume that the particles which make up the cathode ray are a part of all matter.

The Irish physicist, Stoney, an associate of Rutherford, suggested in 1891 that the quantity of electricity present in each of the cathode-ray particles be

41

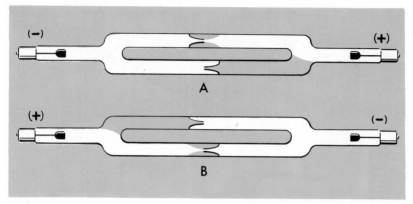

Discharge tube containing small amount of gas to show directional effect of cathode ray. (In A) the glowing gas shows the cathode ray to be "funneled" through the upper branch of the tube; (B) shows the reverse effect of the gas when the position of the negative electrode is reversed. Cathode rays come from the negative electrode.

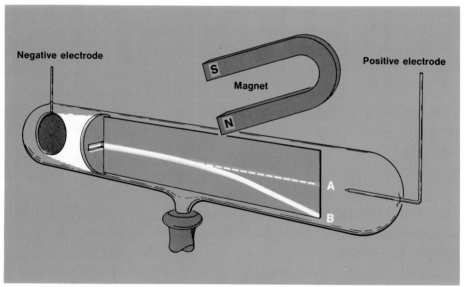

Magnetic deflection of cathode ray from position A to position B when magnetic field is applied. This indicates a negative charge on the moving particles.

FIGURE 2-9   Three cathode-ray tubes and their contributions to the characterization of the electron. (*Continued on opposite page.*)

called an electron. The idea was accepted, and soon the name was applied to the particle itself. For a while there was a move to change the name of the electron to "negatron," but this was never a popular idea.

It will help us to understand such natural phenomena as color, and to rationalize the existence of chemical bonds, if we know something of how electrons are situated about the atom. We will give the fundamentals here and will recall them numerous times in the text to explain chemical phenomena. There are three points to understand.

First, electrons differ in energy, depending upon the different places they occupy in the space around a nucleus.

Second, the electrons which are closest to the nucleus have the lowest total energy of all the electrons in an atom. They are closer to what attracts them—the nucleus. It takes energy for an electron to be moved away from a nucleus, just as it takes energy for us to climb away from what attracts us—the center of earth. The energy put into moving the electron farther from its nucleus is retained in the atom much as a stretched rubber band retains its energy. This means that the electrons in the outermost parts of an atom have the greatest energy just as a person sitting in the top row of seats at a stadium is in a position of higher energy than a person sitting on the first row. Of course, the energy is in the electron-nucleus system, not just in the electron itself.

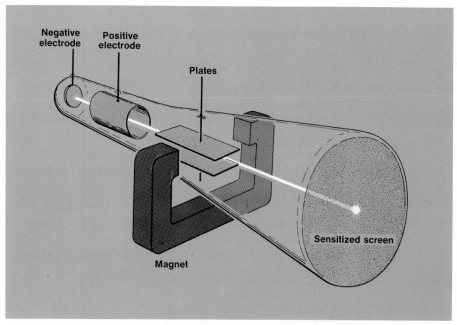

FIGURE 2-9 *(Continued.)* J. J. Thomson experiment. Electric field, applied by plates, and magnetic field, applied by magnet, cancel each other to allow cathode ray (electron beam) to travel in straight line. By means of this tube, the ratio of charge to mass was determined.

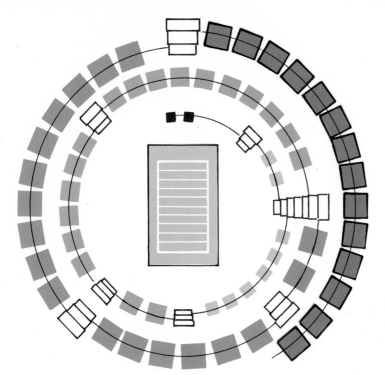

FIGURE 2-10 A stadium has some things in common with the atom as described in the text.

Third, there are discrete units of energy that an electron can have and still maintain stable existence in an atom. This is closely analogous to a book in a bookcase and a spectator in a stadium. A book on a shelf is not likely to move up or down, and in that respect, it is stable. If, for a moment, the book is halfway between one shelf and the next, it is in an unstable position. It is either falling (if it is losing energy), or it is rising (if it has been given energy). A spectator in a stadium is in a stable position (i.e., not likely to go up or down) when seated, but he is not stable when going up or down the steps between tiers. The analogy here is not between the real shelves or tiers, but in the reality of different energy levels existing at various locations and the degree of stability present in them.

Although a very sophisticated theory exists that describes the energy levels of electrons in considerable detail, the main-energy levels will suffice for our purposes. As we go out from the nucleus of an atom, more room is available to electrons. Consequently, each successive main-energy level (moving out from the nucleus), can accommodate more electrons. For example, the first main-energy level can have no more than two electrons, the second main-energy level can have as many as eight, the third, 18, the fourth 32, and so on. The relationship, $2n^2$ (where n is the number of the main-energy level), indicates the maximum number of electrons in a given energy level.

Electrons move about within the atom as the atom absorbs energy (electrons move out) or gives up energy (electrons move in). The arrangement of electrons that is characteristic of a particular atom at its lowest energy level (or content) is called its *ground state*. As electrons in their ground states are added, proceeding from simple electronic structures to complex ones, the first and lowest energy level is filled first ($2n^2 = 2(1)^2 = 2$ electrons), then the second energy level

$(2n^2 = 2(2)^2 = 2 \times 4 = 8$ electrons). There is an exception in the third energy level and each one thereafter. The third energy level can contain $2(3)^2$ or 18 electrons. However, it will only fill to eight electrons before putting two electrons in the fourth level. After the two electrons are put into the fourth level, the third level will add 10 more electrons to achieve its maximum number of 18. The situation becomes a bit more complex as we move into the fourth, fifth, and sixth energy levels, but none of these levels fills to more than eight electrons prior to placing at least two electrons in the next outer energy level.

Now, we shall turn our attention to the structure within specific atoms.

STRUCTURE OF ATOMS

To analyze the structure of an atom, we need to know its atomic number and its mass number. The atomic number, as we have seen, tells us both the number of electrons and the number of protons. The mass number represents the total number of protons and neutrons. Consequently:

Number of neutrons = Mass number—atomic number

The electrons are placed in their orbitals in order, starting with the lowest available orbital. The simplest atom is ordinary hydrogen. Its atomic number is 1, so it has 1 electron and 1 proton. Its atomic mass is 1, so it has no neutrons. In a diagram, this can be shown as:

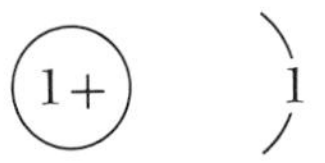

Sodium, Na, has an atomic number of 11 and an atomic mass of 23. Sodium has 11 protons, 11 electrons, and 12 neutrons. An atom of sodium would have the following arrangement:

Chlorine, Cl, has an atomic number of 17 and an atomic mass of 35. A chlorine atom would be arranged like this:

Actually, chlorine has two kinds of atoms; one has a mass of 35 amu and the other a mass of 37 amu. These are *isotopes*. They differ only in the number of neutrons per atom. The 35-amu isotope has already been shown; the 37-amu isotope has this arrangement:

One isotope has 18 neutrons, and the other has 20; this is the only difference between them. The average mass of 35.45 amu in a natural sample of chlorine gas means that there is more of the 35-amu isotope than of the 37-amu isotope.

Electronic structures for argon, Ar, potassium, K, calcium, Ca, and scandium, Sc, illustrate the incomplete filling of the third energy level at the point where electrons are placed in the fourth energy level. The structures are:

Ar   2   8   8

K   2   8   8   1

Ca   2   8   8   2

Sc   2   8   9   2

Recall that the third energy level can have a maximum of 18 electrons, but both potassium and calcium add electrons to the fourth energy level before scandium begins the process of filling the third level to its capacity.

ATOMIC THEORY AT WORK

Throughout this text, we will frequently use the atomic theory to explain various phenomena. Two which we will deal with here are the production of light and laser beams.

Light

According to modern atomic theory, light is emitted when excited electrons move back toward the nucleus into lower energy levels. The sets of energy levels differ by certain discrete amounts of energy called *"quanta."* According to the quantum theory proposed in 1900 by Max Planck (1858–1947; Nobel prize winner in 1918), light and other forms of energy (cosmic, x-ray, heat, sound, electrical, nuclear) are collections of quanta of energy, or *photons*. In the rainbow of colors, the photons decrease in energy content from violet through blue, green, yellow, orange, and red. The particular color emitted in an electron transition depends upon the difference in energy between the two levels. For example, if this energy difference is the same as that of a quantum of blue light, then we see a blue color. Transitions between other energy levels produce other colors.

When a "white light," like sunlight, is passed through a prism or a cloud of raindrops, the light is separated into its assortment of colors, or its *spectrum*. Similarly, if the light from an energized gas is passed through a prism, its light is divided into its own characteristic spectrum. Unlike the sunlight's continuous

Max Planck (1858–1947), a German physicist, explained how energy radiates from a hot object. In 1900 he announced his quantum hypothesis which was later used to explain other natural phenomena by Einstein (photoelectric effect) and Bohr (hydrogen spectral lines). He received the Nobel Prize in 1918.

spectrum (see Color Plate II), the spectrum of an energized gas is a series of separated, distinct lines produced by light of characteristic wavelengths. Each element has its own identifying spectrum, which acts somewhat like a fingerprint for each element. Each line in the spectrum is caused by billions of electrons in billions of "excited" (energized) atoms of the gas making a particular transition toward a lower energy level. The little "bullets" of light gathered together form the ray of colored light.

Lasers

Science-fiction "death ray" guns have a prototype in recently developed light generators called *lasers* (the acronym for **l**ight **a**mplification by **s**timulated **e**mission of **r**adiation). An intense, narrow beam of light is generated in lasers. In a typical light source, the light is emitted in all directions. The laser concentrates the beam in one direction, and all the waves move in unison. All the photons in the emitted light have the same level of energy.

Although laser technology is developing rapidly through the use of many different materials, the principle of the laser can be understood by studying one of the early types. The ruby laser (Figure 2–11), uses a rod machined from a single ruby crystal. Ruby is aluminum oxide in which about 0.05 per cent chromium oxide is distributed. A spiral cathode-ray tube is wrapped around the rod and contains xenon gas. Very intense ultraviolet light from the cathode-ray tube is directed onto the ruby. The chromium atoms in the rod absorb photons from the xenon and, because of a peculiar situation in the crystal, retain most of the quanta. The electrons in a few atoms return to the lowest possible level and in the process emit photons of red light. Because the ends of the rod have been silvered to make them partially reflecting, some of the photons are reflected back through the rod where they encounter the excited chromium atoms, that possess most of the absorbed quanta. This causes all the remaining, excited electrons to make the same jump between energy levels **simultaneously.** If the excited atoms are not stimulated by this reflected light, they maintain their excited level for a few milliseconds before they drop at random to the ground state. The synchronous tumbling of the electrons from the excited state creates an extremely intense beam of photons of the same energy.

47

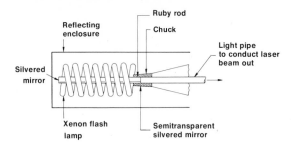

FIGURE 2-11   A simplified scheme of a ruby laser.

The advantages of using an intense beam of energy that can be aimed precisely have stimulated the development of laser technology. Because they are so narrow and do not spread out much with distance from the source, as does ordinary light, laser beams can be reflected from the moon. The laser has been used for cutting holes in diamonds half the diameter of a hair, and for perforating the diamonds that are used for drawing very fine wire. Delicate surgery, such as welding a separated eye retina and chromosome repair, have been accomplished using lasers. A laser eraser has been invented that can correct this mistakes without burning the paper.

QUESTIONS

1. Contrast the Greeks' method and Dalton's method of justifying belief in atoms.

2. Explain what the following terms mean:
   (a) isotopes of an element
   (b) atomic number
   (c) mass number
   (d) an alpha emitter
   (e) element

3. Use a source with atomic numbers and atomic masses to calculate the following:
   (a) the nuclear charge of cadmium, Cd.
   (b) the atomic number of arsenic, As.
   (c) the atomic mass of an isotope of bromine, Br, having 46 neutrons.
   (d) the number of electrons in an atom of barium, Ba.
   (e) the number of protons in an isotope of zinc, Zn.
   (f) the number of protons and neutrons in an isotope of strontium, Sr, with an atomic mass of 88 amu.

4. If the nucleus of a gold atom is 10^{-12} cm in diameter and the diameter of the atom itself is 3×10^{-8} cm, what percentage of the volume of the atom is occupied by the nucleus? The volume of a sphere is $\frac{4}{3}\pi r^3$.

5. If electrons are a part of all matter and electricity is a flow of electrons, why are we not shocked continually by the abundance of electrons around and in us?

6. There are more than 1000 different kinds of atoms and only 105 different kinds of elements. Explain.

7. Protons are to atoms of different elements what _____ are to atoms of different isotopes of the same element.

8. What is a practical application of cathode-ray tubes?

9. How does atomic theory help us to explain

   (a) the generation of light?
   (b) the unique bright-line spectrum for each element?

10. What is a quantum? a photon?

11. Which has more energy per photon, red light or blue light? How is it possible for a beam of red light to contain more energy than a beam of blue light?

12. Write the ground state electronic arrangement of lithium (Li), oxygen (O), cesium (Cs), and arsenic (As).

13. As far as atomic theory is concerned, what is different among the elements?

14. Describe the experiment that led to our concept of a nucleus in the atom.

15. An electron goes from the second energy level to the fifth energy level.

    (a) Does this process require or release energy?
    (b) In this transition, does the electron go toward or away from the nucleus?

SUGGESTIONS FOR FURTHER READING

Reference Works

Ihde, A. J., "The Development of Modern Chemistry," Harper and Row, New York, 1964.
Young, L. B. (Ed.), "The Mystery of Matter," Oxford University Press, New York, 1965.
Jones, M. M., et. al., "Chemistry: A Brief Introduction," W. B. Saunders, Philadelphia, 1969.

Paperbacks

Birks, J. B. (Ed.), "Rutherford at Manchester," W. A. Benjamin, Inc., New York, 1963. (A memorable account of Rutherford's work with articles written by Bohr, Marsden, and others. Also contains reprints of Rutherford's original papers.)
Greenaway, F., "John Dalton and the Atom," Cornell University Press, Ithaca, New York, 1966.
Hochstrasser, R. M., "Behavior of Electrons in Atoms," W. A. Benjamin, Inc., New York, 1964.
Hoffman, B., "The Strange Story of the Quantum," Dover Publications, Inc., New York, 1959.
Jaffe, B., "Crucibles: The Story of Chemistry," Fawcett World Library, New York, 1957.
Lucretius, "The Nature of the Universe," Penguin Books, Inc., New York, 1959.
Moore, R., "Niels Bohr: The Man, His Science and the World They Changed." Alfred A. Knopf, Inc., New York, 1966.

Articles

"Atomic Theory in the Ancient World," *Chemistry*, Vol. 44, No. 4, p. 17 (1971).
Dingal, G. P., "The Elements and the Derivation of Their Names and Symbols," *Chemistry*, Vol. 41, No. 2, p. 20 (1968).
Milliken, Robert, "Electrons—What They Are And What They Do," *Chemistry*, Vol. 40, No. 4, p. 13 (1967).
Szabadvary, F., "Great Moments in Chemistry, Part II: From Thales to Bohr," *Chemistry*, Vol. 42, No. 11, p. 6 (1969).
Wallace, H. G., "The Atomic Theory—A Conceptual Model." *Chemistry*, Vol. 40, No. 10, p. 8 (1967).

3

ATOMS INCOGNITO— COMPOUNDS, MOLECULES, AND IONS

COMPOUNDS AND MIXTURES

Matter is usually found in nature in the form of mixtures. Unique groupings of elements, called *compounds*, are mixed with other compounds or sometimes with a particularly unreactive element. Earth, dust, rock, air, rivers, oceans, food, plants, trees, and you and I are mixtures of many different compounds and elements. The world, in this respect, is like a bargain counter of bead necklaces (Figure 3–1). The whole counter is a mixture of unique groupings of beads (the

FIGURE 3–1   A bargain counter of bead necklaces has groupings and individual units analogous to mixtures, elements, compounds, atoms, and molecules.

necklaces), and perhaps a few individual beads from a broken strand. The individual beads are the atoms in the analogy, and the necklaces are the molecules (a definite number of atoms tied together to form a distinct and separate portion of matter). Necklaces made from the same kind of beads would be molecules of elements. Hydrogen, oxygen, fluorine, nitrogen, and chlorine molecules contain two atoms each. Sulfur molecules ordinarily consist of eight sulfur atoms tied together; phosphorus molecules usually have four phosphorus atoms per molecule. Necklaces made from different kinds of beads are analogous to molecules of compounds. Just as a necklace contains a definite number of red beads, blue beads, round beads, square beads, and so on, a molecule contains a definite number of atoms (for example, two hydrogen atoms and one oxygen atom per molecule of water). For two necklaces to be identical, they must contain the same number and kinds of beads strung in the same order. Every molecule of a given compound contains the same number and kind of atoms arranged in identical order. Of course, the array of atoms, molecules, and ions in nature is more complex than the counter display, but the idea is basically the same.

To make an automobile, or baking soda, or an aspirin tablet, certain compounds have to be extracted from the mixture we call Earth. Sometimes we have to go back further and separate the elements from their compounds. The separation of mixtures into compounds and then into elements is similar to separating the mixture of necklaces on the bargain counter. Very little energy is required to separate the different kinds of necklaces into matching piles. This represents the process of purifying mixtures into separate compounds. It takes more energy to break the strands of necklaces, and to make separate piles of the various kinds of beads. This is analogous to the extra energy it takes to separate compounds into individual elements.

Several ways of separating mixtures are illustrated in Figure 3–2. None of these methods normally separates compounds into elements. A compound can be separated from a mixture if at least one of its properties differs from the properties of the other compounds in the mixture. For example, if the boiling point of the compound differs significantly from the boiling points of the other substances in the mixture, distillation (heating, vaporization, and condensation, with gradually increasing boiler temperatures) will usually separate it. Very pure water is removed from salty ocean water by this process. If one of the compounds is much less soluble in a liquid than the other compounds present, chilling, or evaporation of solvent, followed by filtration might be appropriate. Silt in water or salt from ice water can be removed by filtration through a bed of gravel. Recrystallization depends upon one of the compounds being less soluble, on a percentage basis, in a cold solvent than the other compounds in the mixture. The mixture is dissolved in the hot solvent, and upon cooling the desired nearly-pure compound crystallizes out in solid form. Chromatography depends upon the different tendencies compounds have for sticking to a stationary material as a liquid or gas carries the mixture through or over the stationary substance.

How do we know when we have a pure compound or element, as opposed to a mixture? A substance is pure when, after repeated efforts at purification, the properties of the substance no longer change appreciably (or as far as we

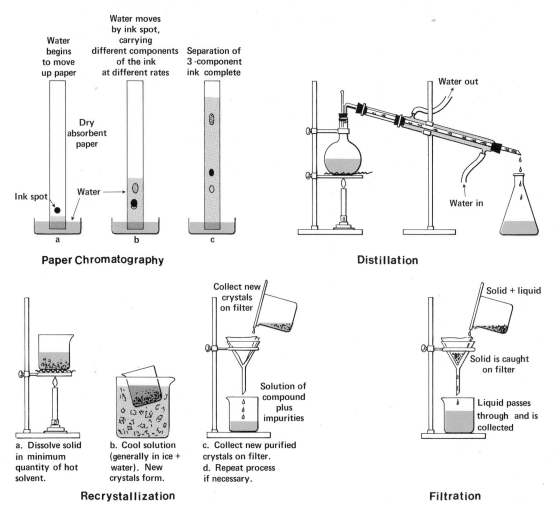

Paper Chromatography

Distillation

Recrystallization

a. Dissolve solid in minimum quantity of hot solvent.

b. Cool solution (generally in ice + water). New crystals form.

c. Collect new purified crystals on filter.
d. Repeat process if necessary.

Filtration

FIGURE 3-2   Four methods of purifying mixtures of elements and compounds. **Paper chromatography.** Owing to the absorbent character of paper, water moves against gravity and carries the ink dyes along its path. If the ink dyes move at different rates, they will be separated in the developed chromatogram. **Distillation.** Sodium chloride dissolves in water to form a clear solution. When heated above the boiling point (indicated by thermometer), water will pass into vapor and into the condenser. Cool water injected into the glass jacket of the condenser circulates over the inner tube causing the steam to liquefy and collect in the receiving flask. In this simple example pure water collects in the receiving flask while the salt remains in the boiling flask. **Recrystallization.** This can be used to separate some solid mixtures. **Filtration.** The separation of a solid from its mixture with a liquid by filtration.

can measure them). The melting point of a solid and the boiling point of a liquid are generally the first criteria to be checked to ascertain the purity of a compound. If the melting point (or boiling point) and **all** other properties remain the same after several attempts at purification, we conclude that the compound may be pure. A compound is recognized by its set of properties and is distinguished from other compounds just as you are recognized by a set of characteristics that distinguish you from other people.

A second distinction between pure compounds and mixtures is that compounds have a definite percentage by weight of each element. This percentage does not change in the process of purification. For example, every sample of *chemically pure* water is 88.8 per cent oxygen by weight and 11.2 per cent hydrogen by weight. The original source of the water makes absolutely no difference. It could come from the great Salt Lake, the Pacific Ocean, the polluted Hudson River, the Okefenokee Swamp, melted snow in Siberia, or a tear; the percentages by weight of hydrogen and oxygen in samples of *purified* water from these sources are always 11.2 and 88.8 per cent, respectively. If another substance has a different ratio of the same elements, it must be a different compound. Hydrogen peroxide, for example, is also a compound of hydrogen and oxygen. The percentages in *pure* hydrogen peroxide are a constant 5.9 per cent hydrogen and 94.1 per cent oxygen by weight.

The constant proportion of weights of elements in compounds was noted in the study of chemistry by Joseph Louis Proust (1754–1826) in 1799. He summarized the general principle that we now know as the *Law of Definite Proportions*, which states that the weight percentage of the elements in a compound is fixed.

The Human Side

Joseph Louis Proust

Proust was a chemist who was associated with such notables as Charles IV of Spain, Napoleon, John Dalton, and Claude Louis Comte Berthollet, an eminent chemist and medical doctor. Young Joseph was raised in an atmosphere pervaded with chemistry, since his father was a pharmacist. During the political upheaval of the French Revolution, Proust lived and worked in Madrid as a chemist under the sponsorship of Charles IV, the King of Spain. When Napoleon's army ousted Charles IV, Proust's laboratory was looted and he lost his position. He returned to France and lived out his life in retirement. Napoleon later offered him a grant to enable him to continue his research, but he was in poor health and declined.

Proust's work which eventually led to the Law of Definite Proportions involved a painstakingly careful analysis of copper carbonate and a number of other compounds. Regardless of how copper carbonate was prepared in the laboratory or how it was isolated from nature, it always contained 5 parts of copper to 4 parts of oxygen to 1 part of carbon by weight. Berthollet (1748–1822) believed that the nature of the final product was determined by the amount of reacting materials one had at the beginning. The running controversy between Proust and Berthollet reached epic proportions. Proust won the argument by showing that Berthollet had made inaccurate analyses and had insufficiently purified his compounds—two great errors in chemistry.

For most substances, Proust's generalization has been verified many times since its formulation. There are, however, several compounds (e.g., copper sulfide) with variable compositions. These are called Berthollide or nonstoichiometric compounds. For most substances, though, the Law of Definite Proportions holds true.

John Dalton saw a very logical way to use his atoms to explain the Law of Definite Proportions. The definite proportions by weight in a compound, according to Dalton, result from the constant ratio of atoms in a compound. For example, carbon monoxide is 42.9 per cent carbon and 57.1 per cent oxygen by weight. If a sample of carbon monoxide is composed of billions of identical molecules each containing one atom of carbon tied to one atom of oxygen, the ratio of the weights of carbon and oxygen in the compound must be the same as that of carbon and oxygen atoms. The atomic mass of C is 12.0 and the atomic mass of O is 16.0. If there are equal numbers of atoms of the two elements in carbon monoxide, the percentage of C by weight must be:

$$\%C = \frac{12}{12 + 16} \times 100\%$$

$$\%C = 42.9\%$$

Formulas from Per cent Analysis and Atomic Masses

The per cent analysis of a relatively pure sample of silicon dioxide (sand is mostly silicon dioxide) is 46.7 per cent silicon and 53.3 per cent oxygen. What is the simplest ratio of silicon atoms to oxygen atoms in a sample of silicon dioxide, or, put another way, what is the chemical formula of silicon dioxide?

The percentage composition of a compound tells us that if we had, for example, 100 grams of silicon dioxide, we should have 46.7 grams of silicon and 53.3 grams of oxygen. A 100 mass unit of any kind would give the same ratio of 46.7 units of silicon to 53.3 units of oxygen. Hence, in 100 amu of silicon dioxide, we should have 46.7 amu of silicon and 53.3 amu of oxygen. If we divide these masses by the atomic masses, we obtain an atomic ratio:

$$?\text{atoms Si} = 46.7 \text{ amu} \times \frac{1 \text{ atom Si}}{28.1 \text{ amu}}$$

$$= 1.66 \text{ atoms Si}$$

$$?\text{atoms O} = 53.3 \text{ amu} \quad \frac{1 \text{ atom O}}{16.0 \text{ amu}}$$

$$= 3.34 \text{ atoms O}$$

The ratio of Si atoms to O atoms in a sample of silicon dioxide is 1.66 to 3.34.

To change this ratio to its equivalent form in whole numbers, we divide each number in the ratio by the smaller number and (if necessary) convert the decimal fractions to whole numbers:

$$\frac{1.66 \text{ atoms Si}}{1.66} = 1$$

$$\frac{3.34 \text{ atoms O}}{1.66} = 2$$

The formula for silicon dioxide is SiO_2. This is called the simplest, or empirical, formula since the ratio of atoms is reduced to the smallest whole number ratio. In the same way the simplest formula for sucrose is $C_{12}H_{22}O_{11}$.

The smallest possible amount of sucrose would have 12 atoms of carbon (C), 22 atoms of hydrogen (H), and 11 atoms of oxygen (O). However, the simplest formula for dextrose (blood-sugar) is CH_2O and benzene is CH, whereas the "true" molecular formulas representing total numbers of atoms of each element in the molecules are, respectively, $C_6H_{12}O_6$ and C_6H_6. In a similar manner, the smallest building blocks of some compounds are *molecules* and in others, the basic units are two or more ions (charged atoms or groups of atoms). The distinction will be made later in this chapter. H_2, N_2, Cl_2, N_2O_4, P_4O_{10}, C_6H_6, and B_2H_6 are molecular formulas, whereas the simplest formulas would be H, N, Cl, NO_2, P_2O_5, CH, and BH_3, respectively.

A formula for a compound represents at least three things: (a) a compound of the indicated composition; (b) a molecule of that compound; or (c) a mole of such molecules. The mass of a molecule can be obtained by simply adding up the masses of the atoms within the molecule. For example, the mass of a molecule of carbon dioxide is 44 amu:

$$1 \text{ atom of C} \times \frac{12 \text{ amu}}{\text{C atom}} = 12 \text{ amu}$$

$$2 \text{ atoms of O} \times \frac{16 \text{ amu}}{\text{O atom}} = 32 \text{ amu}$$

$$\overline{}$$

$$\text{Molecular mass} \quad 44 \text{ amu}$$

Similarly, the molecular mass of H_2SO_4 is the sum of 2 amu for hydrogen (2 atoms $\times$ 1 amu/atom), plus 32.1 amu for sulfur (1 atom $\times$ 32.1 amu/atom), plus 64.0 amu of oxygen (4 atoms $\times$ 16.0 amu/atom) for a total of 98.1 amu. The weight of a mole of H_2SO_4 molecules would be 98.1 grams.

THE PERIODIC TABLE AND COMPOUND FORMATION

The periodic table is helpful to the student of chemistry because it organizes many of the facts and principles that must be used. It is the actual **arrangement** of the elements in the table which contributes most significantly to our knowledge of the elements.

When the elements are arranged in the order of their atomic numbers, a trend in properties occurs that is periodic; that is certain properties are repeated over and over again. This is the *periodic law*. If the elements are arranged in the order of their atomic numbers beginning with hydrogen, a remarkable repetition of chemical properties occurs. Following consecutively the series of Li, Be, B, C, N, O, F, and Ne, is the series of Na, Mg, Al, Si, P, S, Cl, and Ar. What is remarkable about these two series is the coincidence of chemical properties between elements that fall in the same vertical column. For example, notice the similarities in the formulas of the compounds formed with chlorine:

LiCl	$BeCl_2$	BCl_3	CCl_4	NCl_3	Cl_2O	ClF
NaCl	$MgCl_2$	$AlCl_3$	$SiCl_4$	PCl_3	SCl_2	ClCl

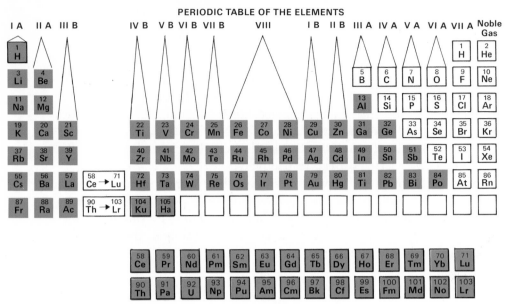

FIGURE 3-3   The periodic table.

Notice that the pattern of 1, 2, 3, 4, 3, 2, 1 chlorine atoms in combination with atoms of the other elements is the same in the two series. This is what we mean by "periodic." Each *period* (horizontal row) shows a trend in properties that repeats itself in the next period.

Each of the elements in a *group* (vertical column) of the periodic table forms compounds of the same general formula as, for example, the chlorides of group II: $BeCl_2$, $MgCl_2$, $CaCl_2$, $SrCl_2$, $BaCl_2$, and $RaCl_2$.

Elements which are found beneath each other in the periodic table usually have the same number of electrons in the highest energy level of each element. All Group IA elements have one high-energy electron; IIA elements have two high-energy electrons; IIIA elements have three high-energy electrons; and so on. For the A groups in the periodic table, the number of the group is the number of highest-energy electrons for each element in the group. The similarity in atomic structure matches the corresponding similarity in chemical properties of the elements in groups.

Dmitri Ivanovich Mendeleev (1834–1907), a brilliant Russian chemist, established the framework of the periodic table. He used atomic weights (masses) to order his elements, **except** where the order conflicted with a similarity in the properties of two elements in a group. In those cases, Mendeleev chose similarity of properties as the true criterion. He was so convinced of the periodic nature of the combining power of elements that he was courageous enough to leave gaps in his table, claiming that the table was not wrong, but that the elements had just not been discovered yet. (Only 63 elements had been discovered when Mendeleev published his first chart in 1869.)

Not only did Mendeleev leave gaps in his chart, but he also predicted the properties of the missing elements based on property-similarities and progressive differences. These predictions also gave clues as to where to search for the "new" elements. This stimulated a flurry of "prospecting" for elements in the 1870's and 1880's. As a result, scientists discovered gallium in 1874, scandium, samarium, holmium, and thulium in 1879, and germanium and dysprosium in 1886. Mendeleev had predicted the properties of gallium, scandium, and germanium, and the properties of the elements fit his predictions closely.

Mendeleev's first periodic table was published in 1869, just a few months before a German chemist, Lothar Meyer (1830–1895), published a similar table based primarily on the periodicity of atomic volumes. Although Meyer is often given partial credit for the discovery of the periodic table, he did not predict undiscovered elements.

The Human Side

Dmitri Ivanovich Mendeleev

In the January 7, 1871 issue of the *Journal of the Russian Chemical Society,* Mendeleev advanced the notion that undiscovered elements would fill in the gaps in his table. This was a remarkable prediction, and equally remarkable is the speed with which the scientific world heard about it. The work of several earlier Russian chemists had been published in the Russian language, which only a few scientists could read. Mendeleev's works were translated immediately into German for all his contemporaries to read.

Mendeleev was the youngest in a family of 14 or 17 children (the records are not clear) and lived in an obscure town in Siberia—certainly not known for intellectual advances—until his late teens. The odds were tremendous against him becoming the most famous chemist in the world, which he indeed became during the 1880's. In 1906, just a few months before his death, he almost received the Nobel Prize in chemistry, missing by one vote. However, his memory will be perpetuated in element 101, prepared synthetically in 1955, and named mendelevium.

Dmitri Mendeleev (1834–1907). Born in Siberia, Mendeleev rose to Professor of Chemistry at St. Petersburg (now Leningrad) and then to director of the Russian bureau of Weights and Measures. Although a prolific writer, a versatile chemist and inventor, and a popular teacher, the fame of this brilliant scientist rests on his discovery of the periodic law.

WHAT HOLDS COMPOUNDS TOGETHER?
THE CHEMICAL BOND

Whatever it may be that holds matter together, we are fairly certain now that it is not the miniature screen-door hooks and eyes proposed by Dalton. All of the evidence we have suggests that there are two methods of holding atoms together. Atoms are either held together by the attraction between unlike charges on charged atoms (ions) or they are joined by the attraction and sharing of negative electrons between two positive centers. In either case, the attraction one atom has for another is electrical and is called a "chemical bond."

IONIC BONDS—UNLIKE CHARGES ATTRACT

How can electrically neutral atoms bond to each other? One way is for electrons to be shifted from one atom to another, producing charged atoms held together by the attraction between their unlike charges. The electrons most readily transferred are those far away from the nucleus in the "peel" of the atom.

The "peel" electrons are called the valence (from Latin *valentia*, meaning strength) electrons. These are the electrons in the highest energy level. For example, sodium has one valence electron (third level), and carbon has four (second level) valence electrons. The valence electrons are the logical ones to be transferred, since they are farthest from the nucleus. We must now ask ourselves questions such as: which kinds of atoms lose electrons? how many do they lose? what happens to the electrons that are lost?

A guiding principle that will help answer these questions becomes apparent when we examine a group of "antisocial," gaseous elements—the noble gases. You are probably familiar with the helium and neon of this group. Helium is used to fill weather balloons and dirigibles, and neon is used in "neon" lights. Argon, krypton, xenon, and radon are less well-known members of the noble gas family of elements. Until recently, compounds of noble gases were unknown. Since 1962, a dozen or so compounds have been made, including XeF_4, XeF_6, $XeOF_4$, XeO_3, and KrF_2, but attempts to prepare stable compounds with He,

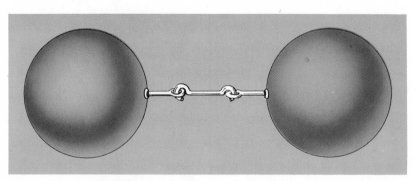

FIGURE 3-4   John Dalton's idea of a chemical bond.

Ne, and Ar, have been unsuccessful so far. By any criterion, this family of elements is relatively inert compared to the other 82 natural elements.

The inertness of the noble gases must be related to their electronic structure. When the electronic arrangements of the noble gases are written out, two features are apparent!

$$
\begin{array}{llllll}
\text{He} & 2 \\
\text{Ne} & 2 & 8 \\
\text{Ar} & 2 & 8 & 8 \\
\text{Kr} & 2 & 8 & 18 & 8 \\
\text{Xe} & 2 & 8 & 18 & 18 & 8
\end{array}
$$

First, all lower energy levels containing any electrons at all are filled to capacity. Second, except for He, the highest numbered energy level contains 8 electrons. These electronic arrangements appear to be particularly stable. Perhaps if other elements achieved these electronic structures, they would also be stable chemically. This seems to be the case for a large group of elements.

The first group (or alkali metal) elements, Li, Na, K, Rb, and Cs, have one more electron than do the noble gases He, Ne, Ar, Kr, and Xe, respectively. If each of these metals loses an electron from each atom, it would have the noble gas electronic arrangement. For example, Li has 3 electrons in the ground-state arrangement:

The loss of one electron gives Li the electronic structure of He. An Li atom with only 2 electrons and 3 protons will have a charge of 1^+. A "charged atom," such as Li^+, or a charged group of atoms such as the sulfate group (SO_4^{2-}), is called an *ion*.

For Mg, Ca, Sr, and Ba to take on the noble gas structure, the atoms of each element must lose 2 electrons each. The loss of 2 electrons would leave 2 protons unneutralized, so each ion would have a charge of 2^+.

As more and more electrons are removed from a single atom, it becomes more difficult to remove the next one. The net positive charge builds up with the loss of each electron and this charge helps to hold the remaining electrons more securely. The difficulty of removing successive electrons from the same atom is emphasized in the ionization energies in Table 3–1. The ionization energy is the quantity of energy required to completely remove an electron from an atom.

The table of ionization energies also indicates how many electrons are relatively easy to remove from an atom. Note that Li, Na, and K have relatively low ionization energies for removal of the first electron, but require a big jump in energy for removal of the second. For Be, Mg, and Ca, the big jump is between the second and third ionization energies. This means that two electrons are relatively easy to remove, but the third is in a set of lower energy and requires considerably more energy to remove it. The electronic structures are consistent with the big jumps in ionization energies. For example, the electronic structure of Ca is

Ca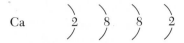

The removal of two electrons wipes out the occupancy of the fourth main energy level. To remove the third electron requires the break-up of pairs of electrons in a lower, main energy level. This takes considerably more energy, and this is exactly what a comparison of the ionization energies of calcium tells us.

The table also shows that electrons are very difficult to remove from elements like oxygen, fluorine, chlorine, and nitrogen. These elements can attain noble gas electronic configurations by *gaining* electrons. For example, F, Cl, Br, and I need one electron each to have the electronic arrangement of Ne, Ar, Kr, and Xe, respectively. An excess of one electron per atom produces a 1^- ion, as in the Cl^- ion. O, S, and Se need to pick up two electrons each to achieve the electronic structure of Ne, Ar, and Kr. An excess of two electrons per atom produces a 2^- ion.

The more electrons added to an atom the more difficult it becomes for each succeeding one to enter. Each electron that comes into an atom after the first

TABLE 3-1   IONIZATION ENERGIES OF SELECTED GASEOUS ATOMS

An electron volt (ev) is the energy acquired by an electron when accelerated by a potential difference of 1 volt. For each element, electrons must be removed to the heavy vertical line in order to attain a noble gas electronic configuration.

ATOMIC NUMBER	ATOM	IONIZATION ENERGIES (ev)							
		1st	2nd	3rd	4th	5th	6th	7th	8th
1	H	13.6							
2	He	24.6	54.4						
3	Li	5.4	75.6	122.4					
4	Be	9.3	18.2	153.9	217.7				
5	B	8.3	25.1	37.9	259.3	340.1			
6	C	11.3	24.4	47.9	64.5	392.0	489.8		
7	N	14.5	29.6	47.4	77.5	97.9	551.9	666.8	
8	O	13.6	35.1	54.9	77.4	113.9	138.1	739.1	871.1
9	F	17.4	35.0	62.6	87.2	114.2	157.1	185.1	953.6
10	Ne	21.6	41.1	64	97.2	126.4	157.9		
11	Na	5.1	47.3	71.7	98.9	138.6	172.4	208.4	264.2
12	Mg	7.6	15.0	80.1	109.3	141.2	186.9	225.3	266.0
13	Al	6.0	18.8	28.4	120.0	153.8	190.4	241.9	285.1
14	Si	8.1	16.3	33.5	45.1	166.7	205.1	246.4	303.9
15	P	10.6	19.7	30.2	51.4	65.0	220.4	263.3	309.3
16	S	10.4	23.4	35.0	47.3	72.5	88.0	281.0	328.8
17	Cl	13.0	23.8	39.9	53.5	67.8	96.7	114.3	348.3
18	Ar	15.8	27.6	40.9	59.8	75.0	91.3	124.0	421
19	K	4.3	31.8	46	60.9	—	99.7	118	155
20	Ca	6.1	11.9	51.2	67	84.4	—	128	147

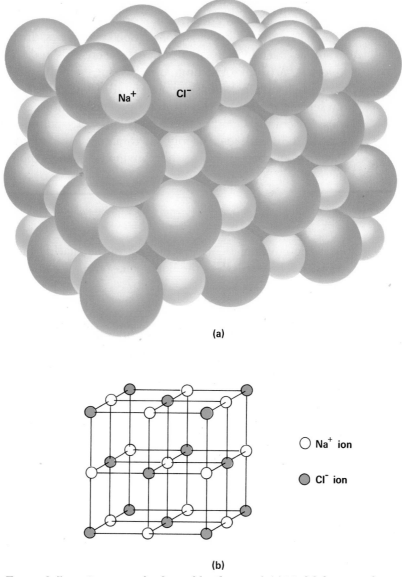

Na^+ Cl^-

(a)

$\bigcirc$ Na^+ ion

$\bullet$ Cl^- ion

(b)

FIGURE 3-5   Structure of sodium chloride crystal, (a) Model showing relative sizes of the ions; (b) ball-and-stick model showing cubic geometry.

must enter against an existing net negative charge. This factor prevents addition of more than three electrons. In ordinary chemical systems, there are no stable 4^-, 5^-, or 6^- simple ions.

With positive and negative ions available, it is very easy to see how a crystal of table salt is held together. The Na^+ ions and Cl^- ions are held in place by electrical attraction. Furthermore, the ratio of sodium ions to chloride ions must be one-to-one if the compound is to be neutral. The simplest formula is found to be NaCl, so the theory is consistent with observations.

The crystalline structure of NaCl is shown in Figure 3–5. Note that each sodium ion is attracted by all six chloride ions around it. Similarly, each chloride

61

ion is attracted by all six sodium ions around it. There is **no unique molecule** in ionic structures; no particular ion is attached exclusively to another ion.

When atoms become ions, their properties are drastically altered. For example, a collection of bromine molecules is red, but bromide 1^- ions are colorless. A chunk of sodium atoms is soft, metallic, and violently reactive with water, but sodium 1^- ions are nonmetallic and stable in water. A large collection of chlorine molecules constitutes a greenish-yellow, poisonous gas, but chloride ions are colorless and nonpoisonous. In fact, sodium and chloride ions in the form of table salt can be put on tomatoes without fear of a violent reaction. When atoms become ions, they obviously change their nature. Ions are atoms incognito.

THE COVALENT BOND

The covalent bond is achieved by the **sharing** of electrons, rather than by transfer. The strength of the bond comes from the nuclei of both atoms attracting the same electrons.

Sharing of Electrons

When two atoms with rather high or nearly equal ionization energies are bonded together, there is little tendency to transfer electrons. Neither atom has the ability to extract an electron from the other. Hence, the atoms may share electrons, forming what is known as a "covalent bond." Substances whose molecules contain nonmetals only are generally bonded covalently. Examples include carbon dioxide (CO_2), table sugar ($C_{12}H_{22}O_{11}$), DNA (deoxyribonucleic acid—our heredity material), and diamond (all C).

The simplest example of electron sharing is found in diatomic molecules of gases such as H_2, F_2, and Cl_2. A hydrogen atom has only one electron. To achieve the helium structure, the hydrogen atom must somehow acquire another electron because He has two electrons. An electron pair can be achieved if two hydrogen atoms share their electrons. If dots are used to represent valence electrons and the element symbol to represent the rest of the atom, the sharing of electrons can be represented by placing the two electrons between the symbols:

$$H:H$$

Since both atoms have a mutual interest in the pair of electrons, the two nuclei are tied together and a chemical bond is formed.

Sharing electrons often, but not always, leads to species in which each element has a noble gas electronic structure. Except for H, this means that there are four pairs of electrons in the valence shell. If we limit the electron-dot representations to include only the valence electrons, and place the shared pairs of electrons between the symbols, the F_2 molecule would be shown as having one pair of shared electrons:

$$2 \; :\!\ddot{F}\!\cdot \; \longrightarrow \quad :\!\ddot{F}\!:\!\ddot{F}\!:$$

Fluorine molecule

Formula	Name	Electron Dot Structure

Single Bonds:

H_2O	Water	H H:O:̈

NH_3	Ammonia	H:N̈:H H

C_2H_6	Ethane	H H H:C:C:H H H

Double Bonds:

CO_2	Carbon dioxide	:Ö::C::Ö:

C_2H_4	Ethylene	H⠀⠀⠀⠀H :C::C: H⠀⠀⠀⠀H

SO_3	Sulfur trioxide	:O: S :O: :O:

Triple Bonds:

N_2	Nitrogen	:N:::N:

CO	Carbon monoxide	:C:::O:

C_2H_2	Acetylene	H:C:::C:H

FIGURE 3-6   Electron-dot structures of some molecules containing single, double, and triple bonds.

Similarly, the nitrogen molecule N_2, would have three pairs of shared electrons:

$$2 \; \cdot \ddot{N} \cdot \longrightarrow \quad :N \colon\colon N \cdot$$

Nitrogen molecule

The pairs of electrons which are not included between the symbols are called "nonbonding electrons." The electron-dot structures of several other molecules are shown in Figure 3–6.

_____

The Human Side

Archibald Couper

In 1858, Archibald Couper (1831–1892) suggested using a dash to represent the chemical bond. This was before electrons were discovered. (The dash has now come to represent an electron pair.) Couper, a Scottish chemist, was a victim of the competitiveness of science. As a child, he had delicate health, and most of his education was at home. He is known for a single scientific paper he wrote in which he paralleled some of August Kekulé's thinking on the bonding in benzene (described later in this chapter). There was some controversy with Kekulé on who had published first. Actually, Couper's paper was more thorough and penetrating. Nevertheless, Kekulé's paper was accepted by his peers. Shortly afterward, Couper suffered first a nervous breakdown, then sunstroke. His scientific career was over before he was 30, although he lived on for 30 years more.

Using a line to represent a pair of bonding (shared) electrons, F_2 and N_2 would be represented as:

$$F\text{---}F \qquad N\equiv N$$

The F_2 molecule is bonded by a single bond and the N_2 molecule by a triple bond.

The rule that there are eight electrons in the valence shell of a bonded atom (often called the octet rule) is not a hard and fast one; indeed, many compounds exist which do not have an octet of electrons around one of the atoms in the molecule. For example, boron trichloride, BCl_3, is a stable compound although boron has only three pairs of electrons in its valence shell.

$$
\begin{array}{cc}
\ddot{:}\!\ddot{Cl}\!\ddot{:} & \\
:\ddot{Cl}\!:\!B & :\ddot{Cl}\!-\!B \\
:\ddot{Cl}\!: &
\end{array}
$$

The fundamental premise of chemical bonding is the tendency for electrons to pair and not the invariable grouping of 8 electrons. In fact, Gilbert Newton Lewis (1875–1946), who is generally credited with developing the octet rule, realized the limitations of his theory when he wrote in 1923, that "the electron pair, especially when it is held conjointly by two atoms, and thus constitutes the chemical bond, is the essential element in chemical structure."

POLAR BONDS

In a molecule like H_2 or F_2, where both atoms are alike, there is equal sharing of the electron pair in the covalent bond. Where two different atoms are bonded together, it is likely that there will be an unequal sharing of the electron pair. The shared pair of electrons is shifted toward the more *electronegative* (electron attracting) element. In any one period of the periodic table, an element with the smallest atom needing the fewest electrons to pair its electrons and complete the bonding shell is the most electronegative element. Electronegativity generally decreases as one moves down or to the left on the periodic table (Figure 3–3).

In a covalent bond between two different atoms, the more electronegative

FIGURE 3–7 Fluorine pulls the negative electron blanket towards itself and exposes the nucleus of hydrogen. The shifting of electrons towards one atom forms a polar bond.

element pulls the electron blanket toward itself. Polar bonds are a type of "intramolecular ionization." The negative portion of the bond is in the region of the more electronegative atom and the positive portion is in the region of the less electronegative atom. For example, in the molecule, HF, the bonding pair of electrons is controlled more by the fluorine atom than by the hydrogen atom (Figure 3–7). Part of the negative "blanket" of charge has been pulled off the hydrogen atom, partially uncovering the positive hydrogen nucleus.

Polar bonds fall between the extremes of pure covalent and ionic bonds as far as separation of charge is concerned. In a pure covalent bond, there is no charge separation; in ionic bonds there is complete separation, and in polar bonds the separation falls somewhere in between.

HYDROGEN BONDS

When hydrogen is attached to a highly electronegative atom like fluorine, oxygen, or nitrogen, the conditions are right for a very important type of inter-

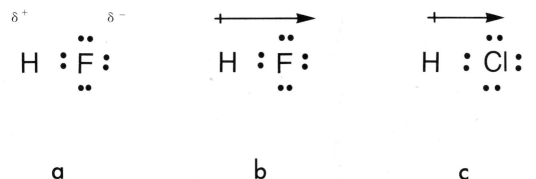

FIGURE 3–8 Polar bonds in HF and HCl molecules. (a) (delta plus-fractional plus charge) and (delta negative-fractional negative charge) are used to indicate poles of charge. In (b) and (c), an arrow is used to indicate electron shift, the arrow having a plus tail to indicate partial positive charge on the hydrogen atom. Note that the longer arrow in HF structure indicates a greater degree of polarity than in HCl.

FIGURE 3–9   Solid paraffin (on the left) sinks in its liquid while solid water (ice, on the right) floats in its liquid. Why the difference? The answer is on the next page.

molecular, positive-to-negative attraction called *"hydrogen bonding"*. The bond is produced by the attraction arising between a slightly positive hydrogen atom on one molecule and an electronegative atom on another molecule. The shifting of an electron pair toward nitrogen, oxygen, or fluorine causes these atoms to take on a partial negative charge.

The hydrogen bond is a "bridge" between two molecules (or between two locations on the same molecule if the molecule is big enough to bend back on itself). Its strength is about one-tenth to one-fifteenth that of an ordinary covalent bond.

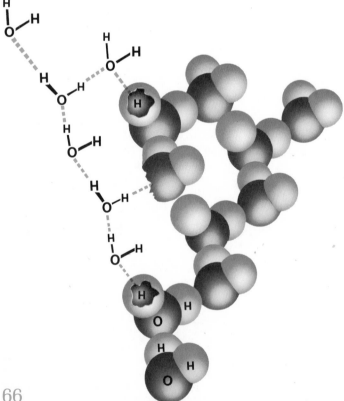

FIGURE 3–10   Hydrogen bonding in the structure of ice.

Hydrogen bonds are found in many substances in the world around us. They are responsible for such phenomena as hard candy getting sticky, cotton fabrics taking longer to dry than nylons, lanolin softening skin, the ultimate shape of proteins and enzymes, and a host of apparent anomalies in the nature of water. Each phenomenon is dealt with in this text, but we shall examine here some of the "strange" things about water that result from H-bonding.

One of the peculiar properties of water is that its solid form, ice, floats. In contrast, most solids will sink in their own liquid form. It is fortunate that ice floats. If ice sank to the bottom of ponds and rivers, the water would continue to freeze all winter until the pond or river became solid. Besides having a devastating effect on aquatic life, the water cycle would be diminished, the lack of water would destroy land plant life, and we would be in severe difficulty.

The reason why ice floats is that water expands on freezing. When rapidly moving water molecules slow down as freezing occurs, **all** water molecules "grab hands" in a vast, interlocking network. This opens up the structure and makes ice less dense. To illustrate, imagine a group of people in a crowded room milling around almost shoulder to shoulder. Now imagine that they were asked to join hands as their hands hung from their sides at a 45° angle. The crowd would become less dense as it assumed this expanded structure, just as water does when it becomes hydrogen-bonded during freezing. Do you think the people could push out the walls of a room just as the freezing of water will burst an iron pipe?

Another "strange" property of water is its relatively high boiling point. Almost all the hydrogen compounds of oxygen's neighbors and family members are gases at room temperature: CH_4, NH_3, H_2S, H_2Se, H_2Te, PH_3, HCl. But H_2O is a liquid. To transform a molecule into the vapor state, the molecule must absorb energy to free itself from other molecules. Since a water molecule is hydrogen-bonded to other water molecules even at the boiling temperature, it must acquire enough energy to break these bonds before it can be set free. (A halfback with three or four tacklers hanging on to him has a similar problem in freeing himself.) When water reaches a temperature of 100°C (212°F), the hydrogen bonds break sufficiently to allow single H_2O molecules to leave the liquid and go into the vapor state. If strong intermolecular bonding, such as hydrogen bonding, is absent, substances generally boil according to their molecular masses. The larger molecular masses require a higher temperature to boil. This is largely explained by the greater number of protons and electrons in the higher molecular-weight molecules that attract electrons and protons, in that order, in other molecules. It is not related to the effect of gravity (weight) on the molecules.

GEOMETRY OF MOLECULES

It is not enough to know that one atom is attached to another, but we should also know their relative position—how the atoms are arranged in space. This is what we mean when we talk about the structure of a substance. The geometry and composition of a molecule determine its structure.

A simple scheme called the "electron pair repulsion theory" enables us to predict the placement of atoms in a molecule.

To understand how atoms are arranged in space about other atoms, all we have to do is imagine that electron pairs behave like charged ping-pong balls tied to a central point. The like charges on the balls would make them repel each other so that they move as far away from each other as possible. The arrangements shown in Figure 3–11 would result if two, three, four, or six ping-pong balls (or electron pairs) were bonded to a central point (atom). When all electron pairs are bonding-electron pairs (that is, two atoms are attached to the pair of electrons), then the structure predicted by the ping-pong balls is very close to the structure of the molecule. Examples of these are $BeCl_2$ (linear), BCl_3 (triangular), $SiCl_4$ (tetrahedral), and SF_6 (octahedral). Electron dot figures show two, three, four, and six pairs of bonding electrons around the central atom in these molecules, respectively. The three atoms of $BeCl_2$ lie along a straight line, and the structure of BCl_3 is triangular, as predicted.

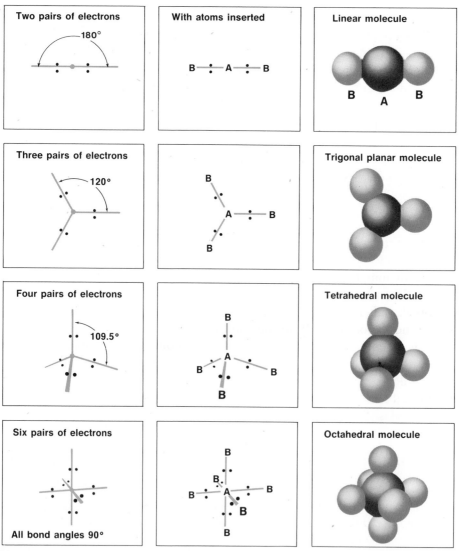

FIGURE 3–11   Arrangements of electron pairs according to the electron-pair repulsion theory.

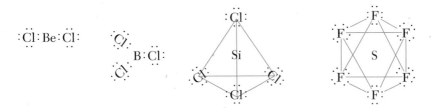

For the $SiCl_4$ molecule, the structure is tetrahedral, and for SF_6, theory and experimental evidence agree again on the octrahedral structure for this molecule. These geometrical figures are readily visualized by looking at Figure 3–11.

At this point you may wonder how the presence of *nonbonding pairs* of valence electrons affect bonding pairs. In brief, a pair of electrons occupies somewhat more volume when it is not involved in bonding. This causes the bonding pairs to move closer together.

Nonbonding pairs of electrons help to explain the structures of ammonia (NH_3) and water (H_2O), shown in Figure 3–12. The bond angles of 106.5° in NH_3 and 104.5° in H_2O are slightly less than the tetrahedral angle of 109° 28′ because the lone, nonbonding pairs of electrons exert more repulsion than the bonding pairs do. Part of the negative charge on the bonding pairs of electrons is diminished by the positive hydrogen nuclei.

The electron-pair repulsion theory also accounts for the structure of organic compounds. Methane, CH_4, has the predicted tetrahedral structure for four atoms attached to a central atom. Ethane, C_2H_6, also has a tetrahedral structure around each carbon atom because each carbon atom has four pairs of bonding electrons (and four atoms) about it (see Figure 3–6). Diamond also is structured tetrahedrally (Figure 3–13). The tying together of all carbon atoms by strong covalent bonds gives diamond its hardness.

Ethylene, C_2H_4, (Figure 3–6) has a double bond and three atoms bonded to each carbon atom. The prediction is a trigonal planar structure around each carbon atom with bond angles of 120°. The double bond squeezes the H—C—H angle slightly so that the angles are not quite 120°.

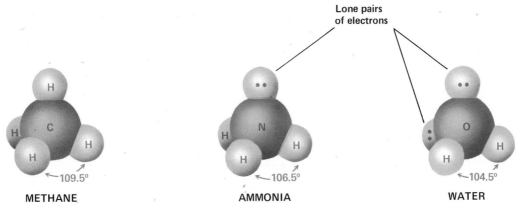

FIGURE 3-12   Those little molecules in water, natural gas (methane) and ammonia have something in common. Methane has only bonding pairs of valence electrons: water and ammonia have bonding and non-bonding pairs of valence electrons. The non-bonding pairs distort the tetrahedral structure slightly.

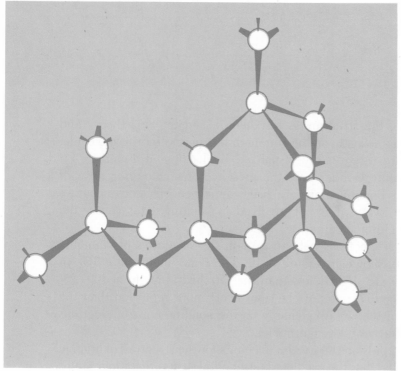

FIGURE 3-13   The tetrahedral arrangement of the carbon atoms in the diamond crystal. Each atom has four nearest neighbors, which are arranged about it at the corners of a regular tetrahedron.

Acetylene, C_2H_2 (Figure 3–6), has a triple bond. Each carbon atom has two groups attached to give a linear structure.

In benzene, C_6H_6, the carbon atoms are joined in a ring with three atoms bonded to each carbon atom. Benzene has a planar structure with 120° bond angles.

If the bonds were alternately single and double around the ring, the molecule would be lopsided because double bonds hold the atoms closer together than do single bonds. Actually, all experimental evidence indicates that the ring is symmetrical. There are no alternating single and double bonds, but rather, each C—C bond is a compromise between a single and double bond. These bonds are intermediate in strength and in length. This structure is often represented

70

by a regular hexagon enclosing a circle, where it is understood that a carbon atom and a hydrogen atom are positioned at each corner.

QUESTIONS

1. What is a chemical bond?

2. When you look at a crystal of table salt, what do you "see"?

3. When you look at a lump of sugar, what do you "see"?

4. One use of calcium chloride is to keep down dust on a road. It is an ionic compound with the formula, $CaCl_2$. Tell what this formula means.

5. Octane is a component of gasoline which, if not burned, is emitted into the atmosphere. It is a covalent compound and has the formula C_8H_{18}. Tell what this formula means.

6. Is Ca^{3+} a possible ion under chemical conditions? Why?

7. Write the electronic configuration (arrangement) for the element potassium (atomic number 19). What will be the electronic configuration when a K^+ ion is formed?

8. How many electrons does a fluorine atom have in its valence shell? How many electrons of fluorine will be involved in bonding?

9. Draw the electron-dot structure for water. Based on bonding theory, why is water's formula not H_3O?

10. Draw electron dot structures for the following molecules:

    (a) NF_3 (c) C_2Cl_2 (e) H_2S (g) N_2H_4
    (b) CCl_4 (d) OF_2 (f) CO (h) CH_3OH

11. The members of the nitrogen family, N, P, As, and Sb, form compounds with hydrogen, NH_3, PH_3, AsH_3, and SbH_3. The boiling points of these compounds are:

$$SbH_3 \quad -17°C$$
$$AsH_3 \quad -55°C$$
$$PH_3 \quad -87°C$$
$$NH_3 \quad -33°C$$

Why doesn't NH_3, the lightest molecule in this list, boil at the lowest temperature?

12. Predict the general kind of chemical behavior (that is, loss, gain, or sharing of electrons) you would expect from atoms with the following electronic configurations:

    (a) 2-8-1
    (b) 2-8-7
    (c) 2-4

13. Show how two fluorine atoms can form a single covalent bond, using electron-dot structures.

14. If covalent bonding is to white as ionic bonding is to black, polar covalent bonding would be represented by _____. On this scale, how could a more polar molecule be distinguished from one that is less polar?

15. Use your chemical intuition and suggest a compound that might be formed by combining boron trichloride, BCl_3, and ammonia, H—N—H.
    |
    H

16. In *Suggestions for Further Reading*, the article "Structures of Noble Gas Compounds" discusses the structure of XeF_6. How many electron pairs would Xe have in its valence shell when bonded to six fluorine atoms? According to electron pair repulsion theory, would this number of electron pairs predict an octahedral structure for the XeF_6 molecule?

17. Name two elements that are found in nature in a relatively pure state.

18. Aluminum does not "rust" and crumble like iron because aluminum forms a thin coating of aluminum oxide on its surface which prevents further combination with oxygen. Use atomic theory to predict the formula for the ionic compound formed between aluminum and oxygen.

19. The first four ionization energies of a certain element are 7.3, 12.8, 78.6, and 135.4 electron volts. How many electrons in the valence shell of this element? In what group in the periodic table is this element located?

20. The smallest basic and stable particle of a simple ionic compound is called a(n) _____ . The smallest basic and stable particle of a covalent compound is called a(n) _____ .

21. Explain why atoms in Groups I and II of the periodic table readily form positive ions, while atoms in Groups IV and V do not.

22. A hypothetical element has the symbol J and six valence electrons. $: \overset{\displaystyle ..}{J} \cdot$

   (a) Write the formula for the expected compound when this element combines with hydrogen.
   (b) If element J were to form an ion, what would be the charge on this ion?
   (c) Element J would be in what group of the periodic table?
   (d) Write the electron-dot formula for the compound formed between J and chlorine.
   (e) What would be the expected structure of the compound between J and Cl?

23. Determine the simplest formulas for compounds with the following percentage by weight:

   (a) 64.86% C, 13.51% H, and 21.63% O.
   (b) 27.48% Mg, 23.66% P, and 48.85% O.
   (c) 38.79% Cl, and 61.21% O.
   (d) 22.9% Na, 21.5% B, and 55.7% O.

SUGGESTIONS FOR FURTHER READING ▬▬▬▬▬▬

Reference Works

Ihde, A. J., "The Development of Modern Chemistry," Harper and Row, New York, 1964.
Pauling, L., "The Nature of the Chemical Bond," Cornell University Press, Ithaca, New York, 1960.

Paperbacks

Hochstrasser, R. M., "Behavior of Electrons in Atoms," W. A. Benjamin, Inc., New York, 1964.
Lagowski, J. J., "The Chemical Bond," Houghton Mifflin Co., Boston, 1966.
Rich, R., "Periodic Correlations," W. A. Benjamin, Inc., New York, 1965.
Ryschkewitsch, G. E., "Chemical Bonding and the Geometry of Molecules," Reinhold Publishing Corp., New York, 1963.

Articles

Benfey, T., "Geometry and Chemical Bonding," *Chemistry*, Vol. 40, No. 5, p. 21 (1967).
Bent, H. A., "The Tetrahedral Atom, Part I: Enter the Third Dimension," *Chemistry*, Vol. 39, No. 12, p. 8 (1966).
Bent, H. A., "The Tetrahedral Atom, Part II: Valence in Three Dimensions," *Chemistry*, Vol. 40, No. 1, p. 8 (1967).
Eyring, H., and MuShik J., "Significant Structure Theory of Water," *Chemistry*, Vol. 39, No. 9, p. 8 (1966).
Greenwood, N. N., "Chemical Bonds," *Education in Chemistry*, Vol. 4, p. 164 (1967).

Kauffman, G. B., "American Forerunners of the Periodic Law," *Journal of Chemical Education,* Vol. 46, No. 3, pages 128 (1969).

Lambert, J. B., "The Shapes of Organic Molecules," *Scientific American,* Vol. 222 No. 1, p. 58 (1970).

Luder, W. F., "Electron Repulsion Theory," *Chemistry,* Vol. 42, No. 6, p. 16, (1969).

"Structures of Noble Gas Compounds," *Chemistry,* Vol. 39, No. 4, p. 17 (1969).

"The Geometry of Liquids," *Chemistry,* Vol. 38, No. 12, p. 20 (1965).

"The Periodic Table, 1869–1969," *Chemistry,* Vol. 42, No. 5, p. 26 (1969).

van Spronsen, J. W., "The Priority Conflict between Mendeleev and Meyer," *Journal of Chemical Education,* Vol. 46, No. 3, p. 136 (1969).

Webb, V. J., "Hydrogen Bond 'Special Agent'" *Chemistry,* Vol. 41, No. 6 p. 16 (1968).

Zimmerman, J., "Mendeleev—His Own Man," *Chemistry,* Vol. 42, No. 11, p. 32 (1969).

4

AS SURE AS DEATH
AND TAXES—
CHEMICAL CHANGE

Change goes on in nature at every moment. Green plants absorb carbon dioxide and water and make sugar and oxygen every second somewhere on earth. Compounds taken from coal and petroleum are changed by chemists into almost a million other compounds. We use some of these, such as plastics, medicine, clothing, food, and many other consumer products, every day. Deterioration and weathering relentlessly change our world (Figure 4–1). The changing of substances (elements and compounds) into other substances is *chemical* change. This is where the action is in chemistry. This is where atoms change partners or lose their identity as ions.

Figure 4-1   Chemical changes produce new substances, often with properties very different from the starting material. The strong steel from which this automobile fender was made has been converted by oxidation to powdered rust. The dollar loss from rusting in the United States is slightly over $30.00 per person per year.

The millions of different chemical changes would make understanding modern chemistry a bewildering, indeed a hopeless, task if we studied the reactions one by one. Fortunately, there are a few guiding principles that can make our task much easier. In this chapter we shall try to define these principles.

REACTANTS BECOME PRODUCTS

When a piece of wood is burned, what happens to it? The two most obvious developments are the heat given off and disappearance of the wood. These represent two of the most fundamental characteristics of chemical change.

> The initial materials are consumed.
> Energy in some form is either given off or absorbed.

In many chemical reactions, the consumption of the initial materials is not as obvious as in the burning of wood. The cooking of an egg or the souring of milk are chemical changes, but it is not entirely obvious that the initial materials have been consumed. The materials, however, have been consumed, and in their place different substances have been formed. In this respect a chemical reaction is like an hour glass containing sand that trickles through it in a given amount of time. At time zero, the upper bulb is filled. As time passes, the sand in the upper bulb empties into the lower bulb. Eventually the sand passes completely into the lower bulb, unless a clod stops up the hole. In a chemical reaction, the reactants (initial materials) change into products. In the process of change, the amount of the reactant diminishes by the same amount that the quantity of product increases.

If the sand were changed into silicon and oxygen as it passed through the hole in the hour glass, the analogy would better approximate a chemical change. Not only do reactants disappear and products form, but there is a definite difference between the materials at the start of the process and those obtained from the reaction. Sometimes we can detect the changes that accompany reactions with our senses; we can see colors changing and precipitates forming, or we may be able to smell gases that are evolved. In other reactions we have to rely on measurements of melting points, boiling points, densities, or any of a host of other properties to show that new substances have been formed.

Returning to the burning of wood, it is fairly obvious that new substances are formed when wood burns. The solid residue, consisting of charcoal and ash, looks quite different from wood. Anything else that was formed must have been in the form of colorless gases because we did not see them. If we run the evolved gases over a reagent that absorbs water (such as calcium chloride), we can show that they contain water vapor. If we include a reagent which is able to absorb carbon dioxide, such as calcium hydroxide, we find that carbon dioxide is one of the products of burning wood.

If we burn the principal component of natural gas, methane (CH_4), in plenty of air, the products are only carbon dioxide and water. There is no ash or solid residue. In this chemical change the methane and the oxygen of the air are consumed, and carbon dioxide and water are formed. We can summarize this in an equation which identifies the products and the reactants:

Methane + oxygen *yields* carbon dioxide + water + heat

MATTER IS CONSERVED IN CHEMICAL CHANGES

Consider the chemical change illustrated in Figure 4–2. You know that a solution of sodium chloride, NaCl (table salt), dissolved in water is clear and colorless. Similarly, a solution of silver nitrate dissolved in water is clear and colorless. When the two solutions are mixed, a white, insoluble substance is produced. The substance is the compound silver chloride, or AgCl. Silver chloride is obviously different from the reactants, sodium chloride and silver nitrate, from the standpoint of solubility. Obviously, a chemical change has occurred, but there is a more basic point to be emphasized here. *The weight of the system does not change during the course of the chemical change.* The weight at the start is the same as the weight at the end of the change.

It was this general type of experimentation that led Antoine Laurent Lavoisier (1743–1794) to the idea that matter is neither created nor destroyed during chemical change. Lavoisier weighed the reactants and products in such chemical changes as the decomposition of mercury oxide (HgO) into mercury and oxygen. Very accurate measurements of the weights of products and reactants showed no change in mass during the course of the reaction.

The Human Side

Antoine Laurent Lavoisier

Lavoisier was born of a wealthy family and was given an excellent education. His father, a lawyer, hoped his son would follow him in that profession, but Lavoisier chose to study and practice chemistry. From the beginning of his chemical researches, he recognized the importance of accurate measurement, especially of weights. Others before him devoted themselves to measurement, but it was Lavoisier who convinced chemists of its importance. It is largely for this reason that Lavoisier is often called the "Father of chemistry."

An examination of all chemical reactions that lend themselves to the kind of quantitative study carried out by Lavoisier led to the *Law of Conservation of Mass.*

Matter is neither lost nor gained during an ordinary chemical reaction.

Dalton's concept of atoms was based on his explanation of the Law of Conservation of Mass. If atoms are indestructible, as Dalton suggested, the appearance of new substances in chemical changes is simply the result of re-arrangement of the atoms. The total weight of the products would have to be

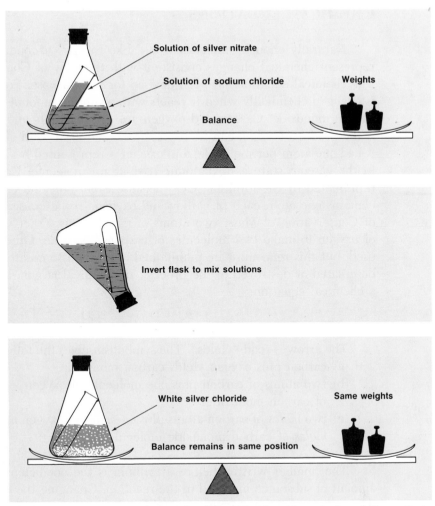

Solution of silver nitrate

Solution of sodium chloride

Weights

Balance

Invert flask to mix solutions

White silver chloride

Same weights

Balance remains in same position

FIGURE 4-2   Mixing a solution of sodium chloride with a solution of silver nitrate produces a new substance, insoluble silver chloride, but the total weight of the matter remains the same.

exactly the same as the weight of the reactants since the same atoms are involved. For example, if a child uses all his blocks in building first a fort, and then reassembles them to form a bridge, the two displays would necessarily weigh the same (See Figure 4–3).

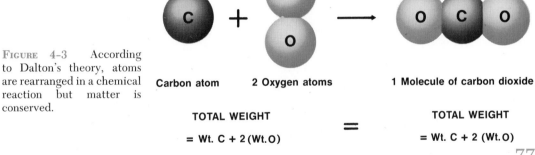

FIGURE 4-3   According to Dalton's theory, atoms are rearranged in a chemical reaction but matter is conserved.

Carbon atom   2 Oxygen atoms

1 Molecule of carbon dioxide

TOTAL WEIGHT

= Wt. C + 2 (Wt. O)

=

TOTAL WEIGHT

= Wt. C + 2 (Wt. O)

77

CHEMICAL EQUATIONS

Naturally enough, chemists would like to have a concise, simple way to represent chemical changes consistent with the Law of Conservation of Mass. The chemical equation has been devised for this purpose. For example, carbon is changed chemically when it reacts with a limited amount of oxygen to form carbon monoxide. Carbon and oxygen are the reactants and carbon monoxide is the product. Carbon can be represented, like most solid elements, as though it had one atom per molecule, and oxygen is represented by "O_2" since it exists in the gaseous state as a diatomic (two-atom) molecule. Carbon monoxide is represented by the formula "CO" because each molecule of carbon monoxide contains one atom each of carbon and oxygen. To be consistent with the Law of Conservation of Mass, two atoms of carbon must react with one molecule of oxygen to make two molecules of carbon monoxide. Other amounts can be used, but this *ratio* must be maintained, otherwise some atoms would have to be created or destroyed in the process. In chemical jargon, this is expressed in a chemical equation:

$$2C + O_2 \longrightarrow 2CO$$

The arrow is read "yields." The symbolism gives the following information:
(a) carbon plus oxygen yields carbon monoxide;
(b) two atoms of carbon plus one molecule of oxygen yield two molecules of carbon monoxide;
(c) two *moles* of carbon atoms plus one *mole* of oxygen molecules yield two *moles* of carbon monoxide molecules.

The number written before a formula is the coefficient, which gives the amount of substance involved in the reaction. Changing the coefficient changes the amount of element or compound involved, whereas changing the subscript would necessarily involve changing from one substance to another. For example, 2CO means either two molecules of carbon monoxide or two moles of CO. However, CO_2 is not carbon monoxide at all; it is another compound of these two elements, called carbon dioxide.

To be consistent with the Law of Conservation of Mass, the mass of reactants must equal the mass of the products. If we add up the atomic masses expressed by each side of the equation, we see the equality:

$$2C \qquad + \qquad O_2 \longrightarrow 2CO$$

$$2 \times 12.0 \text{ amu} + 1 \times 32.0 \text{ amu} = 2 \times 28.0 \text{ amu}$$

$$56.0 \text{ amu} = 56.0 \text{ amu}$$

It is rather obvious that we could use any unit of mass (or weight) in a chemical equation as long as we keep the same numbers. For example, 24 pounds of carbon plus 32 pounds of oxygen will make 56 pounds of carbon monoxide. The unit, pounds, does not change the ratio of the masses of reactants. Tons, grams, pounds, ounces, or any other unit of mass (or weight) can be used in chemical equations as long as the coefficients are multiplied by the formula weights.

As an example of the quantitative use of equations, let us calculate the weight of rocket fuel, liquid hydrogen, that can be obtained from 1.00 ton of water. The balanced equation (no atoms lost or gained) would be:

$$2H_2O \longrightarrow 2H_2 + O_2$$

This equation states that 36 weight units of water (2×18.0) will yield four weight units of hydrogen (2×2.0). We conclude then that we will get four tons of hydrogen from each 36 tons of water used $\left(\dfrac{4 \text{ tons hydrogen}}{36 \text{ tons water}} \right)$. Any weight of water would yield a proportional amount. Hence, one ton of water would yield:

$$1 \text{ ton water} \times \frac{4 \text{ tons hydrogen}}{36 \text{ tons water}} = 0.11 \text{ ton hydrogen}$$

You can imagine how helpful the weight relationships in a chemical equation are to the purchasing agent of a chemical industry. Suppose you are in business to make the detergent, sodium lauryl sulfate. It is prepared from lauryl alcohol, sulfuric acid, and sodium hydroxide as follows:

$$\underset{\textit{Lauryl alcohol}}{C_{12}H_{25}OH} + \underset{\textit{Sulfuric acid}}{H_2SO_4} \longrightarrow \underset{\textit{Lauryl sulfuric acid}}{C_{12}H_{25}OSO_3H} + H_2O$$

$$C_{12}H_{25}OSO_3H + \underset{\textit{Sodium hydroxide}}{NaOH} \longrightarrow \underset{\textit{Sodium lauryl sulfate}}{C_{12}H_{25}OSO_3Na} + \underset{\textit{Water}}{H_2O}$$

In this series of two equations you can see that it takes one formula unit of lauryl alcohol to make one formula unit of sodium lauryl sulfate. In order to turn out 15 tons of detergent a day, how much lauryl alcohol would you need for a day's work? If you add up the formula weights, that of lauryl alcohol is 186.0 and that of sodium lauryl sulfate is 288.1. This means that 186.0 tons of lauryl alcohol will produce 288.1 tons of detergent; therefore, the amount of lauryl alcohol required for each day's operation would be 9.7 tons.

$$\text{Weight of lauryl alcohol} = 15 \text{ tons detergent} \times \frac{186.0 \text{ tons lauryl alcohol}}{288.1 \text{ tons detergent}}$$

$$= 9.7 \text{ tons lauryl alcohol}$$

The purchasing agent can rely on this figure as the minimum amount of lauryl alcohol he should have for a full day's operation. He may also need to estimate an additional amount for waste.

(Other problems of this type are presented and explained in Appendix C. The problems presented here and in the Appendix assume that the reactants change *completely* into products.)

CHEMICAL EQUILIBRIUM

Some chemical reactions appear to stop while some of each reactant still remains. The amount of reactant and product no longer changes with time, and it appears that the reaction has stopped. However, not all activity has stopped, as we shall see later. When the *amount* of reactant and product does not change further, we say that we have arrived at *chemical equilibrium*.

At the beginning of a reaction, we usually are dealing only with reactants. As the reaction proceeds, the amount of reactants decreases while the amount of products increases, until the reaction arrives at equilibrium. At equilibrium there is still some of each kind of reactant remaining in the presence of products. The amount of products may be greater than, equal to, or less than the amount of reactants.

As an example of chemical equilibrium, consider a laboratory experiment in which N_2O_4 (dinitrogen tetroxide) is the reactant and NO_2 (nitrogen dioxide) is the product, and use the equation:

$$N_2O_4 \rightleftharpoons 2NO_2$$

(The double arrow means that the reaction is reversible.) Let us examine what happens if the reaction is carried out in a suitable container in which one mole of N_2O_4 is present initially, as is shown in Figure 4–4. We can follow the progress of the reaction by measuring the color of the mixed gases, since NO_2 is brown and N_2O_4 is colorless. The intensity of the color is thus a measure of the amount of NO_2 present. The brown color eventually develops and then reaches a steady

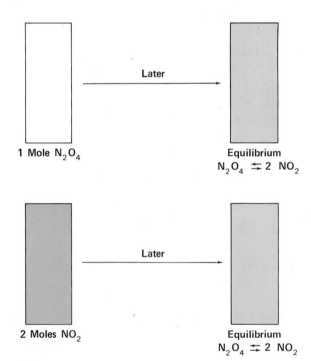

1 Mole N_2O_4

Later

Equilibrium
$N_2O_4 \rightleftharpoons 2\ NO_2$

2 Moles NO_2

Later

Equilibrium
$N_2O_4 \rightleftharpoons 2\ NO_2$

FIGURE 4-4   The equilibrium between N_2O_4 and NO_2 can be approached from either direction because the reactions are easily reversible. The intensity of the color indicates the amount of NO_2 present in the system.

intensity in the reaction vessel. Since the intensity does not change beyond this point, we might be inclined to believe that the reaction was completed. However, if we test the system, we find the vessel contains a mixture of NO_2 and unreacted N_2O_4. Since some N_2O_4 remains, and since the amount of N_2O_4 and NO_2 does not change with time, we say that the reaction has reached a state of equilibrium.

If the experiment is changed by starting with two moles of NO_2, rather than with one mole of N_2O_4, the same shade of brown will eventually appear, provided, of course, that container size and temperature are the same (Figure 4–4). This emphasizes the fact that chemical equilibrium can be attained by approaching from either direction and that chemical reactions are generally reversible.

Further investigations indicate that chemical equilibrium is dynamic in the sense that molecular transformations continue to occur. Although the relative amount of reactants and products does not change with time, experimental studies indicate that product molecules are indeed reverting to reactants and that reactant molecules are becoming products. Both processes, however, occur at the same rate so that the overall amount of products and reactants remains the same. To illustrate, consider the reaction, $N_2 + 3H_2 \rightleftharpoons 2NH_3$. If we allow nitrogen, hydrogen, and ammonia to come to equilibrium, then remove some of the hydrogen (H_2) and replace it with exactly the same number of molecules of deuterium (D_2), we can show the dynamic nature of equilibrium. (Deuterium is a heavier isotope of hydrogen, with an atomic mass of 2 amu.) After a few moments, we analyze the reaction mixture. We find the species HD, NH_2D, NHD_2, and ND_3, as well as the H_2, N_2, and NH_3 present in the original equilibrium mixture. The only way to explain this assortment of molecular species is to assume that hydrogen (D_2 and H_2) is reacting with nitrogen to form ammonia while simultaneously decomposing into nitrogen and hydrogen. This kind of result can be generalized into a fundamental principle of chemical equilibrium.

At equilibrium, the rate of formation of products from reactants is equal to the rate at which products revert back to reactants.

It is the dynamic aspect of chemical equilibrium that we associate with the usual meaning of equilibrium—balance. In chemical equilibrium, there is a balance (or equality) between the forward and reverse-reaction rates. This balance in rates gives rise to constant, relative amounts of reactants and products.

Understanding the principle of equilibrium is important in the industrial laboratory. The chemical engineer needs to know how to shift the equilibrium so that he can obtain more product and reduce his costs. For example, many medicines are made by a series of chemical reactions, each of which proceeds to equilibrium. This means that some of the reactant is not used, and this is reflected in the cost of the product. The job of the engineer is to arrange reaction conditions in order to achieve the greatest conversion.

EFFECT OF TEMPERATURE ON REACTION RATE

The question of how fast or how slowly chemical reactions proceed is one which can be answered quantitatively by considering reaction rates. The rate

FIGURE 4-5   The biochemical processes of decomposition occur more rapidly at higher temperatures. Half the peach shown in this photograph was refrigerated, while the other half was kept warm. The refrigerated half on the right shows little discoloration, while the other one shows the typical sign of decay.

of a reaction is always defined in terms of change in the amount of chemical substances present per unit of time. Thus, if we consider the burning of sulfur to produce sulfur dioxide,

$$S + O_2 \rightarrow SO_2$$

we can discuss the rate of the reaction in terms of the amount of SO_2 formed per minute, or the amount of S or O_2 consumed per minute.

It is possible to alter the rate of a chemical reaction in a predictable manner by changing the temperature. If the temperature is reduced, the rate is almost always decreased. We make use of this principle in cooking foods (a roast will cook at a faster rate at a higher temperature) and in preserving foods (foods spoil less quickly if refrigerated). Figure 4–5 illustrates the effect of temperature on the biochemical reactions of decomposition that take place in fruit.

According to the kinetic molecular theory, when temperature is increased

FIGURE 4-6   Effect of concentration on reaction rate. Steel wool held in the flame of a gas burner is oxidized. It is in contact with air, which is 20 per cent oxygen. When the red hot metal is placed in pure oxygen in the flask, it oxidizes much more rapidly.

molecules move faster. This means that the molecules collide more often, giving them more chances to react. Furthermore, the greater amount of energy loosens up the bonds slightly (which gives the molecules more activation energy), so that when they collide there is, again, a greater chance for reaction. Indeed, the increased amount of energy available at high temperatures is the major factor in increasing the rate of reaction.

EFFECT OF CONCENTRATION ON REACTION RATES

It is possible to alter the rate of a reaction by changing the concentration of the reactants. For example, in the reaction of sulfur and O_2 given previously, if air replaces the oxygen the reaction will proceed at a slower rate since air is one–fifth oxygen and four–fifths inactive nitrogen. Figure 4–6 contrasts the effects of the oxidation of iron when oxygen occurs in relatively small and in relatively large concentrations.

REACTIONS BY CHEMICAL GROUPS

When a survey is made of the known types of chemical reactions, the diversity illustrated by the following examples is found.

$$C + 2F_2 \longrightarrow CF_4$$
Carbon   Fluorine   Carbon tetrafluoride

$$PCl_3 + 3H_2O \longrightarrow H_3PO_3 + 3HCl$$
Phosphorus   Water   Phosphorous   Hydrogen
trichloride   acid   chloride

$$SO_3 + H_2O \longrightarrow H_2SO_4$$
Sulfur   Water   Sulfuric acid
trioxide

$$C + 2S \longrightarrow CS_2$$
Carbon   Sulfur   Carbon disulfide
vapor

$$Fe_2O_3 + 3CO \longrightarrow 2Fe + 3CO_2$$
Iron (III)   Carbon   Iron   Carbon dioxide
oxide   monoxide

These reactions indicate the wide variety of reactions and substances encountered in chemistry. Indeed, the number of known compounds is in the millions, and the number of reactions by which they are produced is in the multimillions. Becoming familiar with chemical reactions is not, however, as hopeless as it may seem at first. There is order to be found in the apparent chaos. Although there seems to be little order in the reactions mentioned here, consider the following reactions:

$$2Na + F_2 \longrightarrow 2NaF$$
Sodium   Fluorine   Sodium fluoride

$$2Na + Cl_2 \longrightarrow 2NaCl$$
Chlorine   Sodium chloride

$$2Na + Br_2 \longrightarrow 2NaBr$$
Bromine   Sodium bromide

$$2Na + I_2 \longrightarrow 2NaI$$
Iodine   Sodium iodide

Note that the group of elements, fluorine, chlorine, bromine, and iodine, all react with sodium in much the same way; that is, products with similar formulas result. Once we have learned how sodium reacts with one of the elements in Group VIIA of the periodic chart, we have some idea of what is likely to happen with other members in the group. This illustrates that the chemical behavior of elements in the same group in the periodic table is similar.

Another such family (Group VIA) includes the group of elements oxygen, sulfur, and selenium, all of which react with calcium in much the same way.

$$2Ca + O_2 \longrightarrow 2CaO$$
Calcium   Oxygen   Calcium oxide

$$2Ca + 2S \longrightarrow 2CaS$$
Sulfur   Calcium sulfide

$$2Ca + 2Se \longrightarrow 2CaSe$$
Selenium   Calcium selenide

Obviously, the study of chemistry is made easier by the grouping of elements and their reactions.

Not only do elements react as members of groups but groups of atoms, such as the SO_4^{2-} ion, may act as a single unit. Consider the following:

$$Mg + H_2SO_4 \longrightarrow MgSO_4 + H_2$$
Magnesium   Sulfuric   Magnesium   Hydrogen
acid   sulfate

$$Ca + H_2SO_4 \longrightarrow CaSO_4 + H_2$$
Calcium   Calcium
sulfate

$$Sr + H_2SO_4 \longrightarrow SrSO_4 + H_2$$
Strontium   Strontium
sulfate

$$Ba + H_2SO_4 \longrightarrow BaSO_4 + H_2$$
Barium   Barium
sulfate

Numerous examples of polyatomic ions are encountered in the study of chemistry. Again, the study is made easier when large groups of facts can be grouped and remembered in terms of a single unifying characteristic.

ENERGY CHANGES IN REACTIONS

When water is decomposed into hydrogen and oxygen, it takes 136,600 calories° to decompose two moles of water molecules.

$$2H_2O(\text{liquid}) + 136,600 \text{ cal.} \longrightarrow 2H_2(\text{gas}) + O_2(\text{gas})$$

When methane burns, 192,000 calories are given off as heat per mole of methane burned.

$$CH_4(\text{gas}) + 2O_2(\text{gas}) \longrightarrow CO_2(\text{gas}) + 2H_2O(\text{gas}) + 192,000 \text{ cal.}$$

When two moles of metallic sodium react with the greenish-yellow gas, chlorine, to give two moles of sodium chloride or table salt, 196,400 calories are emitted as heat (Figure 4–7).

$$2Na + Cl_2(\text{gas}) \longrightarrow 2NaCl + 196,400 \text{ cal.}$$

° A calorie is approximately the amount of heat required to heat one gram of water one degree centigrade.

FIGURE 4-7   Sodium metal burning in pure chlorine to form sodium chloride (salt).

For every chemical change there is a quantitative relationship between the amount of chemicals changed and the amount of energy involved. To illustrate, if 136,600 calories are required to decompose two moles (36 grams) of water, it would take 68,300 calories to decompose one mole (18 grams) of water and 273,200 calories to decompose four moles (72 grams) of water.

Conservation of Energy

At first glance, we might look at the three reactions just cited and conclude that energy seems to be created or destroyed in chemical reactions. A certain amount of energy does appear on the reactant or product side of the equation in each case. The process taking place may be easier to understand, however, if we use the word "transformed" rather than "created" to describe it. If we wish to combine two moles of hydrogen and one mole of oxygen to make water (Figure 4–8), 136,600 calories will be given off—*exactly the amount required* for the decomposition of two moles of water. Evidently, this amount of matter contains more energy when it is in the hydrogen-and-oxygen form than it does when it is in the water form. When the energy is tied up in the hydrogen-and-oxygen form, it can be thought of as *chemical potential energy,* or energy that results from the structure of the atoms and molecules. This is similar to the *mechanical potential energy* that a rock has at the edge of a cliff: it can fall, and so give up its energy.

The fact that chemicals can store a quantity of energy during a chemical

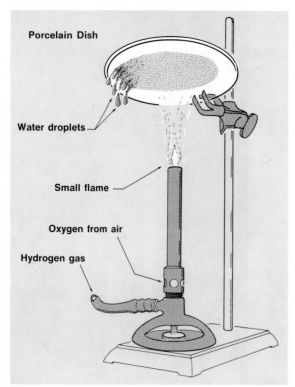

Porcelain Dish

Water droplets

Small flame

Oxygen from air

Hydrogen gas

FIGURE 4-8   Hydrogen and oxygen burning to produce water in the gaseous state. The water is condensed on the cooler porcelain dish.

change, and then release the same quantity of energy when the chemical change is reversed, indicates that energy is neither created nor destroyed in chemical processes, it is merely transformed from one kind of energy into another. Energy can be stored or used but not created or destroyed. This is characteristic of all chemical reactions, so we may write:

Energy is neither created nor destroyed in an ordinary chemical change.

The total amount of chemicals available on earth is finite. Consequently, the total chemical potential energy available is finite. Unfortunately, nearly all the substances in our environment are energy-poor. Only a few are energy-rich, and they are nearly all associated with biochemical processes (Chapters 6 and 10), or are found in fuels, such as oil and coal, that came from biochemical processes.

Man now faces an energy crisis. There is simply not enough chemical potential energy to meet his projected needs. However, when seen in perspective, we realize that chemical potential energy is only a temporary reservoir for the nuclear energy of the sun (Chapter 6). If we are going to have a limitless amount of energy, we will have to tap our nuclear reserves or those of the sun. Then, following the example of nature, we can put the energy on "hold" in chemical structures (stored as food, fuel, manufactured products, and so forth) to make it more usable for our purposes.

What Makes Chemical Reactions Occur?

Some chemical changes, such as the tarnishing of silver, occur without any apparent driving force. In this reaction, black silver sulfide is formed as a result of silver reacting with sulfur-containing compounds in its environment. Under the same conditions, platinum will not tarnish. The natural tendency is for silver to tarnish—not to be "detarnished." We say that the formation of silver sulfide is a spontaneous or natural chemical process.

Spontaneity is not to be equated with rapidity of reaction. Whether a reaction is spontaneous or not tells nothing about how fast the reaction will proceed. Both the slow rusting of iron and the explosive decomposition of nitroglycerin are spontaneous processes at room temperature, but they occur at vastly different rates.

The underlying causes of spontaneous reactions are too complex for a detailed treatment here, although an understanding of a few fundamental points is required for insight into such phenomena as the energy flow in the universe and chemical life processes discussed later in the text. The most fundamental points include the following:

(a) Some chemical reactions are spontaneous; others are not.
(b) Entropy is a measure of disorder in nature.
(c) Spontaneous reactions increase entropy.
(d) Free energy is directly related to possible work that can be done by a chemical change.
(e) The free energy "bank" of the universe is decreased as spontaneous processes occur.

87

(f) The universe appears to be running down, based on observations of free energy and entropy.

Everyone has experienced the increase in entropy (disorder) associated with spontaneous (or natural) processes. What happens to your room during the normal process of living in it? Does it tend to stay in order, or does it become disordered? What is the tendency of an opened bottle of perfume set in a corner of a room? Does it stay ordered by staying separate from other gases in the room or does it mix with the gases and thereby become disordered? Which is the natural process—for an automobile to fall apart or for the separate parts to become ordered into the automobile? In these familiar processes, the natural process increases the disorder.

Chemical reactions that occur naturally or spontaneously are similarly accompanied by an increase in entropy. It is probably not as obvious that the natural tarnishing of silver is accompanied by an increase in entropy, like the spontaneous physical processes mentioned previously. Fortunately, entropy is more than a hazy concept; it is a measurable physical quantity just like the length of a rod or the temperature of a substance. The measured values of the entropy change of a chemical reaction are relied upon when spontaneity is in doubt and when it is desirable to predict whether a reaction is spontaneous or not.

There is a difficulty in using measured values of entropy that limits its use in predicting the spontaneity of a chemical process. Both the entropy change of the reacting chemicals and the entropy change of the affected surroundings must be considered before it is certain that a natural reaction has occurred. This

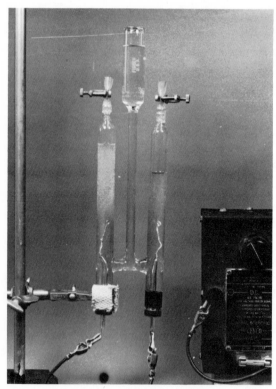

FIGURE 4-9   Electrical energy is required to decompose water into hydrogen (right tube) and oxygen (left tube). Note that there are about two volumes of hydrogen produced for each volume of oxygen.

difficulty is the primary reason entropy is not used very frequently as a criterion for a spontaneous reaction.

The free energy change of a reaction is more difficult to picture than entropy, but it is easier to measure for most chemical reactions, partially because only the free energy change of the reaction itself must be measured and not the surroundings. The free energy change is therefore a more useful criterion for determining the spontaneity of a chemical reaction. The free energy change of a reaction is a measure of its capacity to produce useful work, such as operating an engine, charging a battery, or bringing about a nonspontaneous reaction.

The formation of water from hydrogen and oxygen has a free energy change of -56.7 kcal/mole of water at $25°C$. The minus sign means the reaction can occur in and of itself in some finite time—it is a spontaneous process. The decomposition of water into hydrogen and oxygen has a free energy change of $+56.7$ kcal/mole of water at $25°C$. The positive sign means the reaction is nonspontaneous, that is, it will *never* happen in and of itself; some outside energy, such as electrical energy, will have to be added to get the reaction started and must continue to be added as long as the reaction proceeds (Figure 4-9).

Expended free energy eventually evolves as heat, which is a measure of the random motion of molecules. All kinds of energy ultimately become heat. Mechanical work results from atoms and molecules moving generally in one direction, somewhat like a group of marching soldiers. Electricity is an ordered, concerted flow of electrons, sound is a directed wave of energy in matter, and light is an ordered, directed flow of photons. The ordered, directional flow of

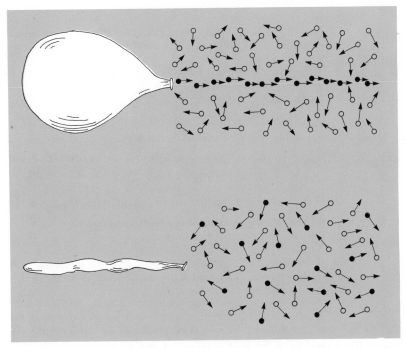

FIGURE 4-10   Gas let out of a balloon. The molecules exiting from the balloon have more free energy and less entropy than the surrounding molecules. The directed motion soon dissipates into randomness.

the submicroscopic species in these forms of energy is capable of doing work, thereby losing the free energy that is transferred eventually into the random motion of molecules (Figure 4–10). The quantity of energy is conserved, but its quality is not.

Apparently, the universe (at least that part which we can observe) is running down. Chemical reactions occur spontaneously only if the entropy of the *universe* is increased, and concurrently, the free energy of the reaction decreases. To be sure, we can put order into a small or isolated system, such as a charged battery or a house of bricks. But remember that the greater system that supplied the energy—the sun—is running down; only so many batteries can be charged and so many houses can be built, before the usable, free energy will be gone. In its place will be scattered matter, moving randomly with relatively little work potential.

QUESTIONS

1. State some ways in which energy plays a role in chemical reactions.

2. Give an example of a chemical reaction with a fast rate and one with a slow rate.

3. List three characteristics of all chemical reactions and give an example of each.

4. Using the periodic chart, select elements which can be expected to have chemical properties similar to those of

   (a) Ca (calcium), atomic number 20
   (b) Sn (tin), atomic number 50
   (c) S (sulfur), atomic number 16

5. In 1968, fire in an Apollo spacecraft cabin killed three astronauts. The fact that pure oxygen composed the cabin atmosphere contributed to the fire. How?

6. The recycling of many by-products in our society, such as paper and glass, involves the principles of reversibility. Outline the ways in which paper and glass may be recycled for further use.

7. Fires have been started by water seeping into bags in which quicklime was stored. Explain why a fire might be started in this way.

$$CaO + H_2O \longrightarrow Ca(OH)_2 + 15.6 \text{ kcal.}$$
   (Quicklime) *(Slaked lime)*

8. When a chemical reaction reaches equilibrium, what does this mean with respect to:

   (a) the relative amounts of reactants and products
   (b) the cessation of chemical change

9. If you were arranging the elements, A, B, C, D, and so on, in a row according to the pattern of the periodic chart, and if you came upon an element G, which had properties similar to A, where would you place it?

10. A fundamental assumption in physical science theory is that molecules move faster as temperature is increased. How would you use this concept to explain why chemical reactions proceed faster as temperature is increased?

11. Why do reactions proceed at a faster rate when the concentrations of the reactants are increased?

12. The natural tendency of chemical reactions is to go toward equilibrium. What does this mean? Give an example.

13. Read the following sentence then give a chemical reaction using the symbols which the chemist employs to convey the same information. "One nitrogen molecule containing two nitrogen atoms per molecule reacts with three hydrogen molecules containing two hydrogen atoms per molecule, to produce two ammonia molecules containing one nitrogen and three hydrogen atoms per molecule."

14. Why is it necessary to balance a chemical equation before it can serve as the basis for a calculation of weight relationships?

15. What is meant by free energy? How is it useful?

16. What is meant by entropy? How is it useful?

17. Balance the following equations.

$$N_2 + H_2 \longrightarrow NH_3$$
$$Fe + H_2O \longrightarrow Fe_3O_4 + H_2$$
$$NH_4NO_3 \longrightarrow N_2O + H_2O$$
$$Cu(NO_3)_2 \longrightarrow CuO + NO_2 + O_2$$
$$TiCl_4 + H_2O \longrightarrow TiO_2 + HCl$$
$$Ag + H_2S + O_2 \longrightarrow Ag_2S + H_2O$$
$$C_2H_6 + O_2 \longrightarrow CO_2 + H_2O$$

18. Contrast the meaning of the expressions 2H and H_2.

19. Which situation has the higher entropy in each of the following pairs?

    (a) an automobile or its separate parts
    (b) the arrangement of cars at the start or finish of a race
    (c) the molecules in a sample of water vapor or the molecules in ice
    (d) a room after it has been cleaned and straightened or before housekeeping was started

20. Some have said it is impossible for us to fight pollution in a universe tending toward disorder. Considering energy that comes from the sun, what do you think of this statement?

21. Much of the sulfur dioxide emitted into the air as a pollutant comes from burning sulfur in coal and petroleum. In the process, a small amount of sulfur trioxide is also produced.

$$2S + 3O_2 \longrightarrow 2SO_3$$

According to this reaction, how much sulfur trioxide could be formed if 96 grams of sulfur were burned completely?

22. A major contribution made by Lavoisier to the field of chemistry was his destruction of the phlogiston theory. You might enjoy reading about the techniques Lavoisier used to discredit this idea. See the references at the end of this chapter.

SUGGESTIONS FOR FURTHER READING

References

Angrist, S. W., and Hepler, L. G., "Order and Chaos; Laws of Energy and Entropy," Basic Books, N.Y., 1967.

Articles

Lonsdale, K., "Disorder in Solids," *Chemistry*, Vol. 39, No. 8, p. 26 (1966).
Schaff, J. R., and Westmeyer, P., "Dynamic Nature of Chemical Equilibrium," *Chemistry*, Vol. 41, No. 7, p. 48 (1968).
Szabadvery, F., "Great Moments in Chemistry. Part I: A Visit with Antoine Lavoisier," *Chemistry*, Vol. 42, No. 4, p. 14 (1969).

5

GIVE AND TAKE IN CHEMICAL CHANGE— ELECTRON AND PROTON TRANSFERS

Two major classes of chemical reactions which have far-reaching applications in our lives are those of *electron transfer* and *proton transfer* between molecular or ionic species. *Oxidation* reactions, such as the rusting of iron, the burning of wood or the bleaching of hair, can be explained in terms of the loss of electrons by chemical species. The opposite reaction, for example the reduction of iron from iron ore, is defined as *reduction*. In a general sense, reduction is the gain of electrons by an atom, ion, or molecule. Obviously, oxidation and reduction always occur simultaneously since the gain in electrons has to be at the expense of another substance losing electrons.

Acids such as vinegar or lemon juice and bases such as lime or baking soda are commonly encountered. Reactions of acids with bases can be defined in terms of the transfer of protons from an acid to a base. Analogous to the electron exchange in oxidation-reduction, there must be an *acid* to give up a proton to an accepting *base* in order to have an acid-base reaction.

ELECTRON TRANSFER OXIDATION-REDUCTION

What do the bleaching of clothes, the rusting of iron, the action of a car battery, and the extraction of iron from its ore have in common? On the basis of simple observation, they do not seem to have anything in common: the clothes become whiter or brighter, the iron becomes crumbly and red, the car starts, and the shiny metal comes from dirt. Only through a knowledge of chemistry do we "see" the similarity in these four processes and realize that many other processes are similar to these four.

92   In each of the four processes, electrons are either completely or partially

shifted between atoms. This is one of the major ways substances react and is called *oxidation-reduction*. The details of shuttling electrons in winning iron from its ore, the rusting of iron, and the action of a battery will be discussed later in this chapter. The bleaching of clothes is discussed in Chapter 15.

Let's take a simpler example to introduce some needed definitions. Do you recall having seen a fireworks spectacular with showers of sparks bursting forth spontaneously from a rocket as it arched to its pinnacle? That beautiful shower of sparks is generally sodium or potassium metal burning with oxygen. The initial reactions can be summarized in chemical equations.

$$4Na + O_2 \longrightarrow 2Na_2^+O^{2-} + energy$$

$$4K + O_2 \longrightarrow 2K_2^+O^{2-} + energy$$

In these reactions, sodium and potassium atoms have lost electrons and are said to be *oxidized*. Oxygen has gained electrons and is said to be *reduced*. If an atom has either lost electrons or lost some control over the electrons, we say it has been *oxidized*. If it has gained electrons or gained a greater degree of control over its bonding electrons, we say it has been *reduced*. The species which cause oxidation are *oxidizing agents*. Oxygen is the oxidizing agent in the previous example. By gaining electrons and being reduced, oxygen makes it possible for sodium and potassium to lose electrons and be oxidized. On the other hand, the species which cause reduction are *reducing agents*. Sodium and potassium are the reducing agents in fireworks. By losing electrons and being oxidized, sodium and potassium make it possible for oxygen to gain electrons and be reduced. Perhaps an easy way to remember this is by the acronym, "LEO says GER"—Lose Electrons, Oxidation—Gain Electrons, Reduction.

The term "oxidation" arose because the first oxidizing agent whose chemical behavior was thoroughly studied was oxygen. The phenomenon was named after the element. As the understanding of chemical reactions deepened, it became apparent that the reaction of a metal with oxygen was very similar to its reaction with fluorine, chlorine, or bromine. Thus, in each of the following reactions:

$$4Fe + 3O_2 \longrightarrow 2Fe_2^{3+}O_3^{2-}$$

$$2Fe + 3F_2 \longrightarrow 2Fe^{3+}F_3^-$$

$$2Fe + 3Cl_2 \longrightarrow 2Fe^{3+}Cl_3^-$$

the iron loses valence electrons to the other reactant. After this similarity was noted, the concept of oxidation was generalized to cover all situations in which an atom, ion, or molecule loses electrons. In the same manner the concept of reduction is now used to cover all situations in which an atom, ion, or molecule gains electrons.

Reduction of Ores to Metals

One of the most practical applications of the oxidation-reduction principle is the winning of metals from their ores. The majority of metals are found in

TABLE 5-1   SOME COMMON ORES

METAL	CHEMICAL FORMULA OF COMPOUND OF THE ELEMENT	NAME OF ORE
Aluminum	$Al_2O_3 \cdot xH_2O$	bauxite
Copper	Cu_2S	chalcocite
Zinc	ZnS	sphalerite
	$ZnCO_3$	smithsonite
Iron	Fe_2O_3	hematite
	Fe_3O_4	magnetite
Manganese	MnO_2	pyrolusite
Chromium	$FeO \cdot Cr_2O_3$	chromite
Calcium	$CaCO_3$	limestone
Lead	PbS	galena

nature in compounds in which they exist as positively charged ions. In order to obtain the metal, the ions must be reduced to neutral atoms. This can be done in an electrochemical cell or with a variety of chemicals which can supply the electrons for the reduction process. Some of the more common metals and an example of an ore for each are listed in Table 5–1. Because of its position as the number one metal in our culture, we shall take a closer look at the reduction of iron.

Iron

The sources of most of the world's iron are large deposits of iron oxides in Minnesota, Labrador, Sweden, France, Venezuela, Russia, Australia, and

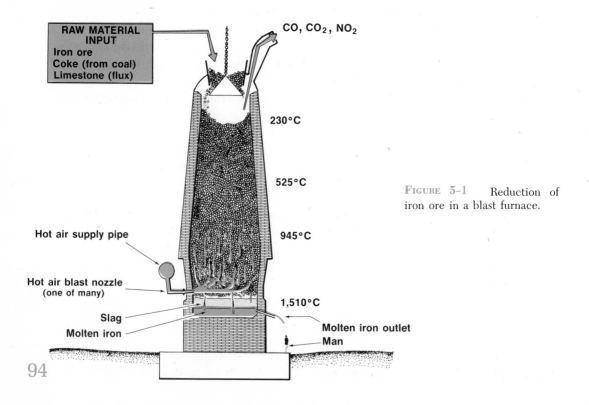

FIGURE 5-1   Reduction of iron ore in a blast furnace.

England. In nature these oxides are mixed with impurities, so the production of iron usually incorporates steps to remove such impurities. Iron ores are reduced to the metal by using carbon, in the form of coke, as the reducing agent (Figure 5–1).

$$2C \ + \ O_2 \ \longrightarrow \ 2CO \ + \text{heat}$$

Carbon   Oxygen   Carbon
monoxide

$$Fe_2O_3 + \ 3CO \ \longrightarrow 2Fe + \ 3CO_2 \ + \text{heat}$$

Iron   Carbon   Iron   Carbon
oxide   monoxide   dioxide

Limestone is added because iron ores usually contain silica (sand, SiO_2) as an impurity. The limestone reacts as follows:

$$CaCO_3 \ \longrightarrow \ CaO \ + \ CO_2$$

Calcium   Calcium   Carbon
carbonate   oxide   dioxide

$$CaO \ + \ SiO_2 \ \longrightarrow \ CaSiO_3$$

Calcium   Silicon   Calcium
oxide   dioxide   silicate

The calcium silicate, or slag, has a melting point low enough to allow it to exist as a liquid in the furnace. Consequently, as the blast furnace operates, two molten layers collect in the bottom. The lower, denser layer is mostly liquid iron which contains a fair amount of dissolved carbon and often smaller amounts of other impurities. The upper, lighter layer is primarily molten calcium silicate with some impurities. From time to time the furnace is tapped at the bottom and the molten iron drawn off. Another outlet somewhat higher in the blast furnace base can be opened to remove the liquid slag.

The iron that is obtained from the blast furnace contains too much carbon for most uses, and so is made into steel. Steel is an alloy of iron with a relatively small amount of carbon (less than 1.5 per cent); it may also contain other metals. In order to convert iron to steel, the carbon content must be decreased; this is done by burning out the excess carbon with oxygen. The total world production of iron and steel is now in excess of one billion tons per year.

Electrolysis

Several metals are either won from their ores or are purified afterward by electrolysis. *Electrolysis* is the term used to designate a chemical reaction carried out by supplying electrical energy.

Figure 5–2 outlines the principal parts of an electrolysis apparatus. Electrical contact between the external circuit and the solution or liquid is obtained by means of *electrodes*, which are often made of graphite or metal. The battery produces a direct current of electrons which flow toward one electrode (the cathode) and away from the other electrode (the anode). The cathode is defined

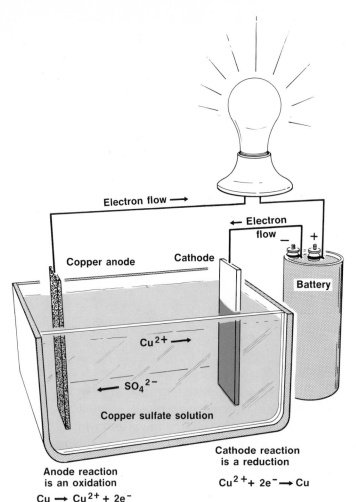

Electron flow →

← Electron
flow

Copper anode   Cathode

Battery

Cu²⁺ →

← SO₄²⁻

Copper sulfate solution

Cathode reaction
is a reduction

$$Cu^{2+} + 2e^- \rightarrow Cu$$

Anode reaction
is an oxidation

$$Cu \rightarrow Cu^{2+} + 2e^-$$

FIGURE 5-2   Electrolysis of a copper sulfate solution. A light bulb in
the circuit shows the current flow.

as the electrode at which reduction takes place; the anode as the electrode at which oxidation takes place.

Figure 5–2 is an illustration of the electrolysis of a copper sulfate solution. This process can be used either to plate an object with pure copper or to purify an impure sample of copper metal.

In the net process, copper metal is transferred from the positive electrode to the negative electrode. If the positive electrode is impure copper to be purified, electrolysis deposits the copper as very pure copper on the negative electrode. If one desires to plate an object with copper, he has only to render the surface conducting and make it the negative electrode; it will become coated with copper, with the copper coating growing thicker as the electrolysis is continued.

Now let us examine how electrolysis transfers copper from the positive electrode to the negative electrode. Electrons flow out of the negative terminal of the battery, through the wire, and into the negative electrode. Somehow this negative charge must be used up at the surface of the electrode.

How? Consider what happens when the electrons build on the negative electrode (cathode). The positive copper ions nearby will be attracted to the surface and take electrons from it. Thus, the Cu^{2+} ions are reduced.

$$Cu^{2+} + 2e^- \longrightarrow Cu$$

In a similar way the negative sulfate ions migrate to the positive electrode (anode). However, it turns out that it is easier to get electrons from the copper metal of the electrode than it is from the sulfate ions. As each copper atom gives up two electrons, it passes into solution:

$$Cu \longrightarrow Cu^{2+} + 2e^-$$

In effect, then, the copper of the positive electrode (anode) passes into solution; the copper ions in solution migrate to the negative electrode (cathode) and plate out as copper metal. The sulfate ions move in such a way as to keep the solution neutral. Large amounts of copper are purified in this way each year. Silver and gold plating can be carried out in a similar fashion.

Aluminum

Seven and one-half per cent of the crust of the earth is aluminum in the form of Al^{3+} ions. However, because of the difficulty of reducing Al^{3+} to Al, only in the past century has man learned to isolate and use this abundant element. Aluminum metal is soft and has a low density. Many of its alloys, however, are quite strong. Hence, it is an excellent choice when a light-weight, strong metal is required. Aircraft could not have developed to their present state without a plentiful supply of cheap aluminum. In structural aluminum, the high chemical reactivity of the element is offset by the fact that a transparent, hard film of aluminum oxide, Al_2O_3, forms over the surface, protecting it from further oxidation:

$$4Al + 3O_2 \longrightarrow 2Al_2^{3+}O_3^{2-}$$

The principal ore of aluminum is bauxite, a hydrated aluminum oxide, $Al_2O_3 \cdot xH_2O$. Because impurities in the ore have undesirable effects on the properties of aluminum, these must be removed before the aluminum is reduced.

Aluminum metal is obtained from the purified oxide by electrolysis in molten cryolite, Figure 5–3. Each aluminum ion must gain three electrons to become an aluminum atom.

$$Al^{3+} + 3e^- \longrightarrow Al$$

Cryolite, Na_3AlF_6, has a melting point of $1006°C$; the molten compound dissolves considerable amounts of aluminum oxide, which in turn lowers its melting point. This mixture of cryolite and aluminum oxide is electrolyzed in a cell with carbon anodes and a carbon cell lining that serves as the cathode on which aluminum is deposited. As the operation of the cell proceeds, the molten aluminum sinks

97

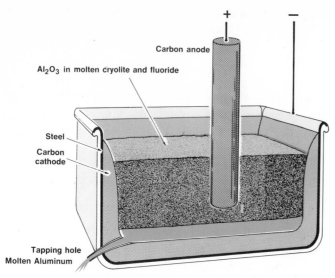

+   −

Carbon anode

Al₂O₃ in molten cryolite and fluoride

Steel

Carbon
cathode

Tapping hole
Molten Aluminum

FIGURE 5-3 Schematic drawing of a furnace for producing aluminum by electrolysis of a melt of Al_2O_3 in Na_3AlF_6. The molten aluminum collects in the carbon cathode container.

to the bottom of the cell. From time to time the cell is tapped and molten aluminum is run off into molds.

The Human Side

Charles Martin Hall

From 1828 when Frederick Wöhler (the "Father of organic chemistry," see also Chapter 7) first isolated aluminum, until 1886 when Charles Martin Hall (1863–1914), an American chemist, and Paul Louis Toussaint Heroult (1863–1914), a French metallurgist, independently discovered an inexpensive, efficient way to win aluminum from bauxite, the price of aluminum never fell below eight dollars a pound. The top of the Washington Monument is composed of a slab of the metal which was then quite expensive. Today aluminum sells for about 30 cents a pound, thanks to the Hall-Heroult process.

There are a number of interesting similarities between Hall and Heroult. They were born in the same year and died in the same year. Both were 23 when they independently discovered the process that made them famous.

Hall's discovery is inspiring to all students of chemistry. At the age of 22 young Hall was inspired by a chance remark of his chemistry teacher to make a tremendously important discovery. His teacher stated that anyone discovering a cheap way of preparing aluminum would grow both famous and rich. Charles took him at his word, made the discovery, and grew rich and famous. He made the discovery in his home laboratory, with homemade batteries. The keys to the discovery involved dissolving the aluminum oxide in cryolite and the use of carbon electrodes.

Incidentally, if you can devise a cheap method of obtaining metals from silicate rocks (in which most of our metals are locked), you, too, will derive riches and fame from your discovery.

FIGURE 5-4 A neighborhood reclamation center for recycling glass, metal, and paper. Each week, four tons of glass and metals are recycled by volunteer workers in West University Place, Texas. (Photo courtesy of Dr. Lee Allbritton.)

Aluminum is used both as a structural metal and as an electrical conductor in high voltage transmission lines. At the present time the world production of aluminum is about 10 million tons per year. Over 500,000 tons of aluminum went into food and beverage cans in the USA in 1972. Plants such as the one shown in Figure 5–4 are being set up over the country to recycle aluminum and other materials.

BATTERIES

One of the most useful applications of oxidation-reduction reactions is in the production of electrical energy. A device that produces an electron flow (current) by a spontaneous chemical reaction is called a "voltaic cell." A series of such cells is referred to as a "battery."

Consider the reaction when a piece of zinc metal is placed in a solution of copper sulfate. The copper is in the form of copper 2^+ ions, Cu^{2+}. The zinc is gradually eaten away, a deposit of copper metal settles to the bottom of the beaker, and the solution gradually loses its blue color. The solution now contains colorless Zn^{2+} ions. During the reaction zinc atoms become zinc ions as copper ions become copper atoms.

$$Zn + Cu^{2+} \longrightarrow Zn^{2+} + Cu + energy$$

The electron transfer takes place between the zinc metal and the copper ions, and the energy liberated simply causes heating of the solution and the zinc strip.

It is evident that copper ions have a greater attraction for electrons than do zinc ions since copper gets the electrons at the expense of the zinc. All spontaneous oxidation-reduction reactions are driven by such unequal attractions for electrons.

99

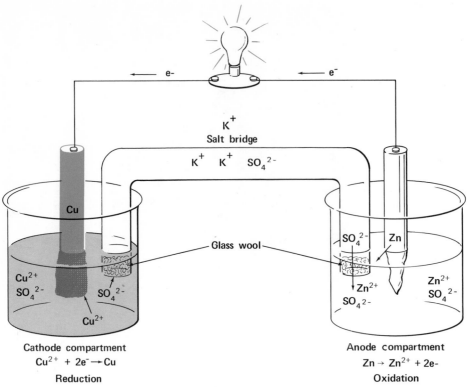

Cathode compartment
$Cu^{2+} + 2e^- \rightarrow Cu$
Reduction

Anode compartment
$Zn \rightarrow Zn^{2+} + 2e-$
Oxidation

FIGURE 5-5   A simple battery involving the oxidation of zinc metal and the reduction of Cu^{2+} ions.

It is possible to design a voltaic cell in which electrons pass from Zn atoms to Cu^{2+} ions indirectly. Figure 5–5 shows a battery built on this principle. The anode reaction is the oxidation of zinc to Zn^{2+} ions.

$$Zn \longrightarrow Zn^{2+} + 2e^-$$

The electrons flow from the Zn electrode through the connecting wire, light the lamp in the circuit, and then flow into the copper cathode where reduction of Cu^{2+} ions takes place:

$$Cu^{2+} + 2e^- \longrightarrow Cu \ (Metal)$$

The copper is deposited on the copper cathode.

This flow of electrons (negative charge) from the anode to the cathode compartment in the battery must be neutralized electrically by corresponding motion of ions. A "salt bridge" for the ions is provided which connects the two compartments. Ions flow through this bridge readily since it is a salt solution (such as K_2SO_4) in a glass tube with porous plugs at either end. During the operation of the battery, a net flow of negative charges just equal to the number of electrons used at the cathode will pass through the bridge to the Zn^{2+} solution. Likewise, the copper solution is kept neutral by negative SO_4^{2-} ions flowing into the salt bridge and positive K^+ ions flowing into the solution. The reaction of

TABLE 5-2   CHARACTERISTICS OF SOME BATTERIES

SYSTEM	ANODE OXIDATION	CATHODE REDUCTION	ELECTROLYTE	TYPICAL OPERATING VOLTAGE PER CELL
Dry cell	Zn	MnO_2	NH_4Cl–$ZnCl_2$	0.9–1.4 volts
Edison storage cell	Fe	Ni oxides	KOH	1.2–1.4
Nickel/Cadmium cell	Cd	Ni oxides	KOH	1.1–1.3
		AgO	KOH	1.0–1.1
Lead storage cell	Pb	PbO_2	H_2SO_4	1.95–2.05
Mercury cell (used in transistor radios)	Zn(Hg)	HgO	KOH–ZnO	1.30

zinc with Cu^{2+} continues until the battery runs down; that is, until equilibrium is attained.

Many different oxidation-reduction combinations are used in voltaic cells. A few of the more popular ones are listed in Table 5–2. The lead storage battery and fuel cells are described in the next chapter.

CORROSION

Many metals undergo corrosion when exposed to moist air over a long period of time. Typically, corrosion is a reaction with the oxygen and water of the air which transforms a metal into its oxide or hydroxide. Corrosion of iron is called "rusting" and leads to the transformation of iron into rust, $(Fe_2O_3 \cdot xH_2O)$. The initial reaction is the oxidation of iron to iron (II) hydroxide:

$$2Fe + O_2 + 2H_2O \longrightarrow 2Fe(OH)_2$$
$$\text{\textit{Moist air}} \qquad \text{\textit{Iron (II) hydroxide}}$$

Iron (II) hydroxide is itself subject to further oxidation in moist air to give iron (III) hydroxide (the Roman numeral refers to the charge on the ions):

$$4Fe(OH)_2 + O_2 + 2H_2O \longrightarrow 4Fe(OH)_3 \text{ (or } 2Fe_2O_3 \cdot 3H_2O)$$

The iron (III) hydroxide loses water readily to form iron (III) oxides with variable amounts of water. The rusting process transforms iron metal back into a compound very similar in composition to its ore and thus undoes all the effort expended in obtaining the metal. The replacement of rusted objects costs several billion dollars a year in the United States alone.

Rusting also occurs when we make an iron object part of an electrical cell or battery. This happens when part of the iron surface becomes wet in contact with air (Figure 5–6). The oxygen in the air will dissolve in the water and remove electrons from the iron:

$$O_2 + 2H_2O + 4e^- \longrightarrow 4OH^- \qquad \text{\textit{Reduction}}$$

$$2Fe \longrightarrow 2Fe^{2+} + 4e^- \qquad \text{\textit{Oxidation}}$$

101

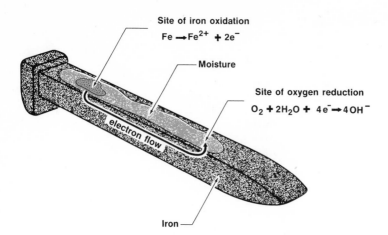

Site of iron oxidation

Fe ⟶ Fe^{2+} + 2e$^-$

Moisture

Site of oxygen reduction

O$_2$ + 2H$_2$O + 4e$^-$ ⟶ 4OH$^-$

electron flow

Iron

FIGURE 5-6 The site of iron oxidation may be different from the point of oxygen reduction owing to the ability of the electrons to flow through the iron. The point of oxygen reduction can be located with an acid-base indicator because of the OH$^-$ ion produced.

Corrosion and rusting are obviously very serious problems which have occupied the attention of scientists for years. There are many ways of preventing or reducing corrosion of a metal object. Three of these are: (a) protective coatings; (b) cathodic protection; and (c) alloying (stainless steel is a corrosion-resistant alloy of iron).

Protective coatings are applied to prevent the access of atmospheric oxygen to the iron surface. Such coatings may be paint, enamel, grease, or another, more resistant metal, such as chromium. This method is successful as long as cracks or holes do not develop in the coating. Galvanized iron contains a surface coating of the more active metal, zinc, which protects the metal according to principles outlined in the next section.

Cathodic protection is a method by which the iron is electrically connected to a more reactive metal, such as magnesium or zinc, which will be oxidized rather than the iron (Figure 5–7). In such a system, the oxygen is reduced at the steel surface via the reaction

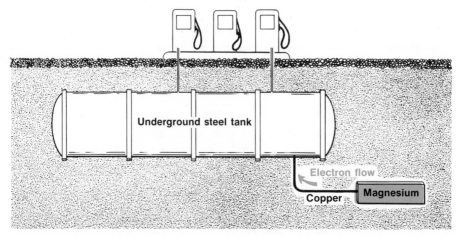

Underground steel tank

Electron flow

Copper **Magnesium**

FIGURE 5-7 Cathodic protection. If magnesium is connected to the steel tank to be protected, the magnesium is more easily oxidized than the iron or copper connecting wire. The magnesium serves as the anode. Hence, the cathode is protected with no points of oxidation occurring on its surface. The anode is the electrode where oxidation occurs; reduction occurs at the cathode. When the magnesium is used up, it is replaced by another block. The replacement is much easier and cheaper than replacing the tank or bridge.

$$O_2 + 2H_2O + 4e^- \longrightarrow 4OH^-$$

but the more reactive metal (magnesium) is the source of the electrons:

$$Mg \longrightarrow Mg^{2+} + 2e^-$$

As a result, the magnesium is corroded rather than the iron. This method is used to protect iron or steel objects buried in the ground, such as pipelines, tanks or bridge supports. Of course, protection ends when the magnesium is depleted.

The cathode is always the electrode where reduction occurs, and the anode is the electrode where oxidation occurs. Cathodic protection is so named because the iron to be protected serves as the cathode, while the more active metal is the anode. Hence, oxidation cannot occur at the iron cathode.

Stainless steels are alloys of iron that contain other metals, such as nickel or chromium. They are made by melting iron and the other elements together in an electric furnace. The resulting alloys are resistant to corrosion because in the presence of oxygen they form a very thin, tough, and impervious adherent layer of metal oxide on their surfaces. This oxide protects the underlying metal from further contact with the oxygen of the air and renders the objects "stainless," or very resistant to corrosion under normal circumstances. A stainless steel alloy which is widely used is the so-called 18–8: 18 per cent chromium, 8 per cent nickel, and the rest iron. Stainless steels are used in many household items; larger amounts are used in the construction of industrial plants for processing food and handling highly corrosive materials.

ACIDS AND BASES

The terms "acid" and "base" have been used by chemists for several hundred years. For example, acids have long been characterized as substances which are sour tasting, corrosive, turn litmus red, and react with substances called bases. Bases, on the other hand, have a bitter taste, make the fingers or skin feel slippery on contact, turn litmus blue, and react with acids. As more and more information was collected on the properties of acids and bases, these simple definitions had to be refined. This has been carried to the point where current definitions of acids and bases are structural in nature.

PROTON TRANSFER AND NEUTRALIZATION

The definitions we shall use here are those first given by J. N. Brønsted and T. M. Lowry, in 1923.

Brønsted acid: A chemical species which can donate hydrogen ions, H^+.
Brønsted base: A chemical species which can accept hydrogen ions.

To illustrate these definitions we consider the reaction between gaseous hydrogen chloride (HCl) and water:

103

$$HCl\ (gas) + H_2O \longrightarrow H_3O^+ + Cl^-$$

Acid   Base   Hydronium   Chloride
                    ion   ion
            (an acid)   (A very weak base)

Hydrochloric acid

The HCl is an acid (since it gives up a hydrogen ion) and the water is a base (since it accepts a hydrogen ion). The reaction is reversible, but when equilibrium is established, almost all the HCl has been converted to $H_3O^{+\circ}$ and Cl^-. This solution is termed an "acid solution" because the hydronium ions (H_3O^+) are donors of hydrogen ions to many common substances.

If the ionic solid sodium oxide, $Na_2^+O^{2-}$, is dissolved in water, a vigorous reaction occurs producing a solution containing sodium ions (Na^+) and hydroxide ions (OH^-). In this process the oxide ion (O^{2-}) reacts with water to form the hydroxide ion:

$$2Na^+ + O^{2-} + H_2O \longrightarrow 2Na^+ + 2OH^-$$

Base   Acid   Sodium hydroxide
                      a strong base solution

The sodium hydroxide solution is commonly termed a base because the OH^- ions contained in it take hydrogen ions from many common substances.

It has long been known that when acids react with bases the properties of both disappear. This process is "neutralization." To get a more precise picture of acid-base neutralization reactions, let us consider what happens when a solution of hydrochloric acid is mixed with a solution of sodium hydroxide. The hydrochloric acid contains H_3O^+ and Cl^- ions; the sodium hydroxide solution contains Na^+ and OH^- ions. When these two solutions are mixed a reaction occurs between H_3O^+ and OH^-.

$$H_3O^+ + Cl^- + Na^+ + OH^- \longrightarrow Na^+ + Cl^- + 2H_2O$$

Strong acid   Strong base   Salt Solution   Water

Hydrochloric   Sodium hydroxide
acid   solution

If we have an equal number of H_3O^+ and OH^- ions, they will react to produce a neutral solution, with the hydronium ions (H_3O^+) donating their protons to the hydroxide ions (OH^-) to form molecules of water. If we have more H_3O^+ ions than OH^- ions, the extra H_3O^+ will make the resulting solution acidic. If we have more OH^- ions than H_3O^+ ions, only a fraction of the OH^- ions will be neutralized, and the extra OH^- ions will make the resulting solution basic.

One water molecule can transfer a proton to another water molecule.

$$2H_2O \rightleftharpoons H_3O^+ + OH^-$$

104    $^\circ H_3O^+$ has the structure $\left[\begin{array}{c} H \\ H:\ddot{O}:H \end{array} \right]^+$

This reaction takes place to only a very small extent, as is indicated by the arrows of unequal length. Since H_3O^+ and OH^- are produced in equal amounts when only water is present, pure water is our reference point. It is generally considered neither acidic nor basic, but is described as neutral.

Strength of Acids and Bases. Concentrations of H_3O^+ and OH^-. The pH Scale.

A chemical species in water solution is commonly called an acid if it donates protons to water and increases the concentration of H_3O^+. Similarly, a base is commonly described as a compound whose addition to water will increase the concentration of OH^-. The concentration of the hydrogen ion times the concentration of the hydroxide is equal to a constant.

$$[H_3O^+][OH^-] = \text{Constant}$$

The brackets, [ ], are used to denote concentration in moles per liter. This equation tells us that as the hydrogen ion concentration increases, that of the hydroxide ion decreases. At room temperature (25°C), the value of the constant is 1×10^{-14}. In a neutral solution, $[H_3O^+] = [OH^-] = 1 \times 10^{-7}$ mole/liter.

In a solution made by dissolving one mole of HCl gas in enough water to make one liter of solution, the hydrogen ion concentration is 1.0 mole/liter since practically all the HCl has lost its protons to the water. Species like HCl, in which proton transfer is virtually complete, are said to be strong acids. By contrast, if one mole of acetic acid (the acid in vinegar) is dissolved in enough water to make a liter of solution, the hydrogen ion concentration is only about one-thousandth as much, or 0.001 mole/liter. Evidently only a small proportion of the acetic acid lost its protons to water. Such an acid is termed weak. Similarly, there are strong bases, such as sodium hydroxide, and weak bases, such as ammonia. In a solution containing one mole per liter of sodium hydroxide, (NaOH), the hydroxide concentration is 1.0 mole per liter; in a solution containing one mole per liter of NH_3, the hydroxide concentration is only about 0.001 mole/liter. Apparently, only a small amount of ammonia undergoes reaction with water to produce hydroxide ions, but this is enough to give it its cleansing action (See Chapter 15).

$$NH_3 + H_2O \rightleftharpoons NH_4^+ + OH^-$$

The concentrations of protons (H_3O^+) in biochemical and laboratory systems are generally less than one mole per liter—quite often as low as 10^{-5} to 10^{-14} mole per liter. In 1909, the Danish chemist, Soren Sorensen (1868–1939), invented a very simple way to express these very low concentrations of the hydrogen ion without use of exponents or many zeros. He devised the pH scale for this purpose; it furnishes a number which describes the acidity of a solution. The pH is defined as follows: $pH = -\log[H_3O^+]$.°

Most solutions have pH's between 0 and 14, although solutions can have

---

° The logarithm of a number is the exponent or power to which 10 must be raised to equal the number. The log of 10^{-7} is -7.

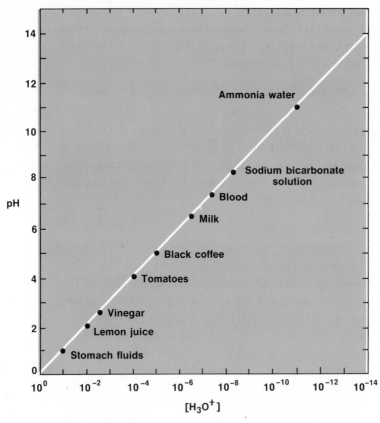

FIGURE 5-8   A plot of pH versus hydrogen ion concentration, $[H_3O^+]$. Note that the pH increases as the $[H_3O^+]$ decreases. The pH values of some common fluids are given for reference.

a pH as low as -1 or as high as 15. A pH of 7 indicates a neutral solution (pure water), a pH below 7 indicates an acid solution, and a pH above 7 indicates a basic solution.

Figure 5–8 graphically displays the relationship between pH and concentration of the hydrogen ion. It also gives the approximate pH values for common solutions. A close examination of Figure 5–8 reveals that basic solutions, such as ammonia water, have hydrogen ion concentrations less than that of pure water.

SULFURIC ACID, H_2SO_4, AN INDUSTRIALLY IMPORTANT STRONG ACID

Sulfuric acid is used in greater quantities than any other chemical produced in the United States. In 1971, over 30 million tons of sulfuric acid were used in the U.S., compared to about 2 million tons of its closest competitor, hydrochloric acid. The acid is used in huge quantities in the manufacture of fertilizers, in the petroleum industry, in automobile batteries, and in the production of steel. It also plays an important role in the manufacture of organic dyes, plastics, drugs, and in many other products. The cost of sulfuric acid, about 1 cent per pound, has not changed much in 300 years, a tribute to man's improving technology.

Most of our sulfuric acid had been made by burning sulfur in air to give sulfur dioxide, SO_2, as the first step.

$$S + O_2 \longrightarrow SO_2(g)$$

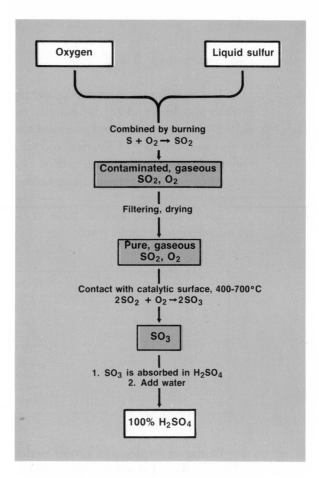

FIGURE 5-9 Flow diagram of contact process for producing sulfuric acid from sulfur.

The source of the sulfur was the huge underground deposits in Louisiana. Much sulfuric acid now comes from by-product sulfur of metallurgical plants and from the removal of SO_2 and much smaller amounts of SO_3 (sulfur trioxide) from stack gases of power plants. Eventually, pollution control will give more H_2SO_4 than we can use.

Regardless of the source, the conversion of the gaseous SO_2 to SO_3 is achieved by passing it over a hot, catalytically active surface such as platinum or vanadium pentoxide:

$$2SO_2 + O_2 \xrightarrow{\text{catalyst}} 2SO_3(g)$$

Although SO_3 can be converted directly in H_2SO_4 by passing it into water, the enormous amount of heat released by the reaction causes the formation of a fog of H_2SO_4. This is avoided by passing the SO_3 first into H_2SO_4 and then diluting the $H_2S_2O_7$ with water:

$$\underset{\substack{\text{Sulfuric} \\ \text{acid}}}{SO_3 + H_2SO_4} \longrightarrow \underset{\substack{\text{Pyrosulfuric} \\ \text{acid}}}{H_2S_2O_7}$$

$$H_2S_2O_7 + H_2O \longrightarrow 2H_2SO_4$$

107

SALTS

Preparation of Salts

The chemical compounds known as salts play a vital role in nature, in animal growth and life, and in the manufacture of various chemicals for human use. Salts are made up of + and − ions. Potassium chloride, for example, is composed of an equal number of K^+ ions and Cl^- ions arranged in definite positions with respect to one another in a lattice (see Figure 3–5). The salt crystal must be electrically neutral; it can have neither an excess nor a deficiency of positive or negative charge.

Let us imagine that we have at our disposal the ions listed below, and let us see what salts could result:

Ion		Possible Salts	Salt Name
Na^+	Sodium	NaCl	Sodium chloride
Ca^{2+}	Calcium	$NaNO_3$	Sodium nitrate
Cl^-	Chloride	Na_2SO_4	Sodium sulfate
NO_3^-	Nitrate	$NaC_2H_3O_2$	Sodium acetate
SO_4^{2-}	Sulfate	$CaCl_2$	Calcium chloride
$C_2H_3O_2^-$	Acetate	$Ca(NO_3)_2$	Calcium nitrate
		$Ca(C_2H_3O_2)_2$	Calcium acetate
		$CaSO_4$	Calcium sulfate

In the examples just given, notice that in order to attain an electrically neutral lattice, it is necessary to balance the charges of the ions. A sodium (Na^+) ion requires just one chloride ion (Cl^-). A sulfate ion (SO_4^{2-}) with two negative charges must have its negative charge balanced by two positive charges. This may be done using two Na^+ ions to give Na_2SO_4 or one Ca^{2+} ion to give $CaSO_4$.

It is possible to form many solid salts of limited solubility by mixing water solutions of different soluble salts with each other. For example, both lead acetate and sodium chloride are soluble in water. If we mix solutions of these salts, we find the insoluble salt, lead chloride, precipitates from the mixture.

$$2Na^+ + 2Cl^- + Pb^{2+} + 2C_2H_3O_2^- \longrightarrow PbCl_2 + 2Na^+ + 2C_2H_3O_2^-$$

Sodium chloride   *Lead acetate in*   *Solid*   *Sodium acetate in*
in solution   *solution*   *lead*   *solution*
                                      chloride

Salts can also be formed as the products of acid-base neutralizations, as in the example:

$$(K^+ + OH^-) + (H_3O^+ + Br^-) \longrightarrow K^+ + Br^- + 2H_2O$$

Potassium hydroxide   *Hydrobromic acid*   *Potassium bromide*
in water   *in water*   *in solution*

QUESTIONS

1. What is the primary reducing agent in the production of iron from its ore?

2. Why is $CaCO_3$ necessary for the production of iron in a blast furnace?

3. What is the difference between iron and steel?

4. How is it possible that oxidation and reduction can occur at different points on a piece of iron?

5. Give examples of three ways by which metals can be protected from corrosion.

6. Describe the contact process for the production of sulfuric acid.

7. Write the equation for a chemical reaction in which water acts as a Brønsted acid, and as a Brønsted base.

8. A solution of hydrochloric acid is electrolyzed. The products are hydrogen at the cathode and chlorine at the anode. Write the reactions occurring at each electrode and state which ions move toward the cathode and which toward the anode.

9. What is the pH of a 0.0001 M solution of HCl?

   (a) What is the hydrogen ion concentration in a solution with pH of 8?

10. In the production of copper from its sulfide ores, large amounts of SO_2 gas are formed as a by-product. What useful product can be prepared from this material? What are the consequences of allowing the gas to escape into the atmosphere? See Chapter 19.

11. Define acid-base reactions in terms of protons, and oxidation-reduction reactions in terms of electrons.

12. What is the difference between a hydrogen ion and a hydronium ion?

13. What can be said about the ability of gold to be oxidized relative to iron? Which metal would you characterize as being more active chemically?

14. Identify the reducing agent, the oxidizing agent, and what is reduced and what is oxidized in the production of iron in the blast furnace.

15. Which represents a spontaneous chemical process, an electrolysis cell or a battery? Why?

16. Explain why you would make both the cathode and the anode out of copper if you wanted to purify copper by electrolysis.

17. Aluminum cannot be reduced to the metal from water solutions of its salts since the water is more easily reduced. What is the liquid used for the commercial reduction of aluminum?

18. Figures 5–2 and 5–5 both have a light bulb in the electrical circuit, but for different reasons. Explain.

19. Corrosion leads to a more stable state of affairs. Is there more order in a given amount iron, oxygen and water, or in an equivalent amount of rust? Does nature prefer order or disorder?

20. Which is more active chemically, iron or magnesium? Use Figure 5–7 to justify your answer.

21. Identify the acid and base in the reaction:

$$H^+Cl^- + Na^+CN^- \longrightarrow HCN + Na^+Cl^-$$

What part does the Na^+ ion play in this reaction?

22. Someone has said that pH is a lazy man's approach to describing the acidity of a solution. Does this seem fair?

23. A solution that has a pH of 3 has ten times the concentration of hydrogen ion as a solution with pH of 4. Can you explain this statement?

24. Why is it safe to put vinegar on turnip greens (vinegar is a solution of acetic acid, water and a few other things) while an equal concentration of sulfuric acid would be damaging?

SUGGESTIONS FOR FURTHER READING

Articles

Eyring, H., and MyShik, J., "Significant Structure Theory of Water," *Chemistry*, Vol. 39, No. 9, p. 8 (1966).
Haggin, J., "New Method of Iron Production." *Chemistry*, Vol. 39, No. 5, p. 24 (1966).
Morris, D. L., "Brønsted-Lowry Acid-Base Theory—A Brief Survey," *Chemistry*, Vol. 43, No. 3, p. 18 (1970).
Schmuckler, J. S., and Mogue, P. H., "Dilute Solutions of Strong Acids: The Effect of Water on pH." *Chemistry*, Vol. 42, No. 9, p. 14 (1969).

6

CHEMICAL AND NUCLEAR ENERGY

To say that man needs energy is to state the obvious. He needs energy from his food to carry out life processes. He needs energy to control his environment when nature provides extreme conditions of heat or cold. He needs energy to provide clothing, transportation and shelter. Modern man requires much more energy, either directly or indirectly, then he can produce with his body. At first, he used the kinetic energy of the wind and falling water, but as his energy demands have increased, he has turned more to the chemical potential energy tied up in molecules and, very recently, to the nuclear potential energy within nuclear structures. The use of these energy sources has now reached a point

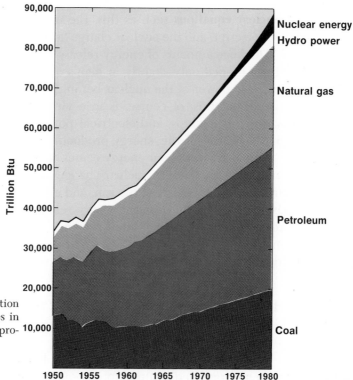

FIGURE 6-1   Total gross consumption of energy resources by major sources in the United States (1950–1965 and projections to 1975 and 1980).

where the people in the United States use an average of 3×10^7 BTU per person per year in their work and daily life. Figure 6–1 shows energy consumption, as well as a projection of needs for the United States.

ENERGY SOURCES

Most of our energy is obtained from chemical reactions in which we burn fossil fuels such as coal, petroleum, or natural gas (Figure 6–1). Typical reactions releasing chemical energy are:

$$C + O_2(gas) \longrightarrow CO_2(gas) + 94 \text{ kilocalories (kcal.)}$$
$$(coal) \quad (air)$$

$$CH_4(gas) + 2O_2(gas) \longrightarrow CO_2(gas) + 2H_2O(gas) + 192 \text{ kcal.}$$

At present, nuclear energy and hydroelectric power provide only a small fraction of our energy needs. However, the energy change associated with a nuclear change may be dramatically larger, atom for atom, than the largest possible energy change for a chemical reaction. For example, when one mole $(6.20 \times 10^{23}$ molecules or 16 grams) of methane, a major component of natural gas, is burned, about 200 kilocalories of heat are liberated. In contrast, when lithium nuclei react with hydrogen nuclei in a nuclear reaction, the energy released per mole of lithium is 23,000,000 kilocalories.

$$^{7}_{3}Li + ^{1}_{1}H \longrightarrow 2^{4}_{2}He + 23,000,000 \text{ kcal}$$

In **nuclear equations** such as this, the mass number of the isotope is given by the superscript, and the nuclear charge (atomic number) is given by the subscript.

The huge amounts of energy released in certain nuclear reactions have thus far been tapped only slightly in man's quest for cheap, controlled energy. The destructive power of the nuclear bomb illustrates the magnitude of the energies involved, but this, of course, is in a virtually uncontrolled form. The nuclear power plants in ships, and electrical power stations may be just the beginning in man's control of this energy-producing giant.

Although most of our energy comes at present from fossil fuels, we can trace even this energy back to nuclear changes. Apparently all life on earth has been nourished by the energy of the sun, and all fossil fuels were produced by various early life forms. The primary energy release of the sun is the result of nuclear changes similar to the one cited previously. Perhaps we will place the concept of energy in proper perspective when we think of chemical energy as only a holding form for the almost unlimited amount of nuclear energy in the universe.

CHEMICAL POTENTIAL ENERGY—
THE ENERGY IN MOLECULAR STRUCTURES

Energy is absorbed when a chemical bond is broken to give isolated atoms. The energy required to break one mole of a particular kind of bond is defined

TABLE 6-1   APPROXIMATE BOND ENERGIES IN KCAL/MOLE

BOND	ENERGY	BOND	ENERGY	BOND	ENERGY
H—H	104	I—I	36	I—Cl	50
C—C	83	H—F	135	O—Cl	49
C=C	146	H—Cl	103	S—Cl	60
C≡C	200	H—Br	88	N—Cl	48
N—N	40	H—I	71	P—Cl	79
N=N	100	H—O	111	C—Cl	79
N≡N	225	H—Se	66	Si—Cl	86
O—O	33	H—S	81	C—O	84
Ȯ—Ȯ in O_2	118	H—N	93	C=O	173
S—S	51	H—P	76	C=O in CO_2	192
F—F	37	H—C	99		
Cl—Cl	58	H—Si	70		
Br—Br	46	Br—Cl	52		

as the bond energy (Table 6–1). It is usually given in kilocalories per mole. The same amount of energy is released when a mole of bonds is formed from isolated atoms.

A given chemical change is *exothermic* (heat is given off) if the formation of new bonds liberates more energy than is required to break the bonds in the reaction. A reaction is *endothermic* if the bonds in the reactants are stronger than those in the products.

Consider again the oxidation of methane, and this time let us focus attention on the energy of the chemical bonds. We can write the equation as:

$$\text{H–C–H} + 2\dot{\text{O}}\text{–}\dot{\text{O}} \longrightarrow \text{O=C=O} + 2\text{H–O–H} + 192 \text{ kilocalories}$$

The energy released in this reaction can be used to heat houses, drive gas turbines or generate electricity. Since energy is released, this means it takes less energy to break four C—H bonds and two Ȯ—Ȯ bonds than is evolved when two C=O bonds and four O—H bonds are formed. The bond energies in Table 6–1 bear this out. It only takes 632 kilocalories to break two Ȯ—Ȯ bonds and

TABLE 6-2   HEATS OF COMBUSTION OF SOME ORGANIC MATERIALS

SUBSTANCE	KILOCALORIES PER GRAM
Methane (principal component of natural gas)	13.2
Gasoline, kerosene, crude petroleum, tallow	9.5–11.5
Lipids	9.0–9.5
Carbon	7.8
Ethyl alcohol	7.1
Proteins	4.4–5.6
Carbohydrates (sugars and starches)	3.6–4.2

These figures correspond to laboratory combustion to yield CO_2, H_2O, and oxides of nitrogen. In the body, proteins are oxidized to CO_2, H_2O, and urea. For this latter process the heat yield is less than the value indicated above. Thus proteins and carbohydrates yield (per gram) about the same energy in the body.

four C—H bonds $2(118) + 4(99) = 632$. The making of bonds produces 828 kilocalories generated by the formation of two C=O bonds and four O—H bonds $2(192) + 4(111) = 828$. This leaves a net of 196 kilocalories ($828 - 632 = 196$), which is in close agreement with the experimental value, 192 kilocalories.

Man's environment on the surface of the earth is made up of relatively stable, low-energy chemical compounds. Yet nature has stored a comparatively large amount of energy in certain chemical structures, such as those found in fossil fuels and other organic materials. The combustion of these materials yields sizable amounts of energy (Table 6–2).

Such chemicals make up nature's chemical potential energy bank, energy that came from the sun and was captured by photosynthesis. The sobering fact is that man is using up in a very short time what it took nature millions of years to produce. It appears that in the future, man will not be able to satisfy his projected energy needs from fossil fuels, photosynthesis, and hydroelectric power. Either he will have to reduce his appetite for energy, learn other methods for the storage of the nuclear energy of the sun, or learn to harness controlled nuclear reactions better on earth.

A WAY TO STORE ENERGY—BATTERIES

In Chapter 5 we saw that some electron transfer reactions are spontaneous and that by the proper construction of a "cell" this type of reaction can be made into a *battery*. Like nature's photosynthesis, we too have a method for storing chemical potential energy.

The lead storage battery in the automobile has positive electrodes filled with lead dioxide, PbO_2, and negative lead electrodes filled with spongy lead; both are dipped in a rather concentrated sulfuric acid solution (Figure 6–2).

When the electrodes are connected through an external circuit, the lead is oxidized:

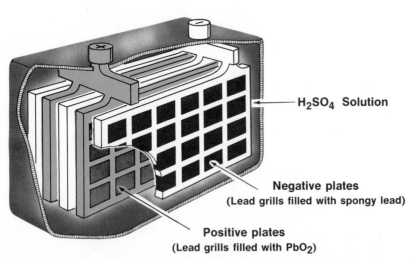

H₂SO₄ Solution

Negative plates
(Lead grills filled with spongy lead)

Positive plates
(Lead grills filled with PbO₂)

FIGURE 6-2   The lead storage battery. (After Lee, G., Van Orden, H. O., and Ragsdale, R. O.: *General and Organic Chemistry*. Philadelphia, W. B. Saunders Co., 1971.)

$$Pb \longrightarrow Pb^{2+} + 2e^-$$

and the PbO_2 is reduced:

$$PbO_2 + 2e^- + 4H^+ \longrightarrow Pb^{2+} + 2H_2O$$

Since lead sulfate ($PbSO_4$) is insoluble, the sulfate ions from the sulfuric acid precipitate the Pb^{2+} ions as soon as they form on both electrodes:

$$Pb^{2+} + SO_4^{2+} \longrightarrow PbSO_4$$

The overall reaction then is

$$PbO_2 + Pb + 2H_2SO_4 \underset{Charge}{\overset{Discharge}{\rightleftharpoons}} 2PbSO_4 + H_2O$$

The double arrows indicate charge and discharge reactions. The voltage (or driving force behind the electron flow) is 2.05 volts. Three such cells in series make a 6-volt battery, and six in series constitute a 12-volt battery.

Note that sulfuric acid is consumed in the discharge of a lead storage battery. Because of this, the density of the liquid in the battery is a measure of the charge in the battery, since sulfuric acid has a greater density than water.

The lead storage battery is but one way to store energy in chemicals. Of course, we think of the car battery as a source of energy, but it could be used for storing the sun's energy when we are getting more than we need at a particular time.

FUEL CELLS

Fuel cells offer an important way to convert the chemical energy stored in fuels directly into electrical energy. In a fuel cell the electrode materials,

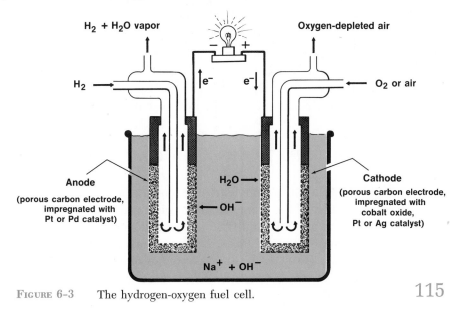

FIGURE 6-3   The hydrogen-oxygen fuel cell.   115

(usually in the form of gases) are supplied continuously to the cell and consumed to produce electricity. A typical fuel cell involves the reaction of hydrogen (the "fuel") and oxygen (the "oxidizer") to form water (Figure 6–3). The electrodes are nonconsumable, porous carbon. At the anode, hydrogen gas is bubbled through carbon that contains a catalyst, such as fine particles of platinum or palladium. At the cathode, oxygen is bubbled through carbon impregnated with cobalt oxide, platinum, or silver as catalyst. A concentrated solution of sodium hydroxide or potassium hydroxide separates the two electrodes.

The catalyst dissociates the H_2 molecules into H atoms which react with hydroxide ions in the solution to form water.

$$H_2 \xrightarrow{\text{Catalyst}} 2H \cdot$$

$$2H \cdot + 2OH^- \longrightarrow 2H_2O + 2e^-$$

Net anode reaction (oxidation)

$$H_2 + 2OH^- \longrightarrow 2H_2O + 2e^-$$

The electrons produced at the anode flow through the external circuit to the cathode. The oxygen, bubbled in at the cathode, uses these electrons to form the hydroxide ions consumed in the anode reaction above:

Cathode reaction $$O_2 + 2H_2O + 4e^- \xrightarrow{\text{Catalyst}} 4OH^-$$

The hydroxide ions complete the cycle by migrating through the solution from the oxygen electrode to the hydrogen electrode. As in all batteries, the electrical power comes from the flow of electrons from one electrode through the external circuit and back to the other electrode. The cell operates between 25° and 60°C and gives an EMF of 0.9 volts.

Interest in fuel cells has been promoted by space travel and in the search for peaceful uses for nuclear energy. For example, sunlight and proper catalysts can be used to decompose water into hydrogen and oxygen that can then be used in fuel cells to produce electrical power. It has been suggested that nuclear power could be made more economical if, during periods of low demand, the power were used to decompose water into hydrogen and oxygen. During peak periods the hydrogen and oxygen would be used in fuel cells to produce electrical energy. Farm tractors have been developed that are powered by fuel cells. Stimulated by the interest in abating air pollution, scientists are making considerable effort to develop automobiles completely powered by fuel cells.

NUCLEAR PARTICLES AND REACTIONS

There are many nuclear reactions that occur in naturally radioactive substances since all the elements above bismuth in atomic number and a few below have one or more naturally occurring radioactive isotopes. Radioactive elements spontaneously give off three different kinds of rays—alpha (α) rays (streams of helium ions, He^{2+}), beta (β) rays (streams of electrons), and gamma (γ) rays (high energy electromagnetic radiation similar to visible and ultraviolet light.)

Let's look at a few examples of nuclear reactions.

The isotope of uranium with atomic mass 238 is an alpha emitter. The atomic number of uranium is 92, which means that uranium-238 has 92 protons and 146 (i.e., 238 − 92) neutrons in the nucleus. When the uranium-238 nucleus gives off an alpha particle, made up of two protons and two neutrons, it necessarily loses 4 units of atomic mass and 2 units of atomic charge. The resulting nucleus would then have a mass of 234 and a nuclear charge of 90. Atoms containing 90 protons in the nucleus are atoms of thorium, not uranium. This spontaneous nuclear reaction then has changed an atom of one element into an atom of another element, and is an example of the transmutation of elements. The decomposition of the U-238 nucleus is stated briefly by the nuclear equation:

$$\ce{^{238}_{92}U} \longrightarrow \ce{^{4}_{2}He} + \ce{^{234}_{90}Th}$$

Th-234 is always found with U-238 in natural ore deposits and almost always in just that concentration which was predicted by the speed of the reaction that produces it and another (below) which consumes it.

Thorium-234 is also radioactive. However, this nucleus is a beta emitter. This poses an interesting question: How can a nucleus containing protons and neutrons emit an electron? It has been established that an electron and a proton can combine outside of the nucleus to form a neutron. Therefore, the reverse process is proposed to occur in the nucleus. A neutron decomposes, giving up an electron and changing itself into a proton.

$$\ce{^{1}_{0}n} \longrightarrow \ce{^{1}_{1}p} + \ce{^{0}_{-1}e}$$

Since the mass of the electron is almost zero compared to that of the proton and neutron, the nucleus would maintain nearly the same mass but it would now carry one more positive charge (a proton instead of one of the neutrons). This nucleus is no longer thorium, since thorium can have only 90 protons in the nucleus; it is now a nucleus of element 91 (protactinium, Pa). The reaction is:

$$\ce{^{234}_{90}Th} \longrightarrow \ce{^{234}_{91}Pa} + \ce{^{0}_{-1}e}$$

Gamma radiation may or may not be given off simultaneously with alpha or beta rays, depending upon the particular nuclear reaction involved. Since gamma rays involve no charge and essentially no mass, it is evident that the emission of a gamma photon cannot alone account for a transmutation event.

HALF-LIFE

Uranium decay is extremely slow compared to thorium decay. Each of these, however, and every other radioactive decay reaction, is found to have a characteristic period of time required for half of the radioactive materials originally present to undergo transformation. This period of time, *the half-life*, is independent of the amount of radioactive material present, and is determined only by the composition and structure of the nucleus. For example, in the reaction

117

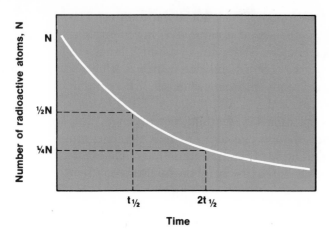

FIGURE 6-4 Half-life. The rate of decay for a radioactive atom depends in a very special way on the number of those atoms present. The rate is such that in a given period of time—the half-life for the species—one-half of the original number of atoms will be gone regardless of the number present at the start. In this graph the number of atoms remaining is plotted with time. $t_{\frac{1}{2}}$ is the half-life, and at the end of one $t_{\frac{1}{2}}$ period the original number, N, is reduced to $\frac{1}{2}$N. After two of these periods, the number is reduced to one-half of $\frac{1}{2}$N, or $\frac{1}{4}$N.

just mentioned, half the thorium will remain after 24 days. In another 24 days one-half of that half ($\frac{1}{4}$) will be left. This process continues indefinitely with one-half the thorium-234 that remains decaying each 24 days. Some half-lives are extremely long and others are extremely short. The half-life for the U-238 alpha decay is 4.5 billion years. As one would expect, relatively large amounts of U-238 can be found in nature, while only traces of Th-234 occur (Figure 6–4).

The radioactive decay of thorium-234 into protactinium-234 is the second

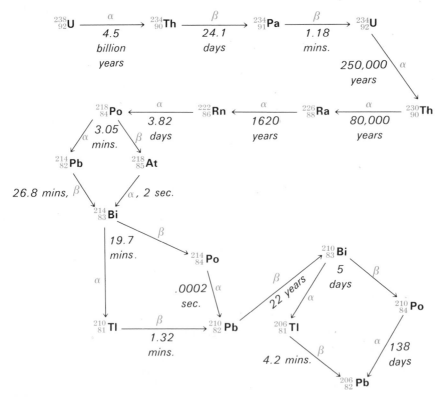

FIGURE 6-5   The uranium radioactive decay series. The times given are half-lives.

step in a series of nuclear decays that starts with U-238 and after 14 periodic decays results in a stable (nonradioactive) isotope of lead, Pb-206. This decay series is called the uranium series (Figure 6–5).

DATING THE UNIVERSE

If the assumption is made that a uranium ore originally contained uranium compounds and no other members of the uranium decay series, one can measure the present concentrations of the various isotopes in the rock and thereby determine the age of the rock. This is possible since the half-life for each decay reaction is known. For example, 1.0 gram (0.0042 mole) of uranium-238, after one half-life of 4.5 billion years, would leave 0.50 g (0.0021 mole) of U-238, and in the process would produce 0.43 g (0.0021 mole) of lead-206. Now if one finds in a uranium ore the ratio of 0.50 g of uranium to 0.43 g of lead, it would follow that the rock is 4.5 billion years old. Ages determined in this fashion for various rocks taken from different parts of the world show that the rocks are approximately 3 billion years old. Some meteorites have been determined to be 4.5 billion years old and some rocks from the moon are equally old. In general, radioactive dating techniques determine the age of the planets in the solar system to be approximately 4.5 billion years.

ARTIFICIAL NUCLEAR REACTIONS

After it was realized that the nuclei of some of the heavier isotopes were unstable, scientists wondered whether nuclear reactions could be initiated with other nuclei that were apparently stable. In order to explain the stability of any nucleus, one must postulate the existence of short-range, attractive, nuclear forces between positive particles (protons) and neutral particles (neutrons). Such forces must be stronger than the electrostatic forces that would tend to make the positive particles fly apart. It seemed reasonable to believe that two nuclei might possibly react to form new nuclear species if they could be brought so close together that the short-range nuclear forces could be operative. In order to achieve this, the two nuclei would have to approach each other with sufficient kinetic energy to overcome the electrostatic repulsion. In such a case, one could postulate an unstable compound nucleus that would emit particles, or energy, or both, in seeking a new and more stable structure.

In 1919, Ernest Rutherford was successful in producing the first artificially radioactive isotopes. He placed nitrogen gas in a cloud chamber and directed alpha particles into the chamber. He found that a nuclear reaction occurred:

$$^{14}_{7}\text{N} + ^{4}_{2}\text{He} \longrightarrow ^{18}_{9}\text{F} \longrightarrow ^{17}_{8}\text{O} + ^{1}_{1}\text{H}$$

This experiment was the basis of the discovery of the proton, $^{1}_{1}\text{H}$.

Following Rutherford's original transmutation experiment, there was considerable interest in bombarding stable nuclei with high energy particles to discover new nuclear reactions. As you might guess, numerous reactions were found. For example, beryllium can be converted to carbon when subjected to an alpha ray bombardment:

$$\ce{^9_4Be + ^4_2He} \longrightarrow \ce{^{13}_6C} \longrightarrow \ce{^{12}_6C + ^1_0n}$$

Although the carbon-12 produced in this reaction is stable, the neutron is given off with sufficient energy to provoke additional nuclear reactions in nuclei with which it collides. This nuclear reaction was used by James Chadwick in 1932 to prove the existence of the previously postulated neutron.

TRANSURANIUM ELEMENTS— THE BEGINNINGS OF NUCLEAR ENERGY

The heaviest known element prior to 1940 was the natural element, uranium. The invention of the cyclotron and other devices to obtain high-energy particles made it possible to bombard heavy nuclei with these particles to obtain even more massive nuclei, the *transuranium elements*, with atomic numbers greater than 92.

In 1940, McMillan and Abelson prepared element 93, the synthetic element neptunium (Np). The experiment involved directing a stream of high-energy deuterons ($^2_1$H) onto a target of uranium-238. The initial reaction involved the conversion of uranium-238 to uranium-239.

$$\ce{^{238}_{92}U + ^2_1H} \longrightarrow \ce{^{239}_{92}U + ^1_1H}$$

Uranium-239 has a half-life of 23.5 minutes, converting spontaneously to the new element, neptunium, by the emission of beta particles.

$$\ce{^{239}_{92}U} \longrightarrow \ce{^{239}_{93}Np + ^0_{-1}e}$$

Neptunium itself, however, is unstable, with a half-life of 2.33 days; it converts into a second new element, plutonium.

$$\ce{^{239}_{93}Np} \longrightarrow \ce{^{239}_{94}Pu + ^0_{-1}e}$$

Plutonium-239, like its parent atom, is an alpha emitter, but the half-life of Pu-239 is 23,100 years. Because of the relative values of the half-lives, very little neptunium could be accumulated, but the plutonium could be obtained in larger quantities. Plutonium-239 is important as fissionable material, and nuclear power stations can be run using it as fuel (see Fission Reactions, the next section). In this way it supplements our supply of naturally-occurring uranium-235 as fission fuel. The names of the two synthetic elements were taken from the mythological names, Neptune and Pluto.

Although Neptune and Pluto are the last of the known planets in the solar system, their namesakes are not last in the list of elements. The rush of transuranium experiments that followed these earlier experiments produced additional new elements: americium (Am), curium (Cm), berkelium (Bk), californium (Cf),

TABLE 6-3   NUCLEAR REACTIONS USED TO PRODUCE TRANSURANIUM ELEMENTS

ELEMENT	ATOMIC NUMBER	REACTION
Neptunium, Np	93	$^{238}_{92}U + ^{1}_{0}n \longrightarrow ^{239}_{93}Np + ^{0}_{-1}e$
Plutonium, Pu	94	$^{238}_{92}U + ^{2}_{1}H \longrightarrow ^{238}_{93}Np + 2^{1}_{0}n$
		$^{238}_{93}Np \longrightarrow ^{238}_{94}Pu + ^{0}_{-1}e$
Americium, Am	95	$^{239}_{94}Pu + ^{1}_{0}n \longrightarrow ^{240}_{95}Am + ^{0}_{-1}e$
Curium, Cm	96	$^{239}_{94}Pu + ^{4}_{2}He \longrightarrow ^{242}_{96}Cm + ^{1}_{0}n$
Berkelium, Bk	97	$^{241}_{95}Am + ^{4}_{2}He \longrightarrow ^{243}_{97}Bk + 2^{1}_{0}n$
Californium, Cf	98	$^{242}_{96}Cm + ^{4}_{2}He \longrightarrow ^{245}_{98}Cf + ^{1}_{0}n$
Einsteinium, Es	99	$^{238}_{92}U + 15^{1}_{0}n \longrightarrow ^{253}_{99}Es + 7^{0}_{-1}e$
Fermium, Fm	100	$^{238}_{92}U + 17^{1}_{0}n \longrightarrow ^{255}_{100}Fm + 8^{0}_{-1}e$
Mendelevium, Mv	101	$^{253}_{99}Es + ^{4}_{2}He \longrightarrow ^{256}_{101}Mv + ^{1}_{0}n$
Nobelium, No	102	$^{246}_{96}Cm + ^{12}_{6}C \longrightarrow ^{254}_{102}No + 4^{1}_{0}n$
Lawrencium, Lr	103	$^{252}_{98}Cf + ^{10}_{5}B \longrightarrow ^{257}_{103}Lr + 5^{1}_{0}n$
Kurchatovium, Ku	104	$^{242}_{94}Pu + ^{22}_{10}Ne \longrightarrow ^{260}_{104}Ku + 4^{1}_{0}n$
Hahnium, Ha	105	$^{249}_{98}Cf + ^{15}_{7}N \longrightarrow ^{260}_{105}Ha + 4^{1}_{0}n$

einsteinium (Es), fermium (Fm), mendelevium (Md), nobelium (No), lawrencium (Lr), kurchatovium (Ku), and hahnium (Ha), the new elements being named after countries, states, cities, and people. Reactions employed in the production of the transuranium elements are given in Table 6–3. As accelerators with greater and greater energy capabilities are produced, elements of even higher atomic number may be produced.

FISSION REACTIONS

Perhaps no manifestation of matter has captured the awe of people to quite the extent that atomic fission has in recent years. Fission, or the splitting of heavy nuclei into middle-weight nuclei, is the basis of the atomic bomb and of nuclear power stations. Since 1938 when Otto Hahn, Fritz Strassmann, Lise Meitner, and Otto Frisch discovered that $^{235}_{92}U$ is fissionable, man has speculated on the possibility of obtaining a major portion of his energy needs from this process. We are already quite far along toward obtaining that goal.

Fission can occur when a thermal neutron (with a kinetic energy about the same as that of gaseous molecules at ordinary temperatures) enters certain heavy nuclei with an odd number of neutrons (^{235}U, ^{233}U, ^{239}Pu). The splitting of the heavy nucleus produces two smaller nuclei, two or more neutrons (an average of 2.5 for ^{235}U), and much energy. Typical nuclear fission reactions are written as follows:

$$^{235}_{92}U + ^{1}_{0}n \longrightarrow ^{141}_{56}Ba + ^{92}_{36}Kr + 3^{1}_{0}n + energy$$

$$^{235}_{92}U + ^{1}_{0}n \longrightarrow ^{103}_{42}Mo + ^{131}_{50}Sn + 2^{1}_{0}n + energy$$

121

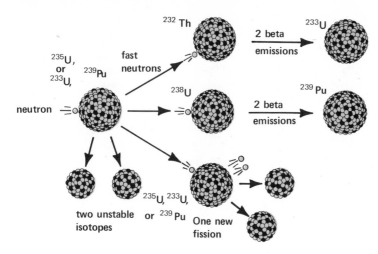

FIGURE 6-6 A chain reaction. A thermal neutron collides with a fissionable nucleus and the resulting reaction produces three additional neutrons. These neutrons can either convert nonfissionable nuclei, such as ^{232}Th, to fissionable ones or cause additional fission reactions. If enough fissionable nuclei are present, a chain reaction will be sustained.

Note that the same nucleus may split in more than one way. The fission products $^{141}_{56}$Ba, $^{92}_{36}$Kr, and so forth, emit beta particles and gamma rays until stable isotopes are reached.

The emitted neutrons can cause fission of other heavy atoms. For example, the three emitted neutrons of the first reaction above could produce fission in three more U atoms; the nine emitted neutrons could produce nine more fissions; the 27 neutrons from these fissions could produce 81 neutrons; the 81, 243; the 243, 729; and so on. This process is called a "chain reaction" (Figure 6–6), and it occurs at a maximum rate when the sample of uranium is large enough for most of the emitted neutrons to be captured by a nucleus before passing out of the sample. Sufficient sample to sustain a chain reaction is called the *critical mass*.

In the atomic bomb the critical mass is kept separated into several smaller subcritical masses until detonation, at which time the masses are driven together. It is then that the tremendous energy is liberated and heats the gases in the vicinity up to five or 10 million degrees. The sudden expansion of gases literally explodes everything nearby and scatters the radioactive fission fragments over a wide area. In addition to the tremendous movement of gases which is, to a lesser degree associated with conventional bombs, there is the tremendous vaporizing heat that make the atomic bomb so devastating.

There is no danger of an atomic explosion in the mineral deposits in earth for two reasons. First, uranium is not found pure in nature; it is found only in compounds which in turn are mixed with other compounds. Second, less than 1 per cent of the uranium found in nature is ^{235}U which is fissionable. The other 99 per cent is ^{238}U which is not fissionable by thermal neutrons.

MASS DEFECT—A SOURCE OF ENERGY

What is the source of the tremendous energy of the fission process? It ultimately comes from mass being converted into energy according to Einstein's famous equation, $E = mc^2$, where E is energy that results from the loss of an amount of mass, m, and c^2 is the speed of light squared. If separate neutrons, electrons and protons are combined to form any particular atom, there is a loss of mass called the *mass defect*. For example, the calculated mass of one ^4_2He atom from the masses of the constituent particles is:

$2 \times 1.007825 = 2.015650$ amu, mass of 2 protons and 2 electrons
$2 \times 1.008665 = \underline{2.017330}$ amu, mass of 2 neutrons
4.032980 amu, calculated mass of ^4_2He atom

Since the measured mass of a ^4_2He atom is 4.002604 amu, the mass defect is 0.030376 amu.

4.032980 amu
$\underline{4.002604}$ amu
0.030376 amu, mass defect

Since the atom is more stable than the separated neutrons, protons and electrons, the atom is in a lower energy state. Hence, the 0.030376 amu lost per atom would be released in the form of energy if the ^4_2He atom were made from protons, electrons and neutrons. The energy equivalent of the mass defect is called the *binding energy*. The binding energy is analogous to the earlier concept of bond energy in that both are a measure of the energy necessary to separate the package (nucleus or molecule) into its parts.

Atoms with atomic numbers between 30 and 63 have a greater mass defect per nuclear particle than very light elements or very heavy ones, as shown in Figure 6–7. This means the most stable nuclei are found in the atomic number range 30 to 63.

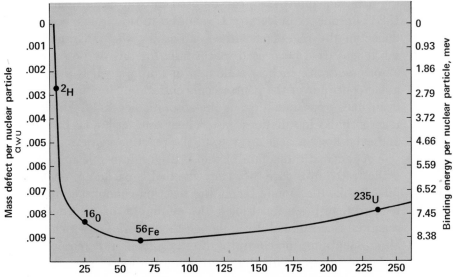

FIGURE 6–7   Mass defect for different nuclear masses. The most stable nuclei center around ^{56}Fe, which has the largest mass defect per nuclear particle.

Because of the relative stabilities, it is in the intermediate range of atomic numbers that most of the products of nuclear fission are found. Therefore, when fission occurs and smaller, more stable nuclei result, these nuclei will contain less mass per nuclear particle. In the process, mass must be changed into energy. This energy obtained from mass gives the fission process its tremendous energy. It takes only about one kilogram of ^{235}U or ^{239}Pu undergoing fission to be equivalent to the energy released by 20,000 tons (20 kilotons) of ordinary explosives like TNT. The energy content in matter is further dramatized when it is realized that the atomic fragments from the 1 kg of nuclear fuel weigh 999 g, so only one-tenth of one per cent of the mass is actually converted to energy. The fission bombs dropped on Japan during World War II contained approximately this much fissionable material.

CONTROLLED NUCLEAR ENERGY

The fission of a uranium-235 nucleus by a slow-speed neutron to produce smaller nuclei, extra neutrons, and large amounts of energy suggested to Fermi and others, not only the possibility of an uncontrolled nuclear explosion, but also that the reaction could proceed at a moderate rate if the number of neutrons could be controlled. If a neutron control could be found, the concentration of neutrons could be maintained at a level sufficient to keep the fission process going but not high enough to allow an uncontrolled explosion. It would then be possible to drain the heat away from such a reactor on a continuing basis to do useful work. In 1942, Fermi, working at the University of Chicago, was successful in building the first atomic reactor, called an "atomic pile."

An atomic reactor has a number of essential components. The charge material (fuel) must be made of or contain significant concentrations of a fissionable isotope such as U-235, Pu-239, or U-233. Ordinary uranium, which is mostly the nonfissionable U-238, can be used since it has a small concentration of the U-235 isotope. A moderator is required to slow the speed of the neutrons produced in the reactions without absorbing them. Graphite, water and other substances have been used successfully for this purpose. A substance that will absorb neutrons, such as cadmium or boron steel, is present in order to have a fine control over the neutron concentration. Shielding, to protect the workers from dangerous radiation, is an absolute necessity. Shielding tends to make reactors bulky installations. Also, a heat-transfer fluid is provided to affect a large and even flow of heat away from the reaction center.

A schematic diagram of the X-10 graphite pile in Oak Ridge, Tennessee, is illustrated in Figure 6–8. Here the fuel is packed in openings in the graphite pile. Unlike the atomic bomb, the reaction cannot lead to an atomic explosion if more than the critical mass (the minimum mass needed for a self-sustaining reaction) is assembled. Even though the material of which the pile is made may absorb neutrons, the control rods offer a final and complete control to stop the fission reactions. With the control rods properly placed, the reaction can be maintained at various safe levels of activity.

124 Once the heat is produced in a nuclear reactor and safety measures are

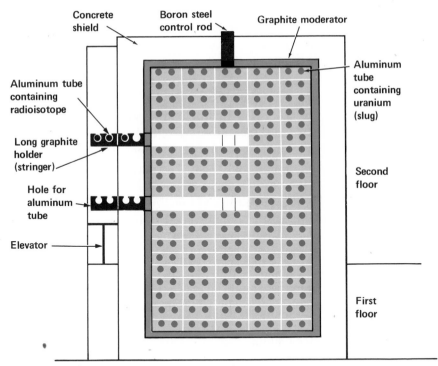

FIGURE 6-8   Schematic diagram of X-10 reactor at Oak Ridge, Tennessee.

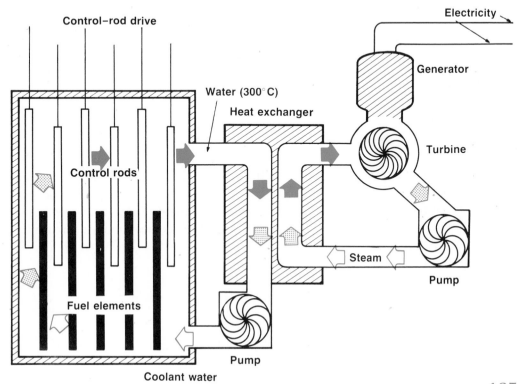

FIGURE 6-9   Electricity produced by atomic energy.

125

employed to protect against radiation, conventional technology allows this energy to be used to generate electricity, to power ships, or to operate any device that uses heat energy. A system for the nuclear production of electricity is illustrated in Figure 6–9.

Numerous nuclear power stations for the generation of electricity are now in operation around the world. However, this source of energy had trouble competing in cost with hydroelectric power, where such power is plentiful, during the two decades after World War II. Such costs have inhibited the development of nuclear power plants. Recent developments indicate that design size is a critical factor, and that with large enough installations, nuclear power may furnish the world's cheapest electricity.

There is not enough U-235 in nature to provide a long-lasting energy supply in terms of increasing human needs. However, U-238 can be converted into fissionable Pu-239 by using a *breeder reactor*. The breeder reactor contains a blanket concentrated in Pu-239 (or U-235). The amount of fissionable material in the blanket not only provides fuel for power generation, but also produces even more fissionable material in other "blankets" containing high concentrations of nonfissionable U-238. In addition to the conversion of U-238 to fissionable Pu-239, the relatively abundant Th-232 can be converted into fissionable U-233 in the breeder reactor. The equations are:

$$^{232}_{90}\text{Th} + ^{1}_{0}\text{n} \longrightarrow \ ^{233}_{90}\text{Th}$$

$$^{233}_{90}\text{Th} \xrightarrow{\text{23 min.}} \ ^{233}_{91}\text{Pa} + ^{0}_{-1}\text{e}$$

$$^{233}_{91}\text{Pa} \xrightarrow{\text{27 days}} \ ^{233}_{92}\text{U} + ^{0}_{-1}\text{e}$$

One of the major problems with the breeder reactor is what to do with the long-lasting radioactive wastes that are produced.

FUSION REACTIONS

When very light nuclei, such as H, He, Li, and so forth, are combined or fused to form an element of higher atomic number, energy must be given off consistent with the greater stability of the elements in this intermediate atomic number range (Figure 6–7). This energy, which comes from a decrease in mass, is the source of the energy released by the sun and by hydrogen bombs. Typical examples of fusion reactions are:

$$4^{1}_{1}\text{H} \longrightarrow \ ^{4}_{2}\text{He} + 2^{0}_{+1}\text{e} + \text{energy}$$

$$^{2}_{1}\text{H} + ^{2}_{1}\text{H} \longrightarrow \ ^{3}_{2}\text{He} + ^{1}_{0}\text{n} + \text{energy}$$

$$3^{4}_{2}\text{He} \longrightarrow \ ^{12}_{6}\text{C} + \text{energy}$$

$$^{12}_{6}\text{C} + ^{4}_{2}\text{He} \longrightarrow \ ^{16}_{8}\text{O} + \text{energy}$$

$$^{3}_{1}\text{H} + ^{2}_{1}\text{H} \longrightarrow \ ^{4}_{2}\text{He} + ^{1}_{0}\text{n} + \text{energy}$$

The first reaction ($^0_1$e is a positron) just mentioned is thought to be the overall reaction which takes place on the sun. It is responsible for the tremendous amount of energy released by the sun—about four times as much per gram of starting materials as fission reactions.

Fusion reactions take place rapidly when the temperature is of the order of 100 million degrees or more. At these high temperatures atoms do not exist as such; instead, there is a plasma of nuclei and of electrons. In this plasma nuclei merge or combine. In order to achieve the high temperatures required for the fusion reaction of the hydrogen bomb, a fission bomb (atomic bomb) is set off first.

One type of hydrogen bomb depends on the production of tritium ($^3_1$H) in the bomb. In this type lithium deuteride (^{6}Li^{2}H, a solid salt) is placed around an ordinary ^{235}U or ^{239}Pu fission bomb. The fission is set off in the usual way. Lithium-6 absorbs some of the neutrons produced and splits into tritium, $^3_1$H; helium, $^4_2$He; and deuterium, $^2_1$H.

$$^6_3\text{Li}^2_1\text{H} + ^1_0\text{n} \longrightarrow ^3_1\text{H} + ^4_2\text{He} + ^2_1\text{H}$$

The temperature reached by the fission of ^{235}U or ^{239}Pu is sufficiently high to bring about the fusion of tritium and deuterium:

$$^3_1\text{H} + ^2_1\text{H} \longrightarrow ^4_2\text{He} + ^1_0\text{n} + 17.6 \text{ mev}$$

A 20-megaton bomb usually contains about 300 lbs of lithium deuteride as well as a considerable amount of plutonium and uranium.

As yet the fusion reactions have not been "controlled." No physical container can contain the plasma without cooling it below the critical fusion temperature. Magnetic bottles, enclosures in space bounded by a magnetic field, have confined the plasma but not for long enough periods (Figure 6–10). Recent developments suggest that these "bottles" may soon hold the plasma long enough

FIGURE 6–10   A controlled fusion experiment in a "magnetic bottle." (From *Chemical and Engineering News*, August 30, 1971.)

so that the fusion reaction will occur—more energy will then be obtained than is fed into the plasma. Other interesting approaches to creating such intense energies use laser and electron beams at their maximum energy output. Controlled fusion energy may be very near. At this point, there is no clear understanding of all the environmental problems that will certainly arise with the production and utilization of such vast amounts of energy. We know there will be no fission-product waste (radioactive materials), but thermal pollution is a real possibility. This subject is discussed in chapter 18.

DIRECT USE OF THE SUN'S ENERGY

Earth's ultimate source of energy is the sun. The sun provides energy for photosynthesis, which gives us food to power our bodies, wood to burn, and the starting materials for coal and petroleum. The sun provides the energy for the water cycle. Without the sun, there would be no natural water power.

Although Earth, a speck in space, receives only about three ten-millionths (0.0000003) of the energy emitted from the sun, the power that the sun supplies to the earth's upper atmosphere is still staggering—about 2×10^{15} kcal/min or, 2.0 cal/cm^2/min. Due to reradiation from the atmosphere and the absorption and scattering of radiant energy by molecules in the lower portions of the atmosphere, the amount of radiation actually reaching the surface of this planet is about 1 cal/cm^2/min. The actual value depends on location, season, and weather conditions. Even this is a large amount of energy. For example, the roof of an average-sized house will receive about 10^8 calories/day when the radiation level is 1 cal/cm^2/min. This is equivalent to the heat energy derived from burning about 150 pounds of coal.

Nature makes use of a tiny fraction of this energy in photosynthetic processes and in the heating of air and water which gives rise to our weather. Man has long dreamed of using the seemingly "free" energy of the sun for his comfort, food preparation, and transportation. Most of the solar energy utilization experiments have involved the concentration of the radiation by optical means and the subsequent boiling of water or a similar fluid. These technical experiments have not led to a widespread use of solar energy, however, due to the high costs involved.

The future for solar energy use is tantalizing, nevertheless. For example, the world's present energy requirements per year are about 10^{18} kcal. An area of desert of about 28,000 square miles with little cloud cover or dust, near the equator, would receive about 1×10^{17} kcal/year of solar radiation. Such deserts exist in northern Chile, and could, if the need were great enough to justify the costs, supply a large portion of the world's energy needs. As an example of how this energy could be used, the solar energy could heat water to steam, which in turn could generate electricity, which could electrolyze water to hydrogen and oxygen. The hydrogen, then, could be piped to where the energy is needed and burned.

Another approach to the direct utilization of solar energy is the *solar battery*. The solar battery converts energy from the sun into electron flow. Solar batteries

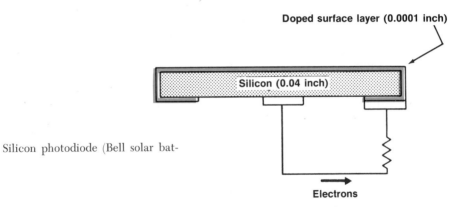

Doped surface layer (0.0001 inch)

Silicon (0.04 inch)

Electrons

FIGURE 6-11   Silicon photodiode (Bell solar battery.)

are efficient and are capable of generating electrical power from sunlight at the
rate of at least 90 watts per square yard of illuminated surface. They are now
used on space ships and communication satellites, and in Israel and other parts
of the Middle East, India and Pakistan, South Africa, and Azerbaijan SSR to
obtain electric power.

One type of solar battery consists of two layers of almost pure silicon (Figure
6–11). The lower, thicker layer contains a trace of arsenic and the upper, thinner
layer a trace of boron. Silicon has four valence electrons and forms a tetrahedral,
diamond-like, crystalline structure. Each silicon atom is covalently bonded to
four other silicon atoms. Arsenic has five valence electrons. When arsenic atoms
are included in the silicon structure, only four of the five valence electrons of
arsenic are used for bonding with four silicon atoms; one electron is relatively

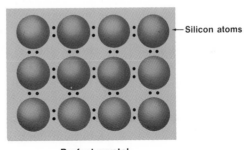

Silicon atoms

Perfect crystal

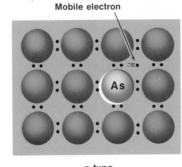

Mobile electron

Positive hole

FIGURE   6-12   Schematic
drawing   of   semiconductor
crystal layers derived from
silicon (after Masterton and
Slowinski).

n-type   **p-type**

129

free to roam (Figure 6–12). Boron has three valence electrons. When boron atoms are included in the silicon structure, there is a deficiency of one electron around the boron atom; this creates "holes" in the boron-enriched layer. Even without sunlight, the "extra" electrons of arsenic diffuse into the holes in the boron. The driving force is the strong tendency to pair electrons. Potentials as high as 1000 volts are incapable of reversing the flow. The negative charge built up in the boron layer would hinder the flow of electrons into that layer and eventually stop the flow. The opposing factors—repulsion between free electrons and the drive to pair electrons—finally bring about an equilibrium.

When sunlight strikes the boron layer, the equilibrium is disturbed. If the wafer is connected, as an ordinary battery would be, electrons flow from what appears to be the negative arsenic layer, through the circuit, and back into the boron layer. The fact that electrons enter the circuit from the arsenic layer can be explained if sunlight unpairs electrons in the boron layer, and the freed electrons are repelled to the arsenic layer and into the circuit. This pulls electrons into the boron layer to maintain the new balance between electron pairing and electron repulsion.

The solar battery can develop electrical potentials of up to half a volt, and up to nine watts of power, from each square foot exposed to the sun. Only about 10 per cent of the energy received on the surface of a solar battery is converted into electrical energy. This is not much, but the advantage of the solar battery is that it has no moving parts, no liquids, no corrosive chemicals—it just keeps on generating electricity indefinitely merely by lying in the sun. The drawbacks of the solar battery are the very large area required for large amounts of power, the high costs of the very pure materials, and the fact that they work only when the sun is shining. Since the first practical use of solar batteries in 1955 to power eight rural telephones in Georgia, they have undergone much development and much more is expected because of their great potential use.

In summary, an almost endless amount of energy is available in the universe for man's use. As yet he has not learned to control it sufficiently well to have cheap energy for all. However, he is making progress. It is evident that chemical energy can only be a holding bank and that fission of heavy nuclei is limited by the amounts of fuels and the discharge of radioactive wastes. If we divert too much of the sun's radiant energy we are not at all sure of the environmental results.

QUESTIONS

1. What is your attitude toward using up the fossil fuels within a few decades? Do we owe future generations a supply of these resources? Would you agree to give up air-conditioning, private cars, and power tools, to mention a few examples, and to limit heating and cooking, so as to share these fuels with your grandchildren?

2. Which theoretically gives greater energy per mole?

   (a) the burning of gasoline
   (b) the fission of ^{235}U
   (c) the fusion of ^{56}Fe

3. Which is the more efficient use of energy?

(a) burning coal to produce electricity which in turn is used to heat homes
(b) burning gas or coal to heat the home directly

4. Give three examples of systems which contain chemical energy that can be used as a source of heat energy.

5. Is the electrical energy where you live produced by burning fossil fuels? If not, what is the energy source? Are there pollution problems connected with the generation of the electrical power?

6. What produces the tremendous energy of a fission reaction?

7. What components would you need to assemble a charged lead storage battery?

8. Why is it difficult to fuse two $^{52}_{24}$Cr nuclei? Explain.

9. What is meant by a chain reaction?

10. How can ages of rocks be determined? What assumptions are made in using this method?

11. Use data given in the chapter to calculate the mass defect of the isotope $^{9}_{4}$Be, which has an atomic mass of 9.01219 amu.

12. What major problem is associated with harnessing the energy from a fusion reaction?

13. (a) If a power plant burns 1000 tons of coal a day, the coal containing 2 per cent sulfur, what is the maximum weight of sulfur that could go into the air each day from the power plant?
(b) If the sulfur is converted to sulfur dioxide according to the equation: $S + O_2 \longrightarrow SO_2$, what weight of sulfur dioxide would be emitted into the air in a day?
(c) Sulfur dioxide is soluble in water. Can you suggest a plan whereby this gas can be removed from fumes from coal burning?

14. In general, how have the synthetic transuranium elements been produced?

15. What does the symbol $^{11}_{5}$B mean?

16. (a) The half-life of ^{218}Po is 3 minutes. How much of a 2-gram sample of this isotope remains after 15 minutes?
(b) Suppose you wanted to buy some of this isotope and it requires 5 hours for it to reach you. How much should you order if you want to use 0.001 gram?

17. Look up the origin of the word "mutation" and explain why the word "transmutation" was an apt choice to describe the changing of one element into another.

18. If radium atoms ($^{226}_{88}$Ra) lose one alpha particle per atom, what element is formed? What is its atomic weight? What is its atomic number?

19. Suppose you were given $1000.00 and told that you could spend half of it the first year, half of the balance the second year, and so on. If you spent the maximum allowed, at the end of what year would you have $31.25 of the original $1000.00 left? How is this problem analogous to the half-life of a radioactive substance? Can you think of ways in which it is not?

20. Estimate your daily bodily energy expenditure. If solar batteries can furnish 20 calories per second per square yard, what size solar battery would you need for your own personal energy needs?

21. Using the bond energies in Table 6–1 calculate the expected heat effects for the reactions

(a) $H_2 + F_2 \longrightarrow 2HF$

131

(b) $2H_2S + 3O_2 \longrightarrow 2H_2O + 2SO_2$
(c) $2C_2H_6 + 7O_2 \longrightarrow 4CO_2 + 6H_2O$
(d) $SiH_4 + 4Cl_2 \longrightarrow SiCl_4 + 4HCl$

SUGGESTIONS FOR FURTHER READING

References

Angrist, S. W., and Hepler, L. G., "Order and Chaos," Basic Books, Inc., New York, 1967.
Ayres, E., and Scarlett, C. A., "Energy Sources—The Wealth of the World," McGraw-Hill Book Co., Inc., New York, 1952.
Chalmers, B., "Energy," Academic Press, New York, 1963.
Daniels, F., "Direct Use of the Sun's Energy," Yale University Press, New Haven, 1964.
Goldsby, R. A., "Energy and Cells," MacMillan, New York, 1967.
Libby, W. F., "Radiocarbon Dating," University of Chicago Press, Chicago, 1955.
Ross, F., "The World of Power and Energy," Lothrop, 1967.

Paperbacks

Seaborg, G. T., "Man-Made Transuranium Elements," Prentice-Hall, Inc., New York, 1963.

Articles

Chappin, G. R., "Nuclear Fission," *Chemistry*, Vol. 40, No. 7, p. 25 (1967).
Clark, H. M., "The Origin of Nuclear Science," *Chemistry*, Vol. 40, No. 7, pp. 8–11 (1967).
"Detecting Forgeries in Paintings," *Chemistry*, Vol. 41, No. 5, p. 5 (1968).
"Element 105," *Chemistry*, Vol. 43, No. 6, p. 20 (1970).
"Energy and Power," *Scientific American*, (entire issue of September, 1971 issue devoted to various facets of this important topic).
Flerov, G. N., and Zvara, I., "Synthesis of Transuranium Elements," *Science*, Vol. 4, No. 7, p. 63, July (1968).
Mills, Ga., H. R. Johnson, and H. Perry, "Fuels Management in an Environmental Age," *Environmental Science and Technology*, January 1971, p. 30.
"Nuclear Power and By-Product Plutonium," *Chemistry*, Vol. 40, No. 3, p. 7 (1967).
"Otto Hahn," *Chemistry*, Vol. 39, No. 12, p. 5 (1966).
"Radiocarbon Dating," *Chemistry*, Vol. 43, No. 7, p. 24 (1970).
Schur, S. H., "Energy," *Scientific American*, Vol. 209, No. 3, p. 111 (1963).
Seaborg, G. T., "Some Recollections of Early Nuclear Age Chemistry," *Journal of Chemical Education*, Vol. 45, p. 278 (1968).
"Sun Powered Furnace," *Chemistry*, Vol. 43, No. 11, p. 23 (1970).

Part 2

A Star Performer— Carbon

7

WHAT IF WE HAD NO CARBON?

Carbon is a paradox. It is only 19th among the elements in abundance, and only about 0.027 per cent of the earth's crust is carbon. But the importance of carbon to life cannot be overemphasized. Consider what the world would be like if all the carbon and carbon compounds were suddenly removed from it. The result would be something like the barren surface of the moon! Many of the everyday things we take for granted would be impossible without the versatile element carbon—no food, fuel, clothing, books, drugs, plastics, or detergents could exist without it. Much of our building materials would be lost, and even we would not exist. If all the carbon were removed from an animal cell, the structure of the cell would be completely destroyed.

Giant industries are based on various aspects of carbon chemistry. Petroleum and coal furnish us with fuels which we burn to make life more comfortable, to provide electric power, and to transport us from place to place. In addition, petroleum and coal yield a wide variety of carbon compounds which end up in such varied applications as double knit slacks, lipstick, and video tape. Some simple bacteria will even feed and grow in petroleum, under proper conditions, yielding proteins which could help solve the world's food problems were it not for the fact that we are rapidly using up our petroleum and coal reserves in other ways. It has been projected that petroleum and coal will be used only as sources for chemicals by the beginning of the next century.

The very large and important branch of chemistry devoted to the study of carbon compounds is called "organic chemistry." The name "organic" is actually a relic of the past when chemical compounds produced from once-living matter were called "organic" and all other compounds were called "inorganic."

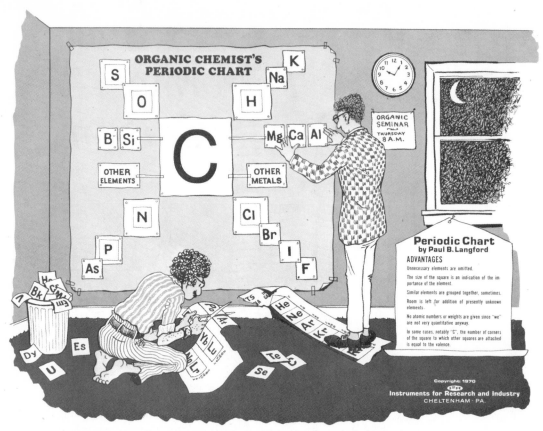

FIGURE 7-1   The organic chemist's periodic chart. Thanks for permission from Paul Langford, David Lipscomb College, and Instruments for Research and Industry, Cheltenham, Pa.

Prior to 1828, it was widely believed that chemical compounds synthesized by living matter could not be made without the living matter—a "vital force" was necessary for the synthesis. In 1828, a young German chemist, Friedrich Wöhler, destroyed the vital force myth and opened the door to the myriad

Frederic Wöhler (1800–1882) was Professor of Chemistry at the University of Berlin and then at Göttingen. His preparation of the organic compound urea from ammonium cyanate, an inorganic source, did much to overturn the theory that organic compounds must be prepared in living organisms. One of the first to study the properties of aluminum, he discovered the element beryllium and is known for many other outstanding contributions to chemistry. He was first to study chemical changes within the body (metabolism).

organic compounds that we now know to exist. Heating a solution of silver cyanate and ammonium chloride, neither of which had been derived from any living substance, Wöhler prepared urea, a biological waste product found in urine of animals.

$$\text{AgCNO} \; + \; \text{NH}_4\text{Cl} \; \longrightarrow \; \text{AgCl} \; + \; \text{NH}_4\text{CNO}$$

Silver cyanate | Ammonium chloride | Silver chloride (Precipitate) | Ammonium cyanate

$$\text{NH}_4\text{CNO} \quad \xrightarrow{\text{Heat}} \quad \text{H}_2\text{N}\overset{\displaystyle O}{\overset{\|}{\text{C}}}\text{NH}_2$$

Ammonium cyanate   Urea

The inevitable questioning and probing of Wöhler's work could not destroy the fact that urea had been prepared from materials not related to living materials. The notion of a mysterious vital force declined as other chemists began to prepare more and more organic chemicals without the aid of a living system. Some form of the "vital force theory" still rears its head from time to time. The theory has been raised again in our time as man tries to prepare "life in a test tube" (see also Chapter 12).

Soon after Wöhler's discovery, it was shown that chemistry could do more than imitate the products of living tissue; it could form unique materials of its own. The preparation of new and different organic compounds through chemical reaction is called *organic synthesis*. As a result of synthesis of new compounds and isolation of those occurring in nature, there are now over *two million different* organic compounds known. Thousands of new ones are added to this list every year.

FIGURE 7-2   Once a new compound has been prepared it must be identified. Here a student is determining the structure of a new compound using a device called a nuclear magnetic resonance spectrometer.

135

Later in this chapter we shall look at a few types of organic syntheses, but to begin our study of carbon chemistry let's try to understand why there are so many compounds of carbon.

WHY SO MANY CARBON COMPOUNDS?

The answer lies in atomic theory. In Chapter 3 we saw that carbon atoms contain four bonding electrons. This means they can bond to four other atoms or three or even two other atoms using single, double, or triple bonding. The intermediate electronegativity value is not large enough to enable a carbon atom to remove electrons completely from metals, but it is sufficient to keep even the most electronegative element, fluorine, from removing an electron completely from carbon. As a result, carbon atoms tend to share electrons with other atoms in covalent bonds.

Carbon atoms have a remarkable tendency to form strong covalent bonds with other carbon atoms, a property not exhibited by many elements. (Other elements sharing this property are oxygen, O_2, nitrogen, N_2, phosphorus, P_4, and sulfur, S_8.) This tendency is so pronounced that compounds are known which contain thousands of these bonds—literally a carbon-to-carbon chain. A molecule of natural rubber is just such a chain.

As we shall see later, the ability of carbon atoms to bond strongly with each other leads to seemingly endless arrays of possible combinations. Other reasons for the multiplicity of carbon compounds are given in the following sections.

CHAINS OF CARBON ATOMS—
THE HYDROCARBONS

Only two elements, hydrogen and carbon, are required to explain the existence of literally thousands of compounds known as "hydrocarbons." The simplest hydrocarbon is CH_4 (methane), the major component of natural gas (Figure 7–3).

Ethane (C_2H_6) is a minor component of natural gas. We are rapidly using our supplies of methane and ethane as they are burned to heat homes and

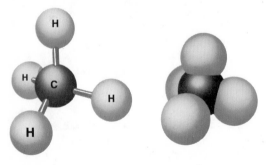

FIGURE 7–3   Methane. (a) Ball-and-stick model showing tetrahedral structure. (b) Model of methane, CH_4, showing relative sizes of atoms.

a   b

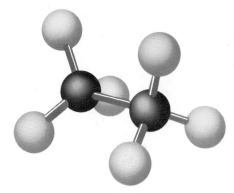

FIGURE 7-4   Ball-and-stick model of ethane.

Ethane

factories and supply power. If the carbon and hydrogen are completely oxidized, the products of combustion of hydrocarbons are carbon dioxide and water.

$$CH_4(gas) + 2O_2(gas) \longrightarrow CO_2(gas) + 2H_2O(gas) + 192,000 \text{ calories/mole}$$

When insufficient air is present, many other products are formed. Some examples of these will be given in the chapter on air pollution.

Atoms are free to twist around a single bond. In ethane, the rotation around the carbon-carbon bond produces many different forms called *conformations*. Two conformations of ethane are shown in Figure 7–5. Since the energy differences between conformations are quite small, it is not possible to isolate them as separate compounds.

The three-carbon structure of the hydrocarbon propane (C_3H_8) is represented in two dimensions as:

$$\begin{array}{ccccc} & H & H & H & \\ | & | & | & \\ H-&C-&C-&C-&H \\ | & | & | & \\ & H & H & H & \end{array}$$

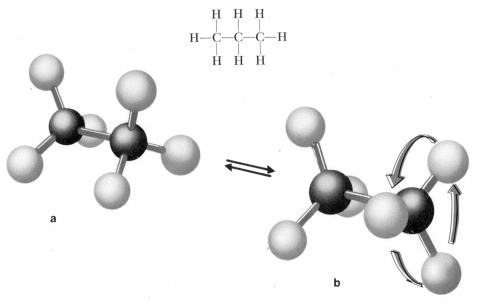

a

b

FIGURE 7-5   Two possible rotational forms of the ethane molecule. The hydrogen atoms in the methyl groups may be in a lined-up position (a), staggered (b), or in any intermediate position.

137

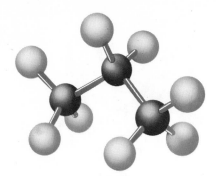

Propane

FIGURE 7-6   Ball-and-stick model of propane.

In Figure 7–6, note that the three carbon atoms in propane do not lie in a straight line because of the tetrahedral bonding about each carbon atom.

It is apparent that these bonding concepts can be extended to a four-carbon molecule and to a limitless number of even larger hydrocarbon molecules. Actually, many such compounds are known.

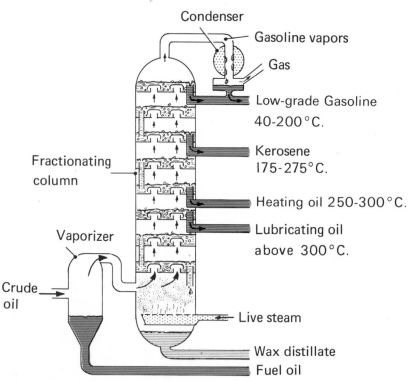

FIGURE 7-7   A diagram of a fractionating column for distilling petroleum. Notice that the higher boiling substances condense out at the lower levels and the lower boiling substances do not condense until the higher, cooler levels. (From Routh, J. I., D. P. Eyman, and D. J. Burton, "A Brief Introduction to General, Organic and Biochemistry," W. B. Saunders Co., Philadelphia (1971).)

138

TABLE 7-1   PETROLEUM FRACTIONS

FRACTION	COMPOSITION	DISTILLATION RANGE
Natural gas	C_1—C_4	Less than 20°C
Liquified petroleum	C_5—C_6	20–60°
Gasoline	C_4—C_8	40–200°
Kerosene	C_{10}—C_{16}	175–275°
Fuel oil, diesel oil	C_{15}—C_{20}	250–400°
Lubricating oils	C_{18}—C_{22}	Above 300°
Asphalt or	C_{20} and above	Non-volatile
petroleum coke	(complicated structures)	

HYDROCARBONS FROM PETROLEUM

Nature provides us with an abundant supply of hydrocarbons in petroleum. The typical petroleum feedstock (crude oil) contains a fantastic number of different compounds. These are separated from each other by fractional distillation in which molecules of roughly the same molecular weight distil in the same temperature range (Figure 7–7).

STRUCTURAL ISOMERS

When given the task of writing the structural formula for butane, C_4H_{10}, we soon discover there are two possible ways of doing it.

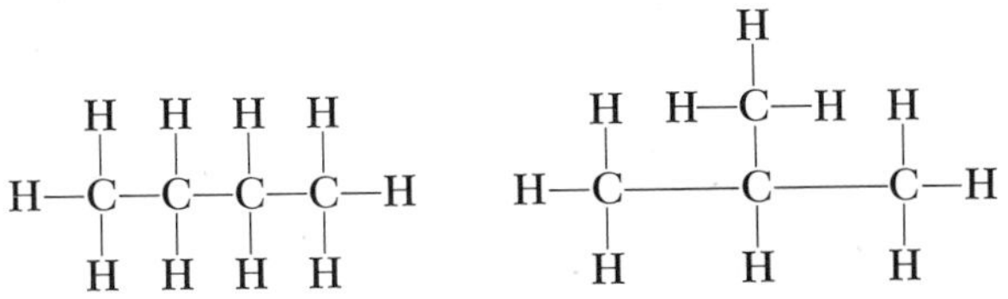

	n-Butane	Isobutane
Melting point	−138.3°C	−160°C
Boiling point	−0.5°C	−12°C
Density	0.610 g/ml	0.557 g/ml

The two formulas represent two distinctly different compounds as their different properties illustrate. These two compounds are *isomers* of each other. Isomers have the same molecular formula but have different structures.

It must be concluded, then, that molecular formulas such as C_4H_{10} are sometimes ambiguous, and that structural formulas are necessary. If no carbon atom is attached to more than two other carbon atoms, the carbon chain is said to be a "straight chain" structure. Actually, as shown in Figure 7–8, the carbon chain is bent (109° 28′) at each carbon atom, but it is called a straight chain because the carbon atoms are bonded together in succession one after the other.

Isobutane is an example of a *branched-chain* hydrocarbon. Isobutane and n-butane are called *structural* isomers because both molecules contain exactly

139

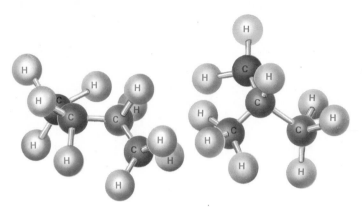

FIGURE 7-8 (a) and (a') Normal butane. (b) and (b') Isobutane.

the same number and kinds of atoms, C_4H_{10}, but the molecules are structured (put together) differently.

Consider the isomeric pentanes with the formula, C_5H_{12}. There are three of these:

	n-Pentane	Isopentane	Neopentane
Melting point	−130°C	−160°C	−20°C
Boiling point	36°C	28°C	9.5°C

Note that only three isomers are predicted by bonding theory. These three isomers of pentane are well known, and no others have ever been found.

Table 7–2 gives the number of isomers predicted by theory for some larger molecular formulas, starting with C_6H_{14}. Every predicted isomer, and no more, has been isolated and identified for the C_6, C_7, and C_8 groups. However, not all the C_{15}'s and C_{20}'s have been separated and put in bottles, but there is sufficient belief in the theory to presume that if enough time and effort were spent, all the isomers could eventually be produced or found in nature. We conclude that structural isomerism helps to explain the vast number of carbon compounds.

TABLE 7–2   STRUCTURAL ISOMERS OF SOME HYDROCARBONS

FORMULA	ISOMERS PREDICTED	FOUND
C_6H_{14}	5	5
C_7H_{16}	9	9
C_8H_{18}	18	18
$C_{15}H_{32}$	4347	—
$C_{20}H_{42}$	366,319	—

NOMENCLATURE

With so many organic compounds, a system of common names quickly fails (the names and formulas of a few simple hydrocarbons are given in Table 7–3). It is evident that a system of nomenclature is needed that makes use of numbers as well as names. Much attention has been given to this problem, and several international conventions have been held to work out a satisfactory system. The

TABLE 7–3   THE FIRST TEN STRAIGHT-CHAIN HYDROCARBONS

NAME	FORMULA	STRUCTURAL FORMULA
Methane	CH_4	
Ethane	C_2H_6	
Propane	C_3H_8	
n-Butane	C_4H_{10}	
n-Pentane	C_5H_{12}	
n-Hexane	C_6H_{14}	
n-Heptane	C_7H_{16}	
n-Octane	C_8H_{18}	
n-Nonane	C_9H_{20}	
n-Decane	$C_{10}H_{22}$	

International Union of Pure and Applied Chemistry has given its approval to a very elaborate nomenclature system (IUPAC System), and this system is now in general use.

Often students become a bit confused when writing and interpreting structural formulas for organic compounds because they fail to realize there are many equivalent ways of writing a single structural formula. For example,

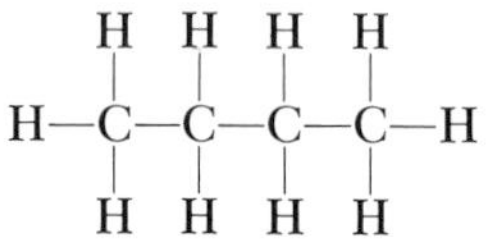

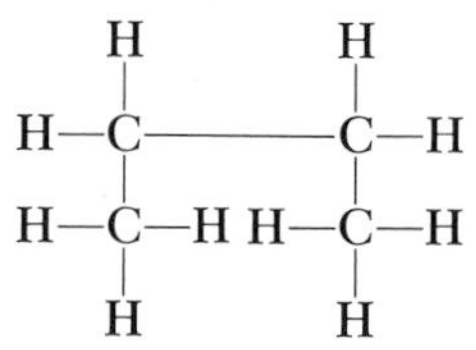

To conserve space, and to emphasize that all the C—C bonds are of equal length, the first method is ordinarily used.

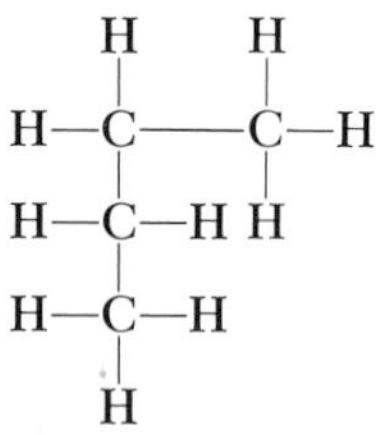

For branched-chain hydrocarbons it becomes necessary to name submolecular groups. We have already named the —CH_3 group the methyl group; this name is derived from methane by dropping the -ane and adding -yl. Any of the 10 hydrocarbons listed in Table 7–3 could give rise to a similar group. For example, the propyl group would be —C_3H_7. In order to illustrate the use of the group names, consider this formula:

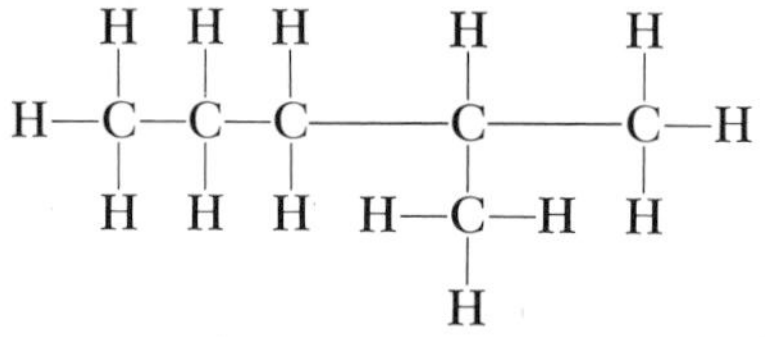

The longest carbon chain in the molecule contains five carbon atoms, hence, the *root name* is pentane. Furthermore, it is a methylpentane (written as one word) because a methyl group is attached to the pentane structure. In addition, a number is needed because the methyl group could be bonded to either the second carbon atom (2) or third carbon atom (3), from one end of the chain.

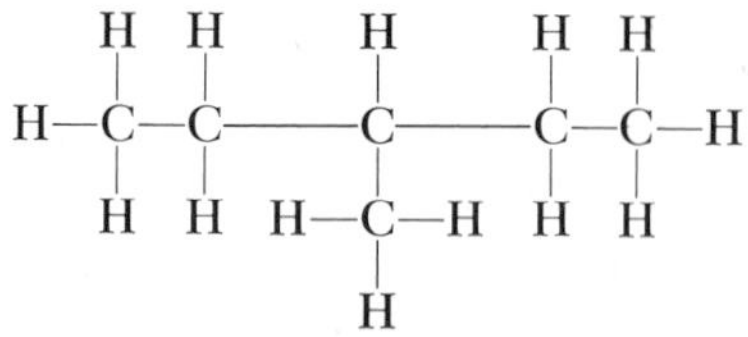

3-Methylpentane 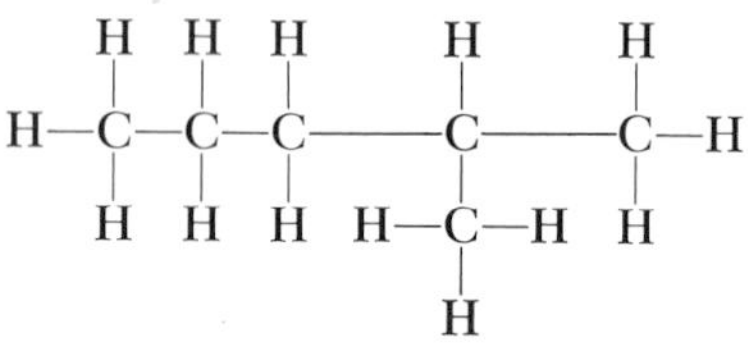   2-Methylpentane

Note that 2-methylpentane is the same as 4-methylpentane since the latter would be the same molecule turned around; *the accepted rule requires numbering from the end of the carbon chain that will result in the smallest numbers.* Therefore, 2-methylpentane is the correct name.

Any number of substituted groups can be handled in this same fashion. Consider the formula for 5,5-diethyl-3,3,4,6-tetramethyloctane:

142

$$CH_3CH_2-\overset{\overset{\displaystyle CH_3}{|}}{\underset{\underset{\displaystyle CH_3}{|}}{C}}-\overset{\overset{\displaystyle H}{|}}{\underset{\underset{\displaystyle CH_3}{|}}{C}}-\overset{\overset{\displaystyle C_2H_5}{|}}{\underset{\underset{\displaystyle C_2H_5}{|}}{C}}-\overset{\overset{\displaystyle H}{|}}{\underset{\underset{\displaystyle CH_3}{|}}{C}}-CH_2CH_3$$

A point of difficulty often arises because the use of common names persists. Consider, for example:

$$CH_3-\overset{\overset{\displaystyle CH_3}{|}}{\underset{\underset{\displaystyle H}{|}}{C}}-CH_3$$

The correct name is methylpropane. However, this compound is more often referred to by its common name, isobutane. Only under special circumstances does the multiplicity of names for a single compound create confusion.

UNSATURATED HYDROCARBONS

Many of the hydrocarbons found in petroleum and others produced by synthetic methods, do not contain as many hydrogen atoms as they would if all the bonds were single electron pairs. Rather, they are found to have some carbon atoms held together by double and triple bonds.

Ethylene is the simplest of a large group of *unsaturated* hydrocarbons, in which there are double or triple bonds.

$$\underset{H}{\overset{H}{>}}C=C\underset{H}{\overset{H}{<}}$$

Ethylene
(*common name*)

The IUPAC name is ethene (*eth* for the C_2 hydrocarbon and *ene* for the double bond). It is possible to add hydrogen (*hydrogenate*) to ethylene using a palladium catalyst, thereby destroying the double bond. As we shall see later, ethylene and other unsaturated hydrocarbons form the basis for a large number of very useful plastics. Compounds with a double bond are often called alkenes or olefins.

$$\underset{H}{\overset{H}{>}}C=C\underset{H}{\overset{H}{<}} + H_2 \xrightarrow[\text{Catalyst}]{\text{Palladium}} H-\overset{\overset{\displaystyle H}{|}}{\underset{\underset{\displaystyle H}{|}}{C}}-\overset{\overset{\displaystyle H}{|}}{\underset{\underset{\displaystyle H}{|}}{C}}-H$$

Unsaturated hydrocarbons can also contain carbon-carbon triple bonds. The simplest of these is ethyne, also known as acetylene (C_2H_2). Notice that the IUPAC name is derived from that of the C_2 hydrocarbon name, *eth*, with *yne* added to the end.

$$H-C\equiv C-H$$

Acetylene
(*common name*)

The combustion of a mole of acetylene (26 grams) gives off 311,000 calories. Gram for gram, acetylene gives off slightly less energy per gram than methane (12,000 calories/gram for acetylene and 13,300 calories/gram for methane). The reactivity of acetylene is useful in synthesizing many important organic and biological chemicals, such as "birth-control" hormones, plastics, and dyes.

HYDROCARBONS FROM COAL— AROMATIC COMPOUNDS

The destructive distillation of coal in the absence of air to produce coke for steel-making leads to a tarry mass known as coal tar. In the early nineteenth century, coal tar was a useless by-product and was often poured into rivers and streams. It was learned however, that upon distillation, coal tar yields more than 250 different compounds (mostly hydrocarbons) and that most of these are useful. In addition to saturated and unsaturated hydrocarbons like the ones discussed previously, coal tar yielded many members of a class of compounds known as aromatic hydrocarbons. These compounds were named aromatics because of the characteristic aroma of benzene, C_6H_6, the simplest aromatic.

The structure of benzene presented many problems for the chemists of the middle nineteenth century. In 1865, Friedrich August Kekulé, a German, postulated a structure for benzene based on a dream he had while wrestling with the structural problem.

> I was sitting, writing at my text-book; but the work did not progress; my thoughts were elsewhere. I turned my chair to the fire and dozed. Again the atoms were gambolling before my eyes. This time the smaller groups kept modestly in the background. My mental eye, rendered more acute by repeated visions of the kind, could now distinguish larger structures, of manifold conformation: long rows, sometimes more closely fitted together; all twining and twisting in snake-like motion. But look! What was that? One of the snakes had seized hold of its own tail, and the form whirled mockingly before my eyes. As if by a flash of lightening I awoke; and this time also I spent the rest of the night in working out the consequences of the hypothesis.
>
> Let us learn to dream, gentlemen . . . then perhaps we shall find the truth . . . but let us beware of publishing our dreams before they have been put to the proof by the waking understanding.°

Kekulé's structure for benzene was a hexagon of carbon atoms in which every other pair of carbons was bonded together by a double bond, with a hydrogen atom bonded to each carbon. He realized that benzene does not react as though it contains double bonds, so he modified his picture of the molecule to one in which the double bonds were constantly shifting. Currently, the more simplified structural formula shown on the right is being used. (The hexagon represents six carbon atoms and six hydrogen atoms.)

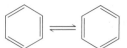

The structure of Kekulé

Modern version

° From Japp, R., "Kekulé Memorial Lecture," *J. Chem. Soc.*, 73, p. 100 (1898).

According to the atomic theory of bonding, after a pair of electrons is shared to form each carbon-to-carbon bond and each carbon-to-hydrogen bond, each carbon atom has one valence electron left over. These electrons form what is called the π (pi) electron cloud of benzene, which is somewhat like a "pie" of electrons. The π electrons of benzene are "delocalized," that is, they are not attached to any particular carbon atoms, but are somewhat free to roam around the ring. In this respect, aromatic rings are "racetracks" for electrons. The π cloud of an aromatic ring is represented by the circle in the hexagon symbol.

BENZENE DERIVATIVES

There are a large number of aromatic compounds that can be *derived* from benzene by replacing one or more of the hydrogen atoms about the ring. These molecules exhibit a wide variety of properties and differ greatly in their chemical reactivity. Some interesting examples of isomerism also are possible. Consider the three different compounds with the formula C_8H_{10} found in some coal tars. There are several names given to these isomers:

1,4-Dimethylbenzene
(*para-xylene*)
m.p. 13.2°C

1,3-Dimethylbenzene
(*meta-xylene*)
m.p. −47.4°C

1,2-Dimethylbenzene
(ortho-xylene)
m.p. −29°C

Each of these isomers has two methyl groups substituted for hydrogen atoms on the ring. The prefixes, para-, meta-, and ortho- are used if there are only two groups on the benzene ring.

Para-xylene is actually a compound of major importance. It is a starting material in the manufacture of the polyester fibers and plastics that are used to make double knits, recording tape, tire cord and many other items. In 1971 the U.S. production of para-xylene was 1.7 *billion* pounds. Much of this production came from petroleum.

POLYNUCLEAR AROMATICS

In addition to benzene and its derivatives, coal tar yields a variety of multi-ring hydrocarbons called *polynuclear* aromatics. Among these are naphthalene (containing two rings of carbon atoms sharing a common edge), anthracene (three rings joined together in a line), and benzo (alpha) pyrene, which was one of the first compounds known to have carcinogenic (cancer producing) properties. Many of the polynuclear hydrocarbons have important uses. Anthracene, for example, is a dye intermediate. (We will discuss dyes in Chapter 15.) Naphthalene is one type of moth repellant.

145

Naphthalene

Anthracene

Benzo (alpha) *pyrene*

OPTICAL ISOMERS

In a preceding section, it was pointed out that structural isomers really represent different compounds because they differ in the way the atoms are joined to one another. However, it is possible for some sets of atoms to form two isomeric molecules, both of which have the same set of bonds as well as the same set of atoms. The only difference in structure is the orientation in space of the atoms. Such a difference gives rise to what is called *optical isomerism*. The existence of optical isomers is another factor explaining why we have so many carbon compounds. It also helps explain why several different sugars have the same formula (Chapter 9) and tells us something about how enzymes work in the body (Chapter 10).

Optical isomerism is possible when a molecular structure is asymmetric (without symmetry). A common example of an asymmetric molecule is one containing a tetrahedral carbon atom bonded to four different atoms or groups of atoms. Such a carbon atom is called an asymmetric atom. Consider the molecule CHIBrCl.

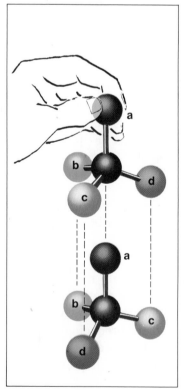

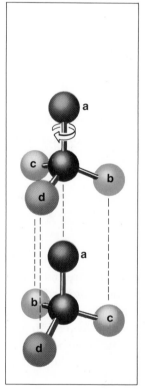

FIGURE 7-9 Optical isomers. Four different atoms, or groups of atoms, are bonded to tetrahedral center atoms so that the upper isomeric form cannot be turned in any way to substitute exactly for the lower structure. These are nonsuperimposable mirror images.

146

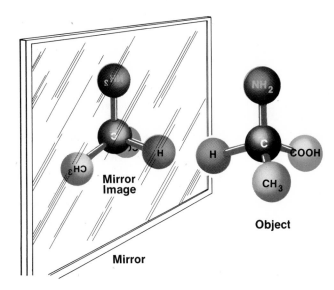

FIGURE 7-10   Optical isomers of the amino acid, alanine.

A study of Figure 7–9 shows that there are two ways to arrange four different atoms in the tetrahedral positions about the central carbon atom. These result in two nonsuperposable mirror-image molecules which are optical isomers.

In Figure 7–10 this mirror-image relationship is shown for isomeric forms of alanine (one of the basic units of proteins), whose molecules contain a tetrahedral carbon atom surrounded by an amine group ($-NH_2$), a methyl group ($-CH_3$), an acid group ($-COOH$), and a hydrogen atom. Note in Figure 7–10 that the carbon atoms in the methyl and acid groups are not asymmetric since these atoms are not bonded to four different groups.

There are a few examples of nonsuperposable mirror images in the macroscopic world. Consider, for instance, right- and left-hand gloves. They are mirror images of each other and are nonsuperposable.

Optical isomers are generally very much alike in their chemical properties. Sometimes, however, one optical isomer will have a chemical effect not shown by its counterpart. The hormone epinephrine (or adrenalin) is one of a pair of optical isomers, and its structure can be represented as

Epinephrine
(Adrenalin)

where C^* designates the asymmetric carbon atom. Only one of the isomers is effective in starting a heart that has stopped beating momentarily or in giving

147

a person unusual strength during times of great emotional stress. The mirror image is ineffective (see also Chapter 10).

Large organic molecules may have many asymmetric carbon atoms within the same molecule. At each carbon atom there exists the possibility of two arrangements of the molecule. The total number of possible molecules, then, increases exponentially with the number of asymmetric centers. With two asymmetric carbon atoms there would be 2^2 or four possible structures; for three centers there would be 2^3 or eight possible structures. It should be emphasized that each of the eight isomers contains the same set of atoms with the same set of chemical bonds. Glucose, a simple blood sugar, contains four asymmetric carbon atoms per molecule. Thus, there are 2^4 or 16 isomers in the family of compounds to which glucose belongs. Obviously, the concept of optical isomers is significant in explaining the vast number of carbon compounds.

$$
\begin{array}{c}
\mathrm{O} \\
\parallel \\
\mathrm{C-H} \\
| \\
\mathrm{H-C^*-OH} \\
| \\
\mathrm{HO-C^*-H} \\
| \\
\mathrm{H-C^*-OH} \\
| \\
\mathrm{H-C^*-OH} \\
| \\
\mathrm{HO-C-H} \\
| \\
\mathrm{H}
\end{array}
$$

Glucose

FUNCTIONAL GROUPS

As pointed out earlier, carbon also forms covalent bonds with a number of other elements. As a result, certain groups of atoms called functional groups appear over and over again in different organic compounds. Consider, for example, the —OH group, called the hydroxyl group. A large number of molecules containing one or more —OH groups have properties which are characteristic of a class of organic compounds called *alcohols*.

The —OH group has the same combining power as an atom of hydrogen. This means that, in any of the hydrocarbons considered thus far, any of the hydrogen atoms, except those on aromatic rings, could be replaced by an —OH group to form an alcohol. A single hydrocarbon molecule can give rise to a number of alcohols if there are different isomeric positions for the —OH group.

When one or more functional groups appear in a molecule, the IUPAC name reveals the group name. For example, the name of an alcohol will use the root of the name of the analogous hydrocarbon to indicate the number of carbon atoms, and the suffix -ol to denote an alcohol. As before, a number is used to indicate the position of the alcohol group.

Some of the major functional groups we will consider in following chapters

TABLE 7-4   SOME FUNCTIONAL GROUPS FOUND IN ORGANIC COMPOUNDS

GROUP	NAME	TYPICAL COMPOUND	COMPOUND NAME	COMMON USE
R—OH	Alcohol (hydroxyl)	$\begin{array}{c} H \\ \| \\ H-C-OH \\ \| \\ H \end{array}$	Methanol (wood alcohol)	Solvent
$\begin{array}{c} O \\ \| \| \\ R-C-H \end{array}$	Aldehyde	$\begin{array}{c} O \\ \| \| \\ H-C-H \end{array}$	Methanal (Formaldehyde)	Preservative
$\begin{array}{c} O \\ \| \| \\ R-C-OH \end{array}$	Acid (carboxyl)	$\begin{array}{c} H\;\;O \\ \|\;\;\| \| \\ H-C-C-OH \\ \| \\ H \end{array}$	Ethanoic Acid (Acetic acid)	Vinegar
$\begin{array}{c} O \\ \| \| \\ R-C-R \end{array}$	Ketone	$\begin{array}{c} H\;\;O\;\;H \\ \|\;\;\| \|\;\;\| \\ H-C-C-C-H \\ \|\;\;\;\;\| \\ H\;\;\;\;H \end{array}$	Propanone (Acetone)	Solvent
R—O—R	Ether	$C_2H_5-O-C_2H_5$	Diethyl ether (Ethyl ether)	Anesthetic
$\begin{array}{c} O \\ \| \| \\ R-O-C-R \end{array}$	Ester	$\begin{array}{c} O \\ \| \| \\ CH_3-CH_2-O-C-CH_3 \end{array}$	Ethyl ethanoate (Ethyl acetate)	Solvent in fingernail polish
$\begin{array}{c} H \\ \| \\ R-N \\ \| \\ H \end{array}$	Amine	$\begin{array}{c} O\;\;H\;\;\;\;H \\ \| \|\;\;\|\;\;/ \\ HO-C-C-N \\ \|\;\;\;\;\backslash \\ H\;\;\;\;H \end{array}$	2-Aminoethanoic acid (Glycine)	Amino acid found in proteins

are listed in Table 7-4. (The symbol R stands for a hydrocarbon group, such as methyl (—CH_3) or ethyl (—C_2H_5), and is shown attached to the designated functional group in order to represent a complete compound.

SOME IMPORTANT COMPOUNDS OF CARBON, HYDROGEN AND OXYGEN

Carbon, hydrogen, and oxygen can be combined to form an enormous number of compounds. Considerable order is introduced into the study of these compounds when they are divided into classes on the basis of the functional groups they contain. The alcohols, and the organic acids and their derivatives (compounds that can be made from them), are very important compounds of carbon, hydrogen, and oxygen since they find such wide application in our everyday lives.

ALCOHOLS

When a hydroxyl (—OH) group is attached to a nonaromatic carbon skeleton (an R-group), the resulting R—OH molecule has properties common to a class

149

TABLE 7-5   SOME IMPORTANT ALCOHOLS

FORMULA	IUPAC NAME	COMMON NAME
$\begin{array}{c} H \\ \| \\ H-C-OH \\ \| \\ H \end{array}$	Methanol	Methyl alcohol (wood alcohol)
$\begin{array}{c} H \quad H \\ \| \quad \| \\ H-C-C-OH \\ \| \quad \| \\ H \quad H \end{array}$	Ethanol	Ethyl alcohol (grain alcohol)
$\begin{array}{c} H \\ \| \\ H-C-OH \\ \| \\ H-C-OH \\ \| \\ H \end{array}$	1,2-Ethanediol	Ethylene glycol (antifreeze)
$\begin{array}{c} H \\ \| \\ H-C-OH \\ \| \\ H-C-OH \\ \| \\ H-C-OH \\ \| \\ H \end{array}$	1,2,3-Propanetriol	Glycerol (glycerin)

of compounds called alcohols. Formulas and names for some important molecules of this type are given in Table 7-5.

Methanol (Methyl Alcohol)

Methanol was originally called wood alcohol because it was obtained by the destructive distillation of wood. It is the simplest of all alcohols and has the formula CH_3OH. In the older method for the production of wood alcohol, hardwoods such as beech, hickory, maple, or birch were heated in the absence of air in a retort (Figure 7–11). Methanol which is 92 to 95 per cent pure can be obtained by fractional distillation of the liquid which results.

In 1923, the price of wood alcohol in the United States was 88¢ per gallon. In that year German chemists discovered how to produce this useful compound synthetically. Methanol is formed when carbon monoxide, CO, and hydrogen are heated at a pressure of 200 to 300 atmospheres over a catalyst of mixed oxides (90 per cent ZnO–10 per cent Cr_2O_3).

$$CO + 2H_2 \xrightarrow[300°C]{ZnO-Cr_2O_3} CH_3OH$$

As a result of this synthetic process, German industrialists were able to sell pure methanol at 20¢ per gallon. Even a high tariff was not able to save the wood distillers in their outdated operations. The synthetic product dominated the market in the United States, and soon the new method was used here.

The production of synthetic methanol in the United States rose to over 4.1 billion pounds in 1969. About one-half of this was used in the production of

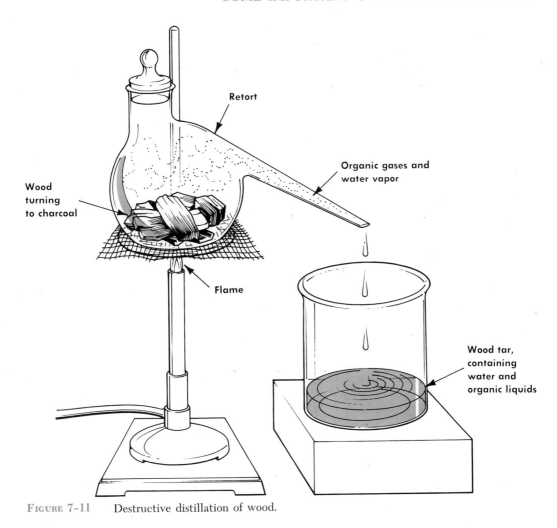

FIGURE 7-11   Destructive distillation of wood.

formaldehyde (used in plastics), 30 per cent in the production of other chemicals, and smaller amounts for jet fuels, antifreeze mixtures, solvents, and as a denaturant (a poison added to ethyl alcohol to make it unfit for beverages). Methanol is a deadly poison; it causes blindness in less than lethal doses. Many deaths and injuries have resulted when this alcohol was mistakingly substituted for ethyl alcohol in beverages.

Ethanol (Ethyl Alcohol)

Ethanol is called grain alcohol because it can be fractionally distilled from the fermented mash made from corn, rice, barley, or other grains that are sources of carbohydrates (Chapter 9). Fermentation is a breakdown of complex organic molecules, such as carbohydrates, brought about by means of enzymes. If the enzyme° *diastase* is mixed with ground grain and water, and the mixture is

---

° Enzymes are complex organic molecules that act as catalysts (Chapter 10).

allowed to stand at 40°C for a period of several days, the starch in the grain is changed into the sugar, maltose.

$$2(C_6H_{10}O_5)_n + nH_2O \longrightarrow nC_{12}H_{22}O_{11}$$

Starch (a carbohydrate)   *Maltose (a sugar)*

The subscript "n" in the formula for starch indicates that starch is made up of many $C_6H_{10}O_5$ units. Brewers call the resulting mixture of maltose and water the wort. The wort is diluted and mixed with yeast and held at a temperature of 30°C for 40 to 60 hours. The living yeast cells secrete two enzymes, maltase and zymase. The maltase causes the sugar, maltose, to decompose into a simpler sugar, glucose.

$$C_{12}H_{22}O_{11} + H_2O \longrightarrow 2C_6H_{12}O_6$$

Maltose   *Glucose*

The glucose, in turn, is converted by zymase to alcohol and carbon dioxide.

$$C_6H_{12}O_6 \longrightarrow 2CO_2 + 2C_2H_5OH$$

Glucose   *Ethanol*

A solution of 95 per cent ethyl alcohol and 5 per cent water can be recovered from the mash by fractional distillation.

Synthetic ethyl alcohol is produced on a large scale for industrial use. The direct hydration of ethylene by addition of water accounts for more than 80 per cent of all ethanol production. Using ethylene obtained from the cracking of petroleum, and a large excess of water, the reaction which occurs is:

$$
\underset{\text{H}}{\overset{\text{H} \quad \text{H}}{\text{H—C}=\text{C—H}}} + \text{HOH} \xrightarrow[300°C]{70 \text{ atm.}} \text{H—}\underset{\text{H}}{\overset{\text{H}}{\text{C}}}\text{—}\underset{\text{H}}{\overset{\text{H}}{\text{C}}}\text{—OH}
$$

Pure ethyl alcohol is 200 proof (exactly twice the per cent). Apart from the alcoholic beverage industry, ethyl alcohol is used widely in solvents and in the preparation of chloroform, ether, and many other organic compounds. Cough syrups often contain upwards of 20 per cent alcohol. The use of ethyl alcohol as a beverage is discussed in Chapter 20.

Ethylene Glycol and Glycerol (Glycerin)

More than one alcohol group (—OH) can be present in a single molecule. Ethylene glycol, the base of permanent antifreeze, and glycerin are examples of such compounds (shown in Table 7-5). Glycerol has many uses in the manufacture of drugs, cosmetics, nitroglycerin, and numerous other chemicals. Perhaps the most important compounds of glycerol are its natural esters (fats and oils) which we shall discuss in Chapter 9.

ORGANIC ACIDS

In Chapter 5 an acid was defined as a species that has a tendency to donate protons. We will now consider a group of organic compounds which are proton donors, the carboxylic acids, characterized by the carboxyl group, $-C\overset{O}{\underset{OH}{\diagup}}$. The electronegative character of the $-C-O$ group tends to remove the negative blanket from the hydrogen atom and gives the hydrogen atom an acidic character. Nearly all the carboxylic acids are much weaker than mineral acids, such as sulfuric and hydrochloric.

Formic Acid

The simplest organic acid is formic acid, in which the carboxyl group is attached directly to a hydrogen atom.

$$H-C\overset{O}{\underset{O-H}{\diagup}}$$

Formic acid

This acid is found in ants and other insects and is part of the irritant that produces itching and swelling after a bite. The sodium salt of formic acid is readily prepared by heating carbon monoxide (CO) with sodium hydroxide (NaOH).

$$CO + NaOH \xrightarrow[\text{6-10 min.}]{200°} \underset{\text{Sodium formate}}{HCOO^-Na^+}$$

If sodium formate is mixed with a mineral acid, formic acid can be distilled from the mixture.

$$\underset{\substack{\text{Sodium} \\ \text{formate}}}{HCOO^-Na^+} + \underset{\substack{\text{Hydrochloric} \\ \text{acid}}}{H_3O^+ + Cl^-} \longrightarrow \underset{\substack{\text{Formic} \\ \text{acid}}}{HCOOH} + \underset{\substack{\text{Sodium} \\ \text{chloride}}}{Na^+Cl^-} + H_2O$$

Acetic Acid

This acid is the most widely used of the organic acids. Vinegar, for example, is an aqueous solution of 4 to 5 per cent acetic acid, plus flavor and colors imparted to vinegars by the constituents of the alcoholic solutions from which they are made. Ethyl alcohol in the presence of certain bacteria and air is oxidized to acetic acid.

$$\underset{\text{Ethyl alcohol}}{CH_3CH_2OH} + \underset{\text{Oxygen}}{O_2} \xrightarrow{\text{Bacteria}} \underset{\text{Acetic acid}}{CH_3COOH} + \underset{\text{Water}}{H_2O}$$

The active bacteria, called "mother of vinegar," forms a slimy growth in a vinegar solution. The growth of bacteria can sometimes be observed in a bottle of commercially prepared vinegar after it has been opened to the air.

Acetic acid is an important starting material for making other chemicals,

153

such as vinyl adhesives and plastics, and is a convenient acidic material when a cheap organic acid is needed.

Fatty Acids

Fatty acids are so named because they can be obtained from fats and oils. A fatty acid is made up of molecules containing a carboxyl group attached to a long hydrocarbon chain. The chains often contain only single carbon—carbon bonds, but may contain carbon—carbon double bonds as well. Two examples are stearic acid and palmitic acid.

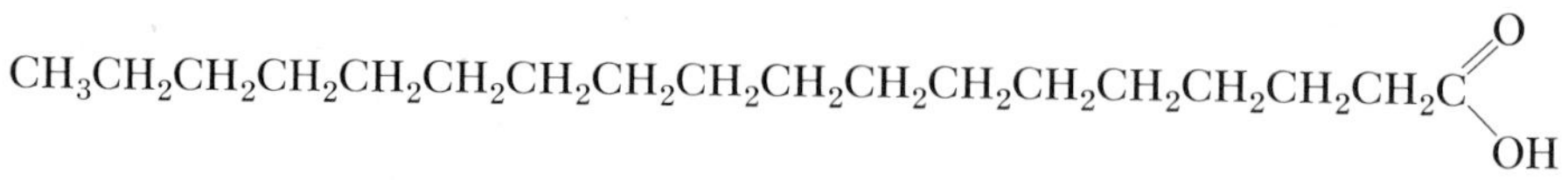

$$CH_3CH_2CH_2CH_2CH_2CH_2CH_2CH_2CH_2CH_2CH_2CH_2CH_2CH_2CH_2CH_2CH_2C\begin{smallmatrix} O \\ \\ OH \end{smallmatrix}$$

Stearic acid, $CH_3{-}(CH_2)_{16}{-}COOH$

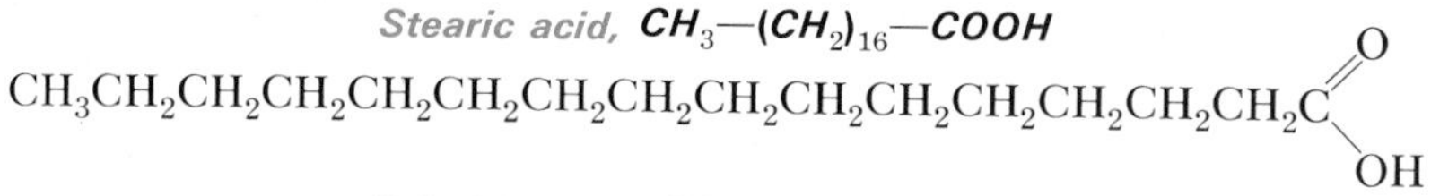

$$CH_3CH_2CH_2CH_2CH_2CH_2CH_2CH_2CH_2CH_2CH_2CH_2CH_2CH_2CH_2C\begin{smallmatrix} O \\ \\ OH \end{smallmatrix}$$

Palmitic acid, $CH_3{-}(CH_2)_{14}{-}COOH$

Stearic acid is obtained by the hydrolysis (reaction with water) of animal fat; palmitic acid results from the hydrolysis of palm oil. These two fatty acids are especially important in the manufacture of soaps.

Special Case of —OH on an Aromatic Ring

Carbolic acid (phenol), which has many uses as a disinfectant, has an —OH group bonded to an aromatic ring, and is a weak acid: Evidently the benzene ring pulls electrons away from the H atom in the —OH group more than is the case in alcohols.

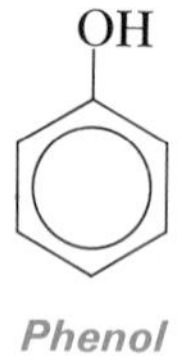

Phenol

The toxic components of poison ivy are related chemically to phenol. The principal culprit is urushiol.

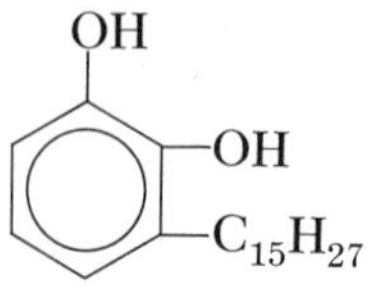

Urushiol

ESTERS

In the presence of certain strong dehydrating agents, organic acids react with alcohols to form a class of compounds called *esters*. For example, when ethyl alcohol is mixed with acetic acid in the presence of sulfuric acid, sweet-smelling ethyl acetate is formed. This reaction is a dehydration in which sulfuric acid acts as a catalyst.

154

TABLE 7-6   SOME ALCOHOLS, ACIDS, AND THEIR ESTERS

ALCOHOL	ACID	ESTER	ODOR OF THE ESTER
CH₃CHCH₂CH₂OH 　　CH₃ Isopentyl alcohol	CH₃COOH Acetic acid	CH₃CHCH₂CH₂—O—C—CH₃ 　　CH₃　　　　‖ 　　　　　　　O Isopentyl acetate	Banana
CH₃CHCH₂CH₂OH 　　CH₃ Isopentyl alcohol	CH₃CH₂CH₂CH₂COOH n-Valeric acid	CH₃CHCH₂CH₂—O—C—CH₂CH₂CH₂CH₃ 　　CH₃　　　　‖ 　　　　　　　O Isopentyl n-valerate	Apple
CH₃CH₂CH₂CH₂OH n-Butyl alcohol	CH₃CH₂CH₂COOH n-Butyric acid	CH₃CH₂CH₂CH₂—O—C—CH₂CH₂CH₃ 　　　　　　　‖ 　　　　　　　O Butyl n-butyrate	Pineapple
CH₃CHCH₂OH 　　CH₃ Isobutyl alcohol	CH₃CH₂COOH Propionic acid	CH₃CHCH₂—O—C—CH₂CH₃ 　　CH₃　　‖ 　　　　　O Isobutyl propionate	Rum

$$CH_3CH_2{-}OH + HO{-}\overset{\overset{\displaystyle O}{\|}}{C}CH_3 \xrightarrow{\;H_2SO_4\;} CH_3CH_2{-}O{-}\overset{\overset{\displaystyle O}{\|}}{C}{-}CH_3 + H_2O$$

Ethyl acetate is a common solvent and is used in fingernail polish remover.

Some of the odors of common fruits are due to the presence of mixtures of volatile esters (Table 7–6). In contrast, higher molecular weight esters often have a distinctly unpleasant odor. Fats and waxes are esters with molecular masses exceeding 700 amu. Both groups of compounds are discussed in later chapters.

This chapter is only an introduction to organic chemistry. In later chapters we shall discuss such topics as detergents, plastics, air pollution, drugs, biochemistry, and consumer products that will build upon your knowledge of organic compounds gained from this chapter.

QUESTIONS

1. Saturated hydrocarbons are so named because they have the maximum amount of hydrogen present for a given amount of carbon. The saturated hydrocarbons have the general formula C_nH_{2n+2}, where n is a whole number. What are the names and formulas of the first four members of this series of compounds?

2. Draw the structural formula for each of the five isomeric hexanes, C_6H_{14}.

3. Write the structural formulas for (a) 2-methylbutane, (b) ethylpentane, (c) 4,4-dimethyl-5-ethyloctane, (d) methylbutane, (e) 2-methyl-2-hexene, (f) 3-methyl-3-hexanol.

4. Give the IUPAC name for:

155

(c) [structural formula]

(d) $H-\overset{\overset{\displaystyle H}{|}}{\underset{\underset{\displaystyle H}{|}}{C}}-C\equiv C-\overset{\overset{\displaystyle H}{|}}{\underset{\underset{\displaystyle H}{|}}{C}}-H$

(e) [structural formula]

5. If cyclohexane has the formula C_6H_{12}, and if it has no double bonds, what would you suggest as its structural formula?

6. What structural feature characterizes aromatic compounds?

7. Indicate the functional groups present in the following molecules:

(a) $CH_3CH_2CH_2COOH$

(b) $CH_3CH_2NH_2$

(c) $CH_3CHCH_2CH_2COOH$
  $\quad\ |$
  $\quad NH_2$

(d) $CH_3-CH-CH_2COOH$
  $\qquad\ |$
  $\qquad OH$

(e) $CH_3-C-CH_2CH_2COOH$
  $\qquad\ ||$
  $\qquad O$

(f) $CH_3-CH-CH_2OH$
  $\qquad\ |$
  $\qquad NH_2$

8. A compound contains 81.8 per cent C and 18.2 per cent H, and has a molecular weight of 44. Can you write a structure for it?

9. Two isomeric compounds, A and B, are found to have molecular weights of 46 and 52.2 per cent C, 13 per cent H, and 34.8 per cent O. Compound A is found to give every evidence of hydrogen bonding to itself in the liquid state, but compound B does not. Suggest structural formulas for A and B on the basis of these data.

10. Write two structural formulas for compounds which can have each of the molecular formulas listed:

(a) $C_5H_{12}O$
(b) C_3H_6O
(c) $C_5H_{10}O_2$

11. Glycine is one of the 21 or so "boxcar" molecules that are bonded together in a "train" to make all of the proteins in us, in animals, and in plants. Is glycine capable of optical isomers?

$$H-\overset{\overset{\displaystyle NH_2}{|}}{\underset{\underset{\displaystyle H}{|}}{C}}-\overset{}{\underset{\underset{\displaystyle O}{||}}{C}}-O-H$$

12. Why are there so many carbon compounds?

13. What are the products in the complete combustion of benzene, C_6H_6? Write the equation.

14. Considering the sources of coal and petroleum, which has the greater chance of having aromatic compounds, you or a tree?

15. There has been considerable interest recently in vitamin C as a possible prevention for the common cold. What are the functional groups in vitamin C (ascorbic acid)?

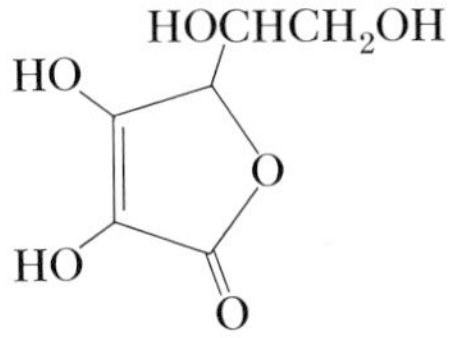

16. Lactic acid is a compound found in muscle tissue and also in sour milk. Show by three-dimensional drawings how two possible optical isomers of this molecule can exist. One is the molecule in muscle, the other in milk.

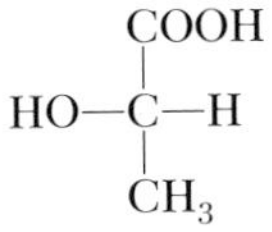

Which functional group makes lactic acid an acid?

17. Draw the structural formula of the ester expected when each of the following acids is reacted with the alcohols given.

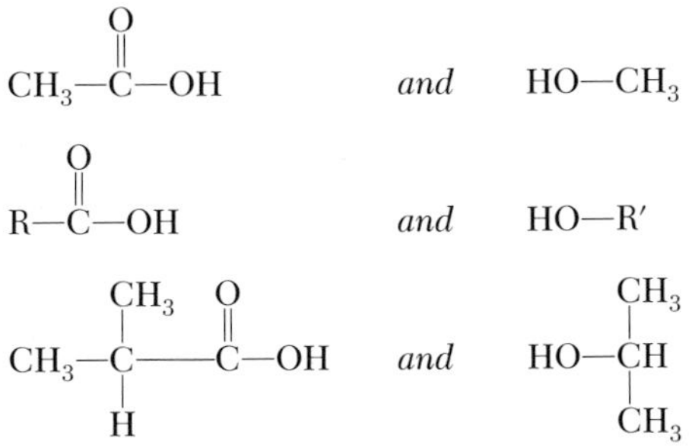

18. A broad generalization about solubility of substances is that "like dissolves in like." Methyl alcohol, CH_3OH, dissolves in water, HOH, but hydrocarbons such as gasoline do not dissolve in water. Suggest a reason for this in terms of the submicroscopic structures of the substances, polar bonds, polar molecules, and hydrogen bonding.

19. Use lines for bonds and draw the structure of $CH_3(CH_2)_3CHOHCH_2CH_3$.

20. Natural gas is a gas at room temperature; gasoline is a liquid at room temperature. Both are nonpolar hydrocarbons. What broad generalization can be made about the molecular weights of the substances comprising natural gas and gasoline?

SUGGESTIONS FOR FURTHER READING ▬▬▬▬▬▬▬▬▬

Reference

Wendland, R. T., "Petrochemicals," Anchor Books, Doubleday and Co., New York, 1969.

Articles

Campaigne, E., "Wöhler and the Overthrow of Vitalism," *Journal of Chemical Education*, Vol. 32, No. 8, p. 403 (1955).

Mills, G. A., "Ubiquitous Hydrocarbons," *Chemistry*, Vol. 44, No. 2, p. 8; p. 12 (1971).

Orchin, M., "Determining the Number of Isomers From A Structural Formula," *Chemistry*, Vol. 42, No. 5, p. 8 (1969).

Rossini, F. D., "Hydrocarbons in Petroleum," *Journal of Chemical Education*, Vol. 37, No. 11, p. 554 (1960).

von Tamelen, E. E., "Benzene, The Story of Its Formula, 1865–1965," *Chemistry*, Vol. 38, No. 1, p. 6 (1965).

Westheimer, F. H., "The Structural Theory of Organic Chemistry," *Chemistry*, Vol. 38, No. 6, p. 13 (1965), and Vol. 38, No. 7, p. 10 (1965).

8

THE GIANTS
ASSEMBLED BY MAN

It is impossible for most people to get through a day without using a dozen or more materials which were completely unknown 30 years ago. Many of these materials are plastics of one sort or another. Examples of these include plastic dishes and cups, combs, automobile steering wheels and seat covers, telephones, pens, plastic bags for food and wastes, plastic pipes and fittings, plastic water-dispersed paints, false eyelashes and wigs, a wide range of synthetic fibers for suits and stockings, synthetic glues, and flooring materials. In fact, these materials are so widely used they are usually taken for granted. We rarely wonder where they come from or how man has achieved the ability to produce materials of this sort for various special uses. All these materials are composed of *giant molecules*. The purpose of this chapter is to examine some of the chemistry and technology that lie behind this flood of indispensable plastic objects.

WHAT ARE GIANT MOLECULES?

Most giant molecules are formed by the bonding together of a large number of smaller molecules. We have already pointed out that diamond is a "molecule" of interlaced carbon atoms. Whereas diamond is analogous to a large group of people bound together by holding hands, the giant molecules in plastics are more like a long train of individual railroad cars hooked together. The smaller molecules are bonded to each other by strong covalent bonds.

Many of the properties of the giant molecules are caused by their rather high molecular masses (or weights). By the late nineteenth century, techniques had been developed that enabled researchers to determine the approximate molecular weights of rubber, cellulose, proteins, and some other natural materials of very high molecular weights. One of these methods, involving the change in the freezing point of water, is illustrated in Figure 8–1. Although this method is not suitable for very high molecular-weight polymers, the basic idea is quite simple: the freezing point of water is depressed if substances are dissolved in it, and the depression of the freezing point is directly proportional to the number

159

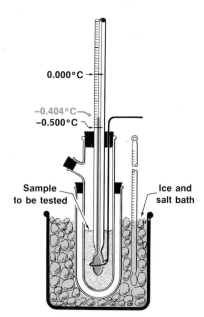

0.000°C →

−0.404°C
−0.500°C

Sample
to be tested

Ice and
salt bath

FIGURE 8–1   Molecular weights of solutes can be determined by the lowering of the freezing point below that of the pure solvent. In this example, water, which freezes at 0°C, dissolves a solute, and the resulting solution freezes at −0.404°C. A salt-ice bath is used in this instance to cool the sample.

of molecules (or moles) of solute added, as long as their total concentration is not too high. One mole of any non-ionized solute dissolved in 1000 grams of water depresses the freezing point by 1.86°C. The results of studies using this approach, usually with organic solvents, and other methods for the determination of molecular weight have led to values of about 12,000 for natural rubber; cellulose-like materials such as starch have molecular weights of 40,000, and the molecular weights of proteins range from a thousand up to a million.

Many chemists were reluctant to accept the concept of giant molecules, but in the 1920's a persistent German chemist, Herman Staudinger (1881–1965; Nobel Prize, 1953), championed the idea and introduced a new term, *macromolecule*, for these giant molecules. Staudinger devised experiments that yielded accurate molecular weights, and, in addition, he synthesized "model compounds" to test his theory. One of his first model compounds was prepared from styrene.

$$H_2C=CH$$

Styrene

Under the proper conditions, styrene molecules use the "extra" electrons of the double bond to undergo a *polymerization* reaction to yield polystyrene, a giant molecule. The word *polymer* means "many parts" (Greek: *poly* = many, *meros* = parts). The molecules of styrene itself are monomers (Greek: *mono* = one); they provide the recurring units in the giant molecule.

The macromolecule polystyrene is represented as a long chain of monomer units bonded to each other. The polymer chain is not an endless one; some polystyrenes made by Staudinger were found to have molecular weights of about

160

600,000, corresponding to a chain of about 5,700 styrene units. The polymer chain can be indicated as

$$R-CH_2-CH-\left(\!CH_2-CH-\!\right)_n CH_2-CH-R$$

where R represents some terminal group, often an impurity, and n is a large number.

Polystyrene is a clear, hard, colorless solid at room temperature. Since it can be molded easily at 250°C, the term *plastic* has become associated with it and similar materials. Polystyrene has so many useful properties that its commercial production, which began in Germany in 1929, today exceeds 2.5 billion pounds per year.

There are two broad categories of plastics. One, when heated repeatedly, will soften and flow; when it is cooled, it hardens. Materials which undergo such reversible changes when heated and cooled are called *thermoplastics;* polystyrene is one example. The other type is plastic when first heated, but when heated further it forms a set of interlocking bonds. When reheated, it cannot be softened and reformed without extensive degradation. These materials are called *thermosetting plastics* and include rigid-foamed polyurethane, a polymer which is finding many new uses as a construction material.

In order to gain a better understanding of polymers it is necessary to look at representative examples of the different types of polymerization processes.

ADDITION POLYMERS

In the previous section it was noted that some polymers, such as polystyrene, are made by adding monomer to monomer to form a polymer chain of great length. Perhaps the easiest addition reactions to understand chemically are those involving monomers containing double bonds. The simplest monomer of this group is ethylene, C_2H_4. When ethylene is heated under pressure in the presence of oxygen, polymers with molecular weights of about 30,000 are formed. In order to enter into reaction, the double bond of ethylene must be broken. This forms reactive sites at either end:

$$\begin{array}{cc} H\ \ H \\ |\ \ \ | \\ C=C \\ |\ \ \ | \\ H\ \ H \end{array} \longrightarrow \begin{array}{cc} H\ \ H \\ |\ \ \ | \\ \cdot C-C\cdot \\ |\ \ \ | \\ H\ \ H \end{array} \quad \textit{Reactive sites}$$

Ethylene

The partial breaking of the double bond can be accomplished by physical means such as heat, ultraviolet light, x-rays, and high energy electrons. The *initiation* of the polymerization reaction can also be accomplished with chemicals such as organic peroxides. These initiators, which are very unstable, break apart into pieces with unpaired electrons. These fragments (called free radicals) are ravenous in trying to find a "buddy" for their unpaired electrons. They react readily

161

with molecules containing carbon—carbon double bonds. Benzoyl peroxide is one commonly used free radical initiator:

Benzoyl peroxide   *(A free radical, R·)*

The odd electron in the free radical which is formed pairs with an electron in ethylene and, in turn, forms another free radical.

$$\text{Peroxide} \longrightarrow R\cdot$$

$$R\cdot + CH_2{=}CH_2 \longrightarrow R{-}CH_2{-}CH_2\cdot$$

Ethylene   *(Another free radical)*

The polymer grows as the resulting free radical reacts with other ethylene molecules to form a long hydrocarbon chain:

$$R{-}CH_2{-}CH_2\cdot + CH_2{=}CH_2 \longrightarrow R{-}CH_2{-}CH_2{-}CH_2{-}CH_2\cdot$$

$$R{-}CH_2{-}CH_2{-}CH_2{-}CH_2\cdot + CH_2{=}CH_2 \longrightarrow$$
$$R{-}CH_2{-}CH_2{-}CH_2{-}CH_2{-}CH_2{-}CH_2\cdot$$

Some time after the chain has begun to form, *termination* of the polymerization process occurs. Occasionally, two long chains may meet and link up their reactive sites.

$$R{-}CH_2{-}CH_2{-}(CH_2{-}CH_2)_n{-}CH_2{-}CH_2\cdot + \cdot CH_2{-}CH_2{-}(CH_2{-}CH_2)_m{-}R$$
$$\longrightarrow R{-}CH_2{-}CH_2{-}(CH_2{-}CH_2)_n{-}CH_2{-}CH_2{-}CH_2{-}CH_2{-}(CH_2{-}CH_2)_m{-}R$$

(In this example n and m are large, and probably different, numbers.)

In addition to this process, the initiator free radicals not used in the initiation process begin to terminate the build-up of some of the polymer chains.

$$R{-}(CH_2{-}CH_2)_n{-}CH_2{-}CH_2\cdot + \cdot R \longrightarrow R{-}(CH_2{-}CH_2)_n{-}CH_2{-}CH_2{-}R$$

Polyethylenes formed under various pressures and catalytic conditions have different molecular structures and hence different physical properties. For example, chromium oxide as a catalyst yields almost exclusively the linear polyethylene shown previously. If ethylene is heated to 230°C at a pressure of 200 atm, irregular branches result. Under these conditions, free radicals undoubtedly attack the chain at random positions, thus causing the irregular branching.

TABLE 8-1   ETHYLENE DERIVATIVES WHICH UNDERGO ADDITION
POLYMERIZATION

FORMULA	MONOMER NAME	POLYMER NAME	USES
$CH_2=CH_2$	Ethylene	Polyethylene	coats, milk cartons, wire insulation, bread boxes
$CH_2=CHCl$	Vinyl chloride	Polyvinyl chloride (PVC)	as a vinyl acetate copolymer in phonograph records, credit cards, rain wear
$CH_2=CH$ $\vert$ C_6H_5	Vinyl benzene (Styrene)	Polystyrene	combs
$CH_2=CH$ $\vert$ $O-C-CH_3$ $\Vert$ O	Vinyl acetate	Polyvinylacetate	latex paint
$CH_2=CH$ $\vert$ CN	Acrylonitrile	Polyacrylonitrile (PAN)	rug fibers
$CH_2=CH-CH=CH_2$	Divinyl (1,3-Butadiene)	BUNA rubbers	tires and hoses
CH_3 O $\vert$ $\Vert$ $CH_2=C-C$ $\diagdown$ $O-CH_3$	Methyl methacrylate	Polymethyl methacrylate (Plexiglas, Lucite)	transparent objects, lightweight "pipes"
$CF_2=CF_2$	Tetrafluoroethylene	Polytetrafluoroethylene (TFE) (Teflon)	Insulation, bearings, nonstick fry pan surfaces

The molecules in linear polyethylene can line up with one another very easily, yielding a tough, high-density compound which is useful in making toys, bottles, and so forth. The polyethylene with irregular branches is less dense, more flexible, and not nearly as tough as the linear polymer, since the molecules are generally farther apart and their arrangement is not as precisely ordered. This material is used for squeeze bottles and other similar applications.

There is a large group of derivatives of ethylene which undergo addition polymerization, usually via a free radical mechanism. Table 8–1 summarizes some pertinent information on these materials.

TAILOR-MADE MOLECULES

When the structure of polypropylene is drawn out to illustrate the three-dimensionality of the carbon—carbon bonds, three unique structures appear (Figure 8–2). In the first structure, called *isotactic*, all the methyl (—CH_3) groups are in identical positions along the polymer chain. When the methyl groups extend in alternate directions away from the chain, a *syndiotactic* arrangement results; finally, there is the possibility of a random, or *atactic*, arrangement.

163

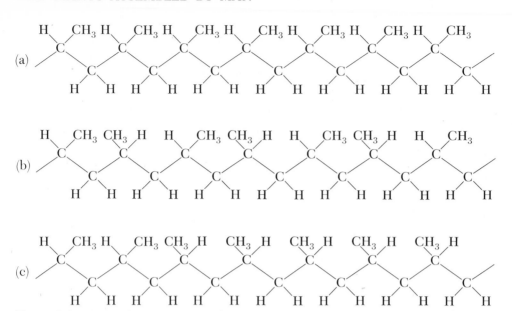

(a)

(b)

(c)

FIGURE 8–2   Three different structures of polypropylene. Each structure imparts different properties to the plastic. See text for discussion.

Polypropylenes of these three types were actually prepared and named by Professor Giulio Natta (who shared the Nobel prize with Karl Ziegler in 1963), in Italy in 1955. Using novel catalysts such as aluminum or iron attached to organic molecules, Natta was the first to control the growth phase of a polymerization reaction. The types of catalysts used by Natta were first used by Professor

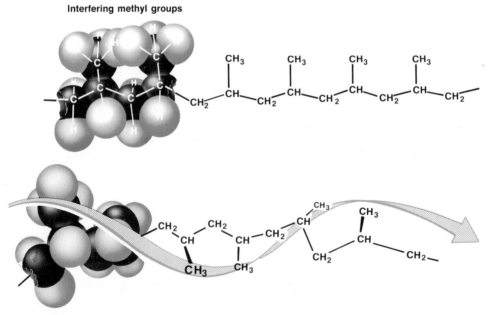

Interfering methyl groups

FIGURE 8–3   (a) Isotactic polypropylene. The bulky methyl groups are close to each other in this arrangement. (b) To eliminate crowding of the methyl groups on one side of the polymer chain, the chain flexes to produce a helical arrangement.

Karl Ziegler in Germany to increase the yields of straight-chain, high-density polyethylene. This type of control is called *stereochemical control*. As a result of this discovery, chemists were able to choose the extent of polymerization (molecular weight) and also the fine structural features of the polymer chain itself.

Natta's three polypropylenes are different in ways that can be related to their structures. The isotactic material (named by Natta's wife) actually has a helical chain structure, owing to the bulkiness of the methyl groups interacting with each other (Figure 8–3); this allows the chains to approach each other closely. Isotactic polypropylene melts at 170°C and can be easily formed into fibers. The atactic polypropylene chains cannot approach as closely because of the randomness of the methyl groups; this irregular structure renders the material rubbery and less dense.

Synthetic "Natural" Rubber

A very interesting application of stereochemical control over polymerization is the manufacture of synthetic rubber. Natural rubber has the composition $(C_5H_8)_n$, and when decomposed in the absence of oxygen, yields isoprene:

$$CH_2{=}\overset{\displaystyle CH_3}{\overset{\displaystyle |}{C}}{-}CH{=}CH_2$$

Isoprene

Natural rubber occurs as latex (a suspension of rubber particles in water) that oozes from rubber trees when they are cut. When the rubber particles are precipitated from the latex, a gummy mass is obtained that is not only elastic and water-repellent but also very sticky, especially when warm. In 1839, after 10 years' work on this material, Charles Goodyear (1800–1860) discovered that heating it with sulfur produced a material that was no longer sticky, but still elastic, water repellent, and resilient.

The Human Side

Charles Goodyear

Before Goodyear, others had tried to mix sulfur with natural rubber gum but with no success. When some of Goodyear's mixture "accidentally" came in contact with a hot stove, some of the portion that wasn't scorched too badly had become dry, flexible rubber. Goodyear began to heat the latex-sulfur mixtures at higher temperatures than anyone else had tried before, and thus discovered vulcanized rubber (after Vulcan, the Roman god of fire). Goodyear patented the process in 1844. He made his first experiments while in a debtor's prison, and died before the process became profitable to him. His debts at the time of his death were estimated to be not less than $200,000 and perhaps as high as $600,000. Since his process was a simple one, he had spent large sums of money protecting his patent.

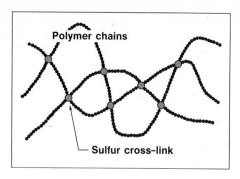 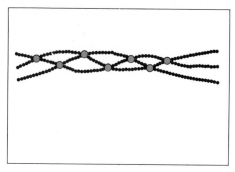

a. Before stretching   **b. Stretched**

FIGURE 8–4   Stretched vulcanized rubber retains its elasticity.

Vulcanized rubber, as Goodyear called his product, contains short chains of sulfur atoms which bond together the polymer chains of the natural rubber. The sulfur chains help to align the polymer chains, so the material does not undergo a permanent change when stretched, but springs back to its original shape and size when the stress is removed (Figure 8–4).

In later years chemists searched for ways to make a synthetic rubber so we would not be completely dependent on natural rubber during emergencies, as during the first years of World War II. In the mid–1920's, German chemists polymerized butadiene (structurally similar to isoprene and obtained from petroleum) to produce Buna rubber, so named because it was made from butadiene (Bu—) catalyzed by sodium (—Na).

The behavior of natural rubber (polyisoprene), it was learned later, is due to the specific arrangement within the polymer chain. We can write the formula for polyisoprene with the CH_2 groups on opposite sides of the double bond (the

Poly-trans-isoprene (the —CH₂—CH₂— groups are trans)

trans arrangement), or with the CH_2 groups on the same side of the double bond (the **cis** from Latin meaning on this side).

Poly-cis-isoprene (the —CH₂—CH₂— groups are cis)

Natural rubber is poly-*cis*-isoprene. However, the *trans* material also occurs in nature, in the leaves and bark of the sapotacea tree, and is known as *gutta-percha*. It is used as a thermoplastic for golf ball covers, electrical insulation and other such applications. Without an appropriate catalyst, polymerization of isoprene yields a solid that is like neither rubber nor gutta-percha.

In 1955, chemists at Goodyear and Firestone discovered, almost simultaneously, how to use stereoregulating catalysts similar to those developed by Ziegler to prepare synthetic poly-*cis*-isoprene. This material is, therefore, structurally identical to natural rubber. Today, synthetic poly-*cis*-isoprene can be manufactured cheaply and is used when natural rubber is in short supply. More than 2.4 million tons of synthetic rubber are produced in the United States yearly.

COPOLYMERS

After examining Table 8–1, one might well wonder what would happen if a mixture of two monomers was polymerized. This type of reaction has been studied in detail and the products are called *copolymers*. If we polymerize pure monomer A, we get a homopolymer, poly A:

—AAAAAAAAAA—

Likewise, if pure monomer B is polymerized, we get a homopolymer, poly B:

—BBBBBBBBBB—

In contrast, if the monomers A and B are mixed and then polymerized, we get copolymers such as

—AABABAAABB—

—AABABABABB—

—BABABBAABA—

In such polymers the order of the units is often completely random; in which case the properties of the copolymer will be determined by the ratio of the amount of A to the amount of B.

A copolymer can have useful properties that are different from and often superior to those of the polymers of its pure constituents. As an example, let's go back to our discussion of rubber and pick up with synthetic rubbers. During

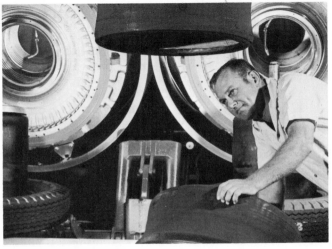

FIGURE 8–5   Ready to take the cure, these "green" truck tires at Goodyear's Danville, Virginia, plant will assume their familiar "doughnut" shape when they are cured, or vulcanized. This involves placing the tires in the huge molds, just behind the technician, and subjecting them to heat and pressure. Note the familiar cured product in the open molds. Goodyear estimated that the industry shipped 28 million truck tires to motor vehicle manufacturers and operators during 1971. (Photo through the courtesy of the Goodyear Tire and Rubber Company, Akron, Ohio.)

World War II it was apparent to our military planners that we would be hard-pressed if our rubber supplies were cut off by Japan. A crash program was begun to develop synthetic rubber which would be as good as natural rubber. The Germans had earlier polymerized styrene, but this is a hard thermoplastic with little elasticity. They had also polymerized butadiene to make the first synthetic rubber, although not a very serviceable one. American chemists found, however, that a 1-to-3 copolymer of styrene and butadiene possessed properties very much like those of natural rubber.

$$CH_2{=}CH + CH_2{=}CH{-}CH{=}CH_2 \longrightarrow$$

$$-CH_2CH{=}CHCH_2CH_2\overset{\displaystyle H}{\underset{\displaystyle}{C}}CH_2CH{=}CHCH_2CH_2CH{=}CHCH_2-$$

Styrene   *Butadiene*

SBR copolymer (Styrene-butadiene rubber)

The double bonds remaining in the polymer chain allow them to undergo vulcanization like natural rubber polymer chains (see Figure 8–4).

SBR rubber is today manufactured on a large scale. About two million tons are used each year in manufacturing automobile and truck tires. A pure form of this polymer has even found its way into our mouths as the replacement for the latex in chewing gum.

CONDENSATION POLYMERS

POLYESTERS

A chemical reaction in which two molecules react by splitting out or eliminating a smaller molecule is called a *condensation reaction*. For example, acetic acid and ethyl alcohol will react, splitting out a water molecule, to form ethyl acetate, an *ester*.

$$CH_3\overset{\displaystyle O}{\overset{\|}{C}}{-}OH + HO{-}CH_2CH_3 \xrightarrow[\text{Catalyst}]{H^+} CH_3\overset{\displaystyle O}{\overset{\|}{C}}{-}OCH_2CH_3 + H_2O$$

Acetic   *Ethyl*   *Ethyl acetate*
acid   *alcohol*

This important type of chemical reaction does not depend upon the presence of a double bond in the reacting molecules. Rather, it requires the presence of two kinds of functional groups on two different molecules. If each reacting molecule has two functional groups, both of which can react, it is then possible for condensation reactions to lead to a polymer. If we take a molecule with two carboxyl groups, such as terephthalic acid, and another molecule with two alcohol groups, such as ethylene glycol, each molecule can react at both ends. The reaction of one acid group of terephthalic acid with one alcohol group of

168

ethylene glycol initially produces an ester molecule with an acid group left over on one end and an alcohol group left over on the other:

Terephthalic acid   *Ethylene glycol*

(An ester)

Subsequently, the remaining acid group can react with another alcohol group, and the alcohol group can react with another acid molecule. The process continues until an extremely large polymer molecule, known as a *polyester*, is produced with a molecular weight in the range of 10,000–20,000.

Poly(ethylene glycol terephthalate)

Polyethylene glycol terephthalate is used in making textile fibers marketed under such names as "Dacron" and "Terylene," and films such as "Mylar." The film material has unusual strength and can be rolled into sheets one-thirteenth the thickness of a human hair. In film form this polyester is often used as a base for magnetic recording tape and for frozen food packaging.

POLYAMIDES (NYLONS)

Another useful condensation reaction is that occurring between an acid and an amine to split out a water molecule and form an *amide*. Reactions of this type yield a group of polymers which perhaps have had a greater impact upon society than any other type. These are the *polyamides*, or nylons.

In 1928, the Du Pont Company embarked upon a program of basic research headed by Dr. Wallace Carothers (1896–1937) who had been hired from the Harvard University faculty. His research interests were high molecular weight compounds, such as rubber, proteins, resins, and the reaction mechanisms that produced these compounds. In February, 1935, his research produced a product known as nylon 66, prepared from adipic acid and hexamethylene diamine.

This material could easily be extended into fibers that were stronger than natural fibers and more chemically inert. The discovery of nylon jolted the American textile industry at almost precisely the right time. Natural fibers were not meeting the needs of twentieth-century Americans. Silk was not durable

169

$$\underset{\text{HO}}{}\overset{\text{O}}{\underset{\|}{C}}-(CH_2)_4-\overset{\text{O}}{\underset{\|}{C}}\overset{}{\underset{\text{OH}}{}} + H_2N-(CH_2)_6-NH_2 \longrightarrow$$

Adipic acid   *Hexamethylene diamine*

$$\underset{\text{HO}}{}\overset{\text{O}}{\underset{\|}{C}}-(CH_2)_4-\boxed{\overset{\text{O}}{\underset{\|}{C}}-\underset{H}{\overset{}{N}}}-(CH_2)_6-\boxed{\underset{H}{\overset{}{N}}-\overset{\text{O}}{\underset{\|}{C}}}-(CH_2)_4-\boxed{\overset{\text{O}}{\underset{\|}{C}}-\underset{H}{\overset{}{N}}}-(CH_2)_6- + xH_2O$$

Nylon 66

The amide groups are outlined for emphasis.

and was very expensive, wool was scratchy, linen crushed easily, and cotton did not lend itself to high fashion. As women's hem-lines rose in the mid-1930's silk stockings were in great demand, but they were very expensive. Nylon changed all that almost overnight. It could be woven into the sheer hosiery women wanted, and it was much more durable than silk. The first public sale of nylon hose took place in Wilmington, Delaware (the hometown of Du Pont's main office), on October 24, 1939. They were so popular they had to be rationed. World War II caused all commercial use of nylon to be abandoned until 1945. Not until 1952 was the nylon industry able to meet the demands of the hosiery industry and to release nylon for other uses as a fiber and as a thermoplastic.

Many nylons have been prepared and tried on the consumer market, but two, nylon 66 and nylon 6, have been most successful. Nylon 6 is prepared from caprolactam, which comes from aminocaproic acid. Notice how aminocaproic acid resembles the first reaction product of adipic acid and hexamethylene-diamine in that it contains an amine group on one end of the molecule and an acid group on the other end.

$$H_2N-(CH_2)_5-\overset{\text{O}}{\underset{\text{OH}}{\overset{\|}{C}}} \xrightarrow{-H_2O} (CH_2)_5\overset{N-H}{\underset{C=O}{<}} \xrightarrow{\text{polymerization}} \text{Nylon 6}$$

Aminocaproic acid   *Caprolactam (An intermediate)*

Figure 8–6 illustrates another facet of the structure of nylon—*hydrogen bonding*. This type of bonding explains why the nylons make such good fibers. In order to have good tensile strength, the chains of atoms in a polymer should be able to attract one another but not so strongly that the plastic cannot be initially extended to form the fibers. Ordinary covalent chemical bonds linking the chains together would be too strong. Hydrogen bonds, with a strength about one-tenth that of an ordinary covalent bond, link the chains in the desired manner. We will see later that this type of bonding is also of great importance in protein structures.

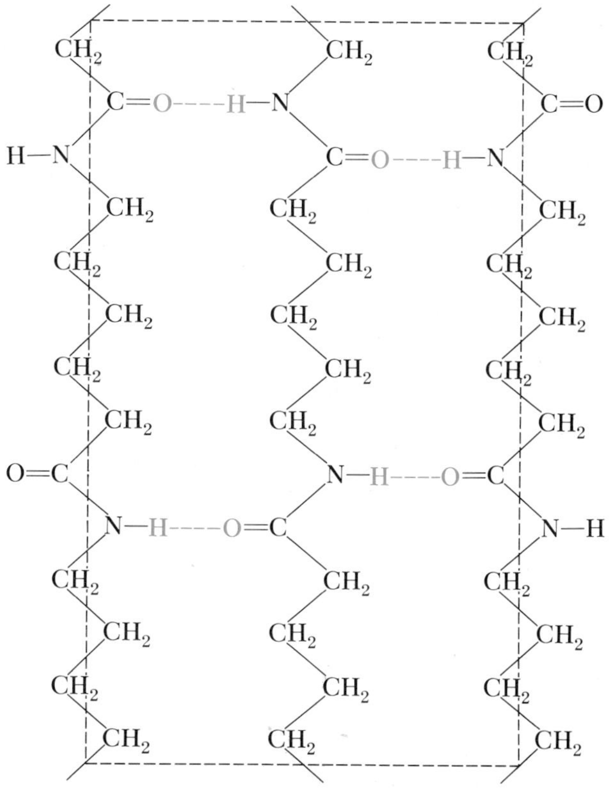

FIGURE 8-6   Structure and hydrogen bonding in nylon 6.

SILICONES

We saw in Chapter 7 that the element silicon can form chains of silicon atoms, but that Si—Si bonds are reactive to water and oxygen. Similarly, Si—H bonds are very reactive. Hence, there are no useful silicon counterparts of most hydrocarbons; however, silicon does form strong bonds with carbon and, especially, with oxygen. A useful and interesting group of condensation polymers are formed as a result of the bonding of silicon, oxygen, and carbon.

In 1945, E. G. Rochow, at the General Electric Research Laboratory, discovered that a silicon-copper alloy will react with organic chlorides to produce a whole new class of compounds, the organo-silanes.

$$2CH_3Cl + Si(Cu) \longrightarrow (CH_3)_2SiCl_2$$

Methyl chloride   *Silicon-copper alloy*   *Dimethyldichloro-silane*

These compounds readily react with water to replace the chlorine atoms with hydroxyl (—OH) groups. In effect, the resulting molecule is like a dialcohol.

$$(CH_3)_2SiCl_2 + 2H_2O \longrightarrow (CH_3)_2Si(OH)_2$$

Two dihydroxy silane molecules undergo a condensation reaction in which

171

a water molecule is split out. The resulting Si—O—Si linkage is very strong; the same linkage holds together all the natural silicate rocks and minerals. Continuation of this condensation process results in polymer molecules with molecular weights in the thousands:

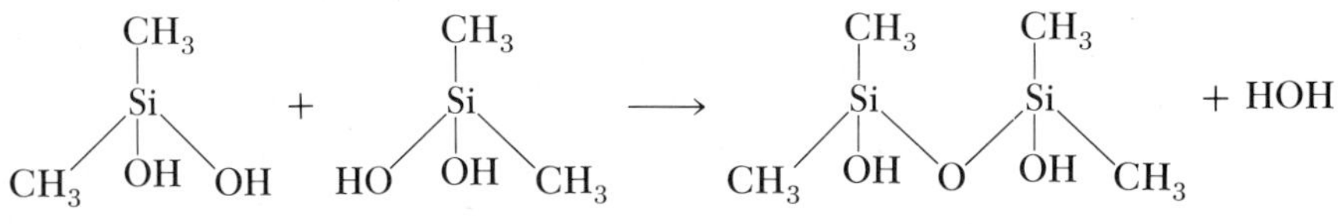

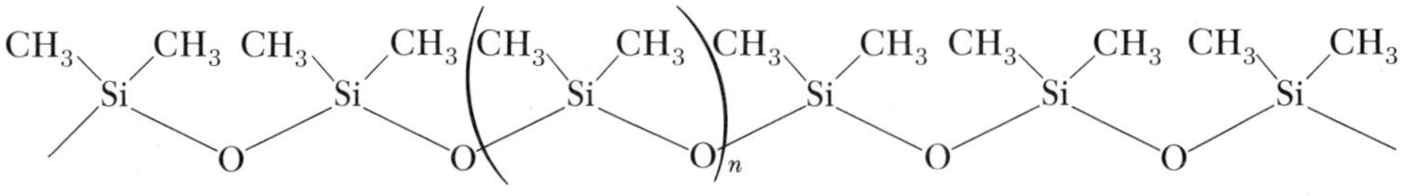

By using different starting silanes, polymers with very different properties result. For example, methyl groups on the silicon atoms result in *silicone oils* which are more stable at high temperatures than hydrocarbon oils and also have less tendency to thicken at low temperatures.

Silicone rubbers are composed of very high molecular weight units bridged together with ethylene or similar groups. Room-temperature-vulcanizing (RTV) silicone rubbers are commercially available. These contain readily hydrolyzable groups which cross link in the presence of atmospheric moisture:

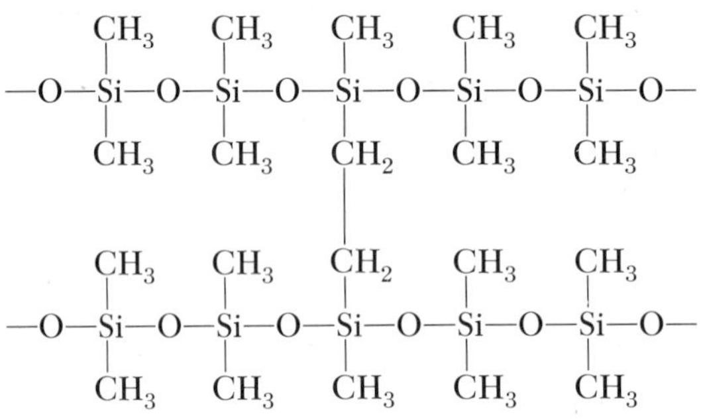

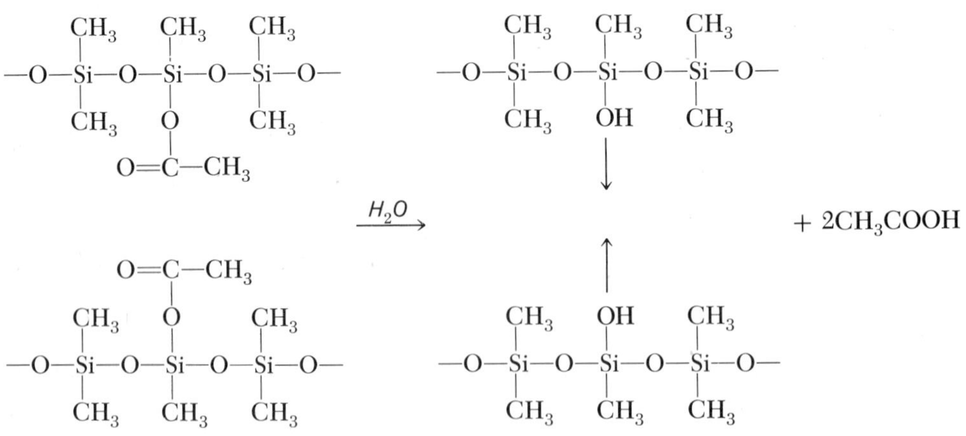

172

Figure 8-7 Examples of the use of silicone in the space program. Soles of lunar boots worn by the Apollo astronauts are made of high-strength silicone rubber. A silicone compound is also used for the air-tight seal of the lunar module hatch from which Astronaut Edwin E. Aldrin, Jr., has just emerged in this photo of the first manned landing on the moon on July 20, 1969.

The —OH groups which are produced are then made to condense, resulting in a cross-linking "cure" which is similar to the vulcanization of organic rubbers.

Over 3,000,000 pounds of silicone rubber are produced each year in the United States. The uses include window gaskets, o-rings, insulation, sealants for buildings, space ships, and jet planes and even some wearing apparel. The first footprints on the moon were made with silicone rubber boots which readily withstood the extreme surface temperatures.

REARRANGEMENT POLYMERS

Some molecules polymerize by rearrangement reactions to yield very useful products. Molecules containing the isocyanate group (—NCO), for example, will react with almost any other molecule containing an active hydrogen atom (such as in an —OH or —NH$_2$ group) in a rearrangement process. An example is the reaction of hexamethylene diisocyanate and butanediol. The urethane linkage

$$\left(\begin{array}{c} -\text{N}-\text{C}-\text{O}- \\ \;\;|\;\;\;\;\| \\ \;\;\text{H}\;\;\;\text{O} \end{array}\right)$$ is a rearrangement of the same atoms in the isocyanate and

alcohol groups and is similar to, but not the same as, the amide bond in nylons.   173

$$OCN(CH_2)_6NCO + HO(CH_2)_4OH \longrightarrow OCN(CH_2)_6\overset{\overset{\displaystyle H}{|}}{N}\overset{\overset{\displaystyle O}{\|}}{C}O\!-\!(CH_2)_4OH$$

Hexamethylene   1,4-Butanediol   Product molecule
diisocyanate

The continued reaction of the other groups gives rise to a polymer chain—a polyurethane.

$$-(CH_2)_6\overset{\overset{\displaystyle H}{|}}{N}\overset{\overset{\displaystyle O}{\|}}{C}O\!-\!(CH_2)_4\!-\!O\!-\!\overset{\overset{\displaystyle O}{\|}}{C}\overset{\overset{\displaystyle H}{|}}{N}\!-\!(CH_2)_6\overset{\overset{\displaystyle H}{|}}{N}\overset{\overset{\displaystyle O}{\|}}{C}O-$$

A portion of polyurethane

A typical polyurethane is structurally similar to a polyamide (nylon), and in Europe it is used similarly to nylon in this country. Polyurethanes have viscosities and melting points that make them useful for foam applications. Foamed polyurethanes are known as "foam rubber" and "foamed plastics."

POLYMER ADDITIVES—TOWARD AN END USE

Few plastics produced today find end uses without some kind of modification. Polyurethanes are a good example. In order for polyurethane to be used as insulation in refrigerators, refrigerated trucks and railroad cars, and as construction insulation, it is foamed.

FOAMING AGENTS

Two methods of producing foamed plastics are commonly used. Hydrocarbons such as pentane or fluorocarbons like trichlorofluoromethane are *physical foaming agents* because they can foam a plastic simply by boiling to produce bubbles in the plastic. Polyurethane can be foamed by dissolving pentane in the liquid under pressure. When this mixture is extruded from an outlet into the atmosphere, the pentane volatilizes, leaving the polyurethane full of small holes. The polyurethane is quickly solidified by cooling. Planks, boards, or logs of foamed polystyrene (styrofoam) can also be made in this way.

Chemical foaming agents produce a gas by chemical reaction, as their name implies. For example, when polyurethanes are formed, it is sometimes possible to make use of gaseous carbon dioxide, generated when an isocyanate group reacts with a water molecule.

$$R\!-\!NCO \quad + H_2O \longrightarrow R\!-\!\overset{\overset{\displaystyle }{|}}{N}\!-\!H + CO_2$$
$$\underset{\displaystyle H}{|}$$

An isocyanate   *An amine*

PLASTICIZERS

Many times a plastic such as polyethylene or polypropylene turns out to be too stiff for its intended application. Chemists have found that the addition

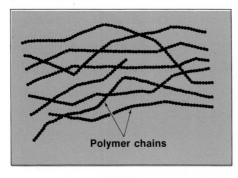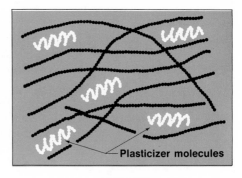

Rigid   **Less rigid**

FIGURE 8-8   Schematic representation of how a plasticizer works.

of certain compounds, called plasticizers, can render a polymer more flexible. The structure of a polymer largely determines its flexibility. The polymer chains are intertwined and somewhat aligned in the solid state. This is enough to prevent adequate flexing of the solid in many instances. Certain compounds such as di-(2-ethylhexyl) phthalate (DOP) and di-(2-ethylhexyl) adipate (DOA), exert a partial dissolving effect by fitting in between the polymer chain and weakening the attractions between chains, thereby increasing the flexibility (Figure 8–8).

In 1968, 1.35 billion pounds of plasticizers were used in plastic formulation. The extent of plastic food wraps for example, would be much more limited without DOA, which has FDA approval. DOP and similar compounds are found in many areas of our environment. Recently, DOP and similar plasticizers were discovered in the heart muscle of cattle, dogs, rabbits, and rats and in the bloodstreams of a group of laboratory workers at a National Laboratory.

DOP   *DOA*

In the case of the synthetic SBR rubber discussed earlier in this chapter, one form of this polymer required ordinary hydrocarbon oil as a plasticizer.

STABILIZERS

Most plastics used in outside locations, such as signs, tarpaulins, indoor-outdoor carpeting, auto seat covers, and toys, must be protected from sunlight, as anyone who has repeatedly parked his plastic-upholstered convertible in the outdoors knows.

Photons of the 290 to 400 nanometer spectral region have sufficient energy

175

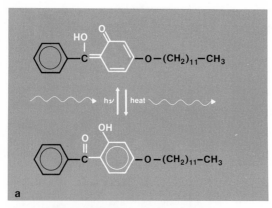

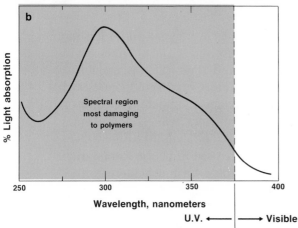

FIGURE 8-9   (a) A possible mechanism by which a molecule can absorb ultraviolet radiation and thus protect a plastic. This particular molecule is almost 100 percent efficient in the process and finds wide use in plastics. (b) Typical ultraviolet absorption by a 2-hydroxybenzophenone-type compound used to protect plastics from sunlight.

FIGURE 8-10   Sunlight (mostly ultraviolet) damages most plastics such as polypropylene webbing. The results are shorter life and higher costs to the consumer. Plastics containing ultraviolet absorbing chemicals have a longer outdoor life.

to break some chemical bonds found in organic molecules. However, there are other molecules that absorb ultraviolet light (Figure 8–9) and liberate the energy as heat, thus preventing broken bonds in the polymer structure. These may be added directly to the plastic (about 0.1 per cent by weight) that is to be subjected to sunlight. Unfortunately, many plastic formulations intended for outdoor use do not contain sufficient ultraviolet stabilizer to prevent decay (Figure 8–10).

THE FUTURE OF POLYMERS

As we have seen in this chapter, the development and use of synthetic polymers is quite recent. Polyethylene, for example, was not discovered until 1933, yet by 1969, its production in the United States amounted to billions of pounds. Chemists are constantly synthesizing new polymers and finding appli-

$$+ \; H_2N-CH_2-CH_2-NH_2 \; \xrightarrow{-H_2O}$$

*Diacid
anhydride*

1,2-diaminoethane

A polyimide

FIGURE 8-11   Preparation of a polyimide (above) and an example of how a polyimide film can withstand, for a short period, the flame of a blowtorch.

177

cations for them. The space age has brought with it the need for new polymers, especially in electronics and as special coatings which can withstand high temperatures without breaking down. Among the newcomers are the polyimides, prepared from the polycondensation of a diacidanhydride and a diamine. Some of these polymers have very high service temperatures (Figure 8–11).

Because polymers are used so extensively in the world today, the problem of waste disposal is inevitable. Engineers have envisioned plants in which solid wastes from cities would undergo first a magnetic separation to remove iron and steel objects, then a ballistic separation based on density, since glass and aluminum objects are more dense than plastics. The plastics thus separated would be treated in two ways. If suitable separation methods could be developed, thermoplastics could be reprocessed into new items (e.g., if all the nylon could be separated from polystyrene). Thermosetting plastics could not be treated this way, however, because breaking the cross-linking would cause complete molecular degradation. If separation and reuse were not feasible, combustion units built near cities could actually use plastics as fuels since they are mostly carbon and hydrogen. There is a danger, though, in that some plastics contain elements which could create massive pollution if released into the atmosphere. An example is polyvinylchloride, which on burning yields hydrogen chloride, a very corrosive gas. However, the combustion products could be recycled as raw materials for other chemical syntheses.

Hopefully, these and similar problems will be solved as man begins to understand more fully how to use what he has on this planet, and how to live in greater harmony with nature.

QUESTIONS

1. Explain how polymers could be prepared from

(a) $CH_3-\overset{\overset{\displaystyle H}{|}}{C}=\overset{\overset{\displaystyle H}{|}}{C}-CH_3$

(b) $HO-\overset{\overset{\displaystyle O}{||}}{C}-CH_2-CH_2-\overset{\overset{\displaystyle O}{||}}{C}-OH$

(c) $\underset{\underset{\displaystyle OH}{|}}{CH_2}-\underset{\underset{\displaystyle OH}{|}}{CH}-\underset{\underset{\displaystyle OH}{|}}{CH_2}$

(d) $H_2N-CH_2-\bigcirc-CH_2-NH_2$

2. Beginning with petroleum, outline the steps involved in the preparation of polyethylene.

3. Draw the monomer structural formula for

(a) polyethylene
(b) neoprene
(c) polystyrene
(d) Teflon

4. How many moles of ethylene gas are required to manufacture a 100 gram polyethylene bottle?

5. What are the monomers used to prepare the following polymers?

(a) $-CH_2CH_2CH_2CH_2CH_2CH_2CH_2CH_2CH_2-$

(b) $-\underset{\underset{\displaystyle CH_3}{|}}{C}HCH_2\underset{\underset{\displaystyle CH_3}{|}}{C}HCH_2\underset{\underset{\displaystyle CH_3}{|}}{C}HCH_2-$

$$\text{(c)} \quad -CH_2\overset{\overset{H}{|}}{C}CH_2\overset{\overset{H}{|}}{C}CH_2\overset{\overset{H}{|}}{C}CH_2\overset{\overset{H}{|}}{C}-$$

6. Write equations showing the formation of polymers by the reaction of the following pairs of molecules:

(a)
$$\text{COOH} \text{(on benzene ring)} \text{COOH}$$ and $HOCH_2CH_2OH$

(b) $HOOCCH_2CH_2COOH$ and $H_2NCH_2CH_2NH_2$

(c)
$$CH_2OH$$
$$H\overset{|}{C}OH$$
$$CH_2OH$$
and
$$\text{(benzene ring with } C{-}OH \text{ and } C{-}OH\text{)}$$

7. A specimen of polystyrene is found to consist of molecules with an average molecular weight of 12,000 amu. What is the average number of styrene monomers in each unit?

8. You are given two specimens of plastic, A and B, to identify. One is known to be nylon, and the other poly(methylmethacrylate). Analysis of A shows it to contain C, H, and O, while B contains C, H, O, and N. What are A and B?

9. Bakelite, one of the first commercially important plastics, is a condensation polymer of formaldehyde and phenol. The oxygen in the formaldehyde reacts with two hydrogens on two molecules of phenol to form water. The hydrogens are either *ortho* or *para* on the phenol. Draw a representative portion of Bakelite.

10. What structural features are necessary in a molecule for it to undergo addition polymerization?

11. What is meant by the term macromolecule?

12. Give an example of a thermoplastic and a thermosetting plastic and compare their properties.

13. Orlon has a polymeric chain structure of

$$-CH_2-\overset{\overset{|}{}}{CH}-CH_2-\overset{\overset{|}{}}{CH}-CH_2-\overset{\overset{|}{}}{CH}-$$
$$\quad\quad\quad CN \quad\quad\quad CN \quad\quad\quad CN$$

What is the Orlon monomer?

14. Illustrate how the presence of too much peroxide initiator could cause an addition polymerization to stop prematurely.

15. What is a copolymer? Give an example of one.

16. What feature do all condensation polymerization reactions have in common?

17. Give an example of the possibilities that exist if a trifunctional acid reacts with a difunctional alcohol.

179

18. What are the starting materials for nylon 66?

19. Why does nylon dry so quickly?

20. Name an additional hazard to firemen when a building containing large quantities of PVC water pipe burns.

21. Suggest a major difference in the bonding of thermosetting and thermoplastic polymers. Which is more likely to have an interlacing of covalent bonds throughout the structure? Which is more likely to have weak bonds between large molecules?

22. How does vulcanization render rubber elastic?

23. Explain how a plasticizer can render a polymer more flexible.

24. How do you suggest we dispose of waste plastic?

25. Write the reaction for the formation of a silicone from $(C_2H_5)_2SiCl_2$.

SUGGESTIONS FOR FURTHER READING

Books

Kaufman, M., "Giant Molecules," Doubleday, New York, 1968.
Mark, H. F., "Giant Molecules," Time Incorporated, New York, 1966.
"Nylon: The First 25 Years," The Du Pont Company, Wilmington, 1963.

Articles

Billmeyer, F. W., Jr., "Measuring the Weight of Giant Molecules," *Chemistry*, Vol. 39, No. 3, p. 8 (1966).
Franer, A. H., "High Temperature Plastics," *Scientific American*, Vol. 221, No. 1, p. 96 (1969).
Mark, H. F., "The Nature of Polymeric Materials," *Scientific American*, Vol. 217, No. 3, p. 98 (1967).
Natta, G., "How Giant Molecules are Made," *Scientific American*, Vol. 197, No. 3, p. 98 (1957).

Part 3
Life and Chemistry

9

GIANT MOLECULES IN BIOCHEMISTRY

The goal of biochemistry is to develop a chemically based understanding of living cells of all types. This includes the determination of the kinds of atoms which are present, the investigation of how they are joined together to form the larger structural units present in cells, and the study of the chemical reactions by which living cells furnish the energy required for the life processes of growth, movement, and reproduction. Furthermore, biochemistry seeks an understanding of how basic chemical structures and processes are combined to produce the complicated and highly specialized cells found in most plants and animals.

In this chapter we shall examine (a) carbohydrates, (b) amino acids and proteins, (c) fats, and (d) waxes. The first three groups of biochemicals are vital parts of the ongoing life processes and waxes are produced by living cells to improve their environment.

CARBOHYDRATES

Carbohydrates are composed of the three elements carbon, hydrogen, and oxygen. Three structural groups are prevalent in carbohydrates: alcohol (—OH), aldehyde $\left(-\overset{O}{\overset{\|}{C}}H\right)$, and ketone $\left(-\overset{O}{\overset{\|}{C}}-\right)$. The carbohydrates can be classified into three main groups: monosaccharides (Latin: *saccharum*, sugar), oligosaccharides, and polysaccharides. Monosaccharides are simple sugars that cannot be broken down into smaller units by mild acid hydrolysis. (Recall that hydrolysis is a reaction in which a chemical bond is broken and water is added—H to one part

181

and OH to the other—at the point of rupture.) Hydrolysis of a molecule of an oligosaccharide yields two to six molecules of a simple sugar; complete hydrolysis of a polysaccharide produces many monosaccharide units.

Carbohydrates are synthesized by plants from water and atmospheric carbon dioxide. The process is called photosynthesis since it is a synthetic reaction that occurs when energized by light. It is an endothermic process; consequently, carbohydrates are energy-rich compounds. These compounds serve as important sources of energy for the metabolic processes of plants and animals. Glucose, $C_6H_{12}O_6$, along with some of the other simple sugars, are quick energy sources for the cell. Polysaccharides, such as starch, store large amounts of energy that can be used by the cell only when the complex unit is broken down into monosaccharides.

Some complex carbohydrates are also used by cells for structural purposes. Cellulose, for example, partially accounts for the structural properties of wood.

MONOSACCHARIDES

Approximately 70 monosaccharides are known; 20 of these simple sugars occur naturally. Unlike many organic compounds these sugars are very soluble in water owing to the numerous —OH groups present which can form hydrogen bonds with water.

The most common simple sugar is D-glucose. The monosaccharide requires three structures for its adequate representation (Figure 9–1).

Structure (a), in which the carbon atoms are numbered for later reference, depicts the "straight-chain" structure with the aldehyde group (—CHO) in position 1. The properties of a water solution of D-glucose cannot be explained by this structure alone. Most of the molecules at any given time exist in ring

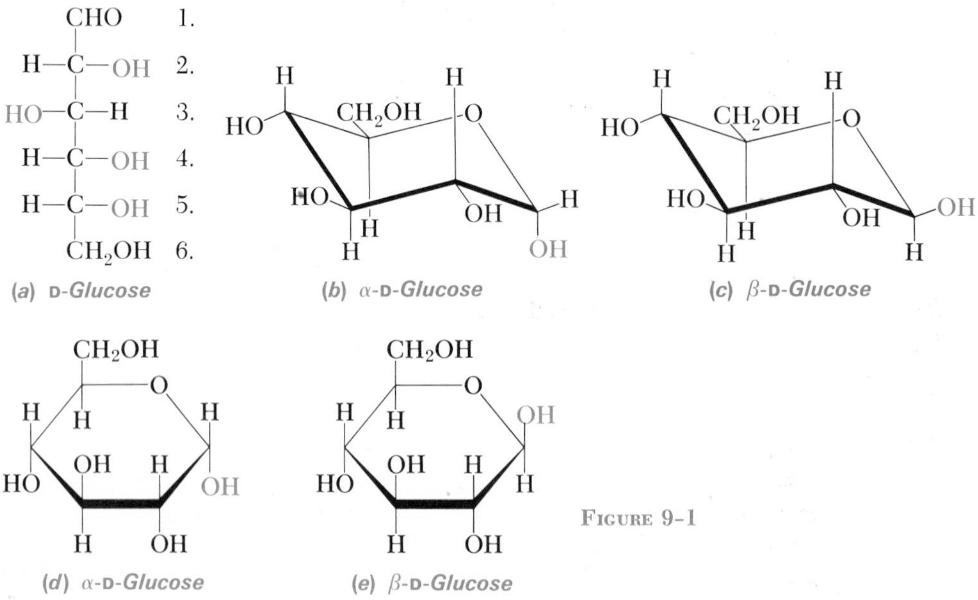

(a) D-*Glucose* (b) α-D-*Glucose* (c) β-D-*Glucose*

(d) α-D-*Glucose* (e) β-D-*Glucose*

FIGURE 9–1

form, structures (*b*) and (*c*), which results when carbon 1 bonds to carbon 5 through an oxygen atom. Both ring structures are possible since the OH group on carbon 1 may form in such a way to point either along the plane of the molecule or out of the plane. It should be emphasized that a solution of D-glucose contains a mixture of the three structures in a dynamic state of equilibrium. The equilibrium is shifted towards the ring forms; very limited amounts of the open chain form are present.

Examination of structure (*a*) in Figure 9–1 shows that four of the carbon atoms (2 through 5) are asymmetric (four different groups attached to each carbon). Consequently, the structural conditions are met for optical isomers. Since there are four asymmetric carbon atoms, there are a total of 16 (that is, 2^4) isomers in this group of simple sugars.

Because it is very sweet but does not crystallize readily D-glucose is used in the manufacture of candy and in commercial baking. This simple sugar, also called dextrose, grape sugar, and blood sugar, is found widely in fruits, vegetables, blood, and tissue fluids. A solution of D-glucose is fed intravenously when a readily available source of energy is needed to preserve life. As will be discussed later, many polysaccharides, including starch, are composed of glucose units and serve as a source of this important chemical upon hydrolysis of their complex structures.

Another very important monosaccharide is D-fructose. Its structure is given in Figure 9–2.

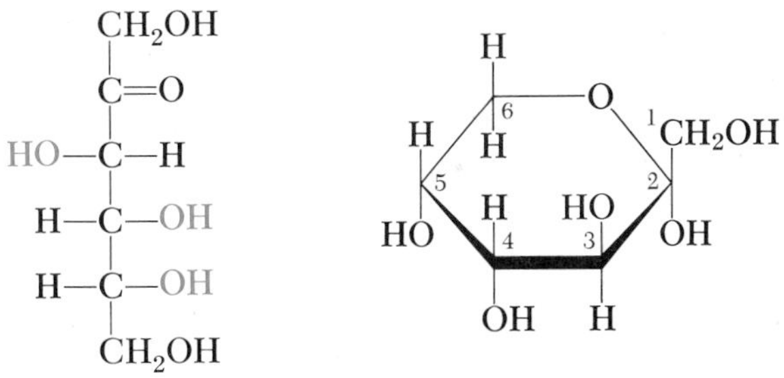

FIGURE 9–2

(a) *Ketone structure*

(b) *α-Ring structure (6-member ring)*

OLIGOSACCHARIDES

The most important oligosaccharides are the disaccharides (two simple sugar units per molecule). Examples include the following widely used sugars:

sucrose (from sugar cane or sugar beets), which consists of a glucose unit and a fructose unit,

maltose (from starch), which consists of two glucose units, and

lactose (from milk), which consists of a glucose unit and a galactose (an isomer of glucose) unit.

The formula for these disaccharides, $C_{12}H_{22}O_{11}$, is not simply the sum of two monosaccharides, $C_6H_{12}O_6 + C_6H_{12}O_6$. A water molecule must be added (hy-

183

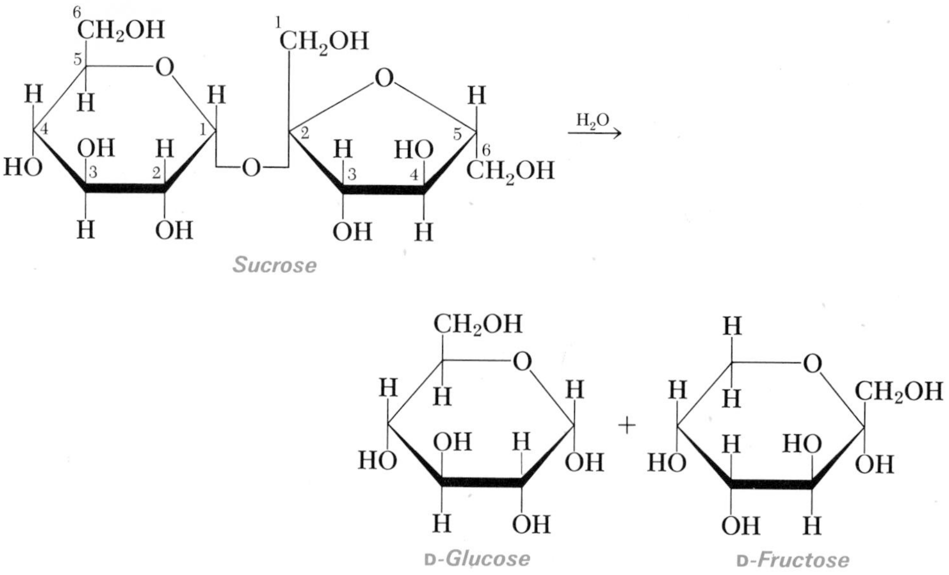

(Note: D-Fructose forms a six-membered ring when isolated and a five-membered ring in the sucrose structure.)

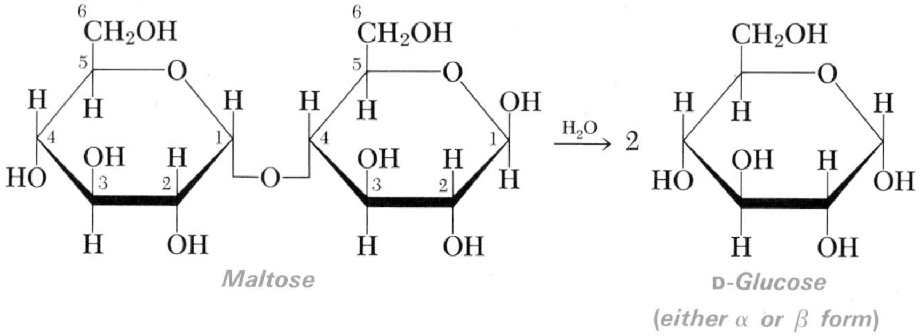

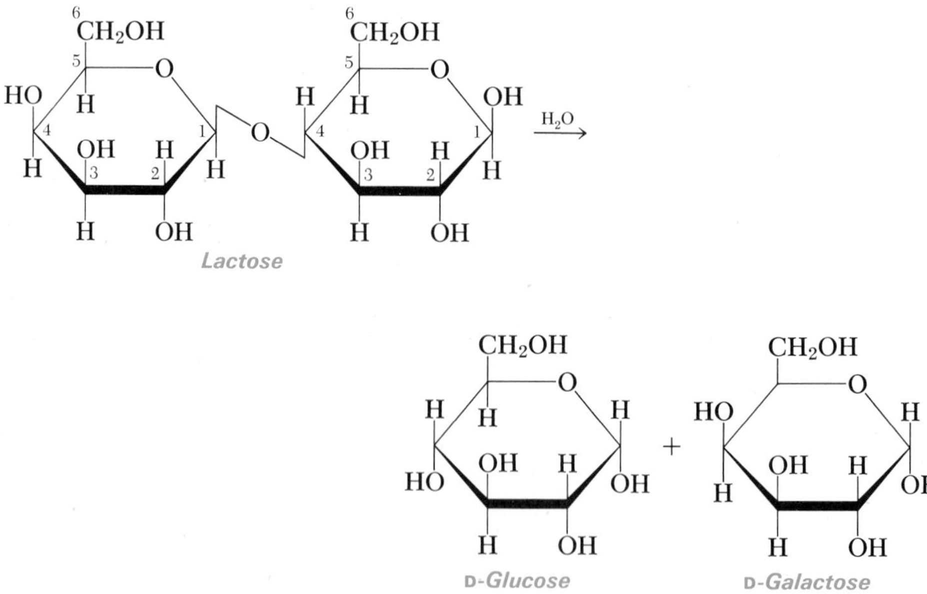

FIGURE 9–3   Hydrolysis of disaccharides (sucrose, maltose, and lactose).

drolysis) to obtain the monosaccharides. The structures for sucrose, maltose, and lactose, along with the hydrolysis reactions, are given in Figure 9–3.

The disaccharides are important as foods; sucrose is produced in a very high state of purity on an enormous scale. The annual production of this food amounts to over 40,000,000 tons. Originally produced in India and Persia, sucrose is now used universally as a sweetener. About 40 per cent of the world sucrose production comes from sugar beets and 60 per cent from sugar cane. Sugar provides a high caloric value (1794 kcal per pound); it is also used as a preservative in jams, jellies, and other foods.

POLYSACCHARIDES

There is an almost limitless number of possible structures in which monosaccharide units (monosaccharide molecules minus one water molecule at each bond between units) can be combined. Molecular weights are known to go above 1,000,000. Apparently nature has been very selective in that only a few of the many different monosaccharide units are found in polysaccharides.

Starches and Glycogen

Starch is found in plants in protein-covered granules. These granules are disrupted by heat, and a part of the starch content is soluble in hot water. Soluble starch is *amylose* and constitutes 10 to 20 per cent of most natural starches; the remainder is *amylopectin*. Amylose gives the familiar blue-black starch test with iodine solutions; amylopectin turns red in contact with iodine.

Structurally amylose is a straight chain of D-glucose units, each one bonded to the next, just as the two units are bonded in maltose (Figure 9–3). While there are amylose structures in excess of 1000 glucose units, molecular weight studies indicate the average "molecule" contains fewer than 500 units. The structure of amylose is illustrated in Figure 9–4.

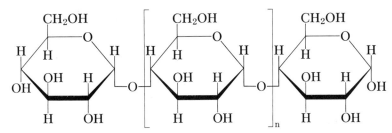

FIGURE 9–4   Amylose structure. n represents a large number of D-glucose units; the type of polysaccharide determines the value of n.

Amylopectin is made up of branched D-glucose units (Figure 9–5). Its molecular weight is generally higher than that for amylose. Partial hydrolysis of amylopectin yields mixtures called *dextrins*. Complete hydrolysis, of course, yields D-glucose. Dextrins are used as food additives, mucilages, paste, and in finishes for paper and fabrics.

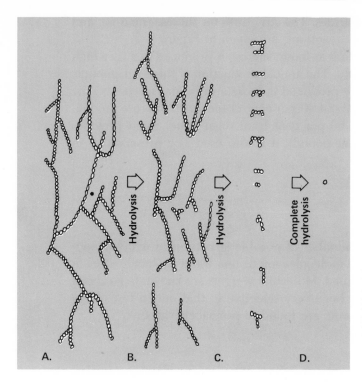

FIGURE 9–5 A, Partial amylopectin structure; B, dextrins from incomplete hydrolysis of A; C, oligosaccharides from hydrolysis of dextrins; D, final hydrolysis product: D-glucose. Each circle represents a glucose unit.

Glycogen serves as an energy reservoir in animals as does starch in plants. Glycogen has essentially the same structure as amylopectin (branched chains of glucose units) except that it has an even more highly branched structure.

Cellulose

Cellulose is the most abundant polysaccharide in nature. Like amylose, it is composed of D-glucose units. The difference between the structure of cellulose and that of amylose lies in the bonding between the D-glucose units. In cellulose, all of the glucose units are in the β-ring form in contrast to the α-ring form in amylose. Review the ring forms in Figure 9–1 and compare the structures in Figures 9–4 and 9–6.

The different structures of starch and cellulose account for their difference

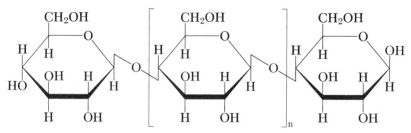

FIGURE 9–6 The structure of cellulose. The value of n may go as high as 10,000 in the natural product.

186

in digestibility. Human beings and carnivorous animals do not have the necessary enzymes (biochemical catalysts) to break down the cellulose structure as do numerous microorganisms. Cellulose is readily hydrolyzed to D-glucose in the laboratory by heating a suspension of the polysaccharide in the presence of a strong acid. Unfortunately, at present this is not an economically feasible solution to man's growing need for an adequate food supply.

Paper, rayon, cellophane, and cotton are essentially cellulose. Figure 9–7 relates the structure of cotton fiber to the chains of glucose units at the submicroscopic level.

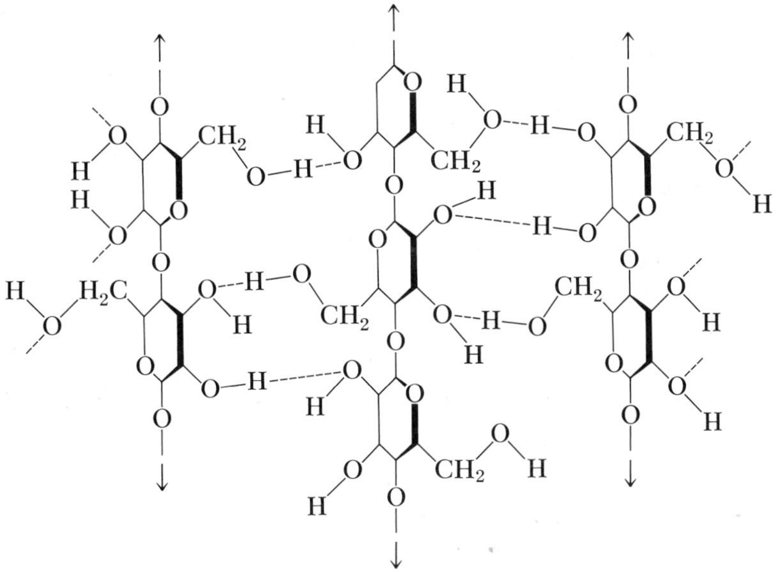

FIGURE 9–7   The properties of cotton, about 98 per cent cellulose, can be explained in terms of this submicroscopic structure. A small group of cellulose "molecules," each with from 2000 to 9000 units of D-glucose, are held together in an approximately parallel fashion by hydrogen bonding (· · · ·). When several of these *chain bundles* cling together in a relatively vast network of hydrogen bonds, a *microfibril* results; the microfibril is the smallest microscopic unit that can be seen. The macroscopic *fibril* is a collection of numerous microfibrils. The absorbent nature of cotton is readily explained in terms of the smaller water molecules being held by hydrogen bonding in the numerous capillaries that exist between the cellulose chains.

PROTEINS, AMINO ACIDS, AND THE PEPTIDE LINKAGE

Proteins occur in all the major regions of living cells. These compounds serve a wide variety of functions including motion of the organism, defense mechanism against foreign chemicals, metabolic regulation of cellular processes, and cell structure. The close relationship between protein structures and life was first noted by the German chemist, G. T. Mulder, in 1835. He named these compounds proteins from the Greek *proteios*, meaning first, indicating this to be the starting point in the chemical understanding of life.

Proteins are macromolecules with molecular weights ranging from 5000 to several million. Like the polysaccharides, these macrostructures are composed

187

of recurring submicroscopic units, the fundamental units in the case of proteins being *amino acids*. Proteins and amino acids are made primarily from four elements; carbon, oxygen, hydrogen, and nitrogen. Other elements occur in trace amounts, the one most often encountered being sulfur.

AMINO ACIDS

The complete hydrolysis of a typical protein yields a mixture of about 20 amino acids. Some proteins lack one or more of these acids, others have small amounts of other amino acids characteristic of a given protein, but the 20 given in Table 9–1 dominate. In a few instances, one amino acid will constitute a major fraction of a protein (the protein in silk, for example, is 44 per cent glycine), but this is not common.

As the name suggests, amino acids contain an amine group ($-NH_2$) and an acid (carboxyl) group ($-COOH$). In all of the amino acids listed in Table 9–1, the amine group and the acid group are bonded to the same carbon atom. Of these acids, 18 have the general formula:

TABLE 9-1   COMMON AMINO ACIDS

All of the amino acids except proline and hydroxyproline have the general formula:

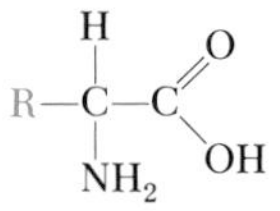

in which R is the characteristic group for each acid. The R groups are as follows.

1. Glycine $-H$
2. Alanine $-CH_3$
3. Serine $-CH_2OH$
4. Cysteine $-CH_2SH$
5. Cystine $-CH_2-S-S-CH_2-$
6. Threonine° $-CH-CH_3$
   with OH
7. Valine° $CH_3-CH-CH_3$
8. Leucine° $-CH_2-CH-CH_3$
   with CH_3
9. Isoleucine° $-CH$ with CH_3 and CH_2-CH_3

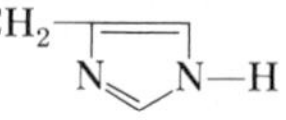

The structures for the other two are:

19. Proline H_2C——CH_2 / H_2C N $CHCO_2H$ / H

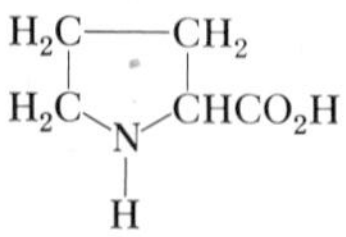

10. Methionine° $-CH_2-CH_2-S-CH_3$
11. Aspartic acid $-CH_2CO_2H$
12. Glutamic acid $-CH_2-CH_2-CO_2H$
13. Lysine° $-CH_2-CH_2-CH_2-CH_2-NH_2$
14. Arginine $-CH_2-CH_2-CH_2-NHCNH_2$ (NH)
15. Phenylalanine° $-CH_2-$ 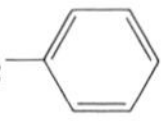
16. Tyrosine $-CH_2-$  $-OH$
17. Tryptophan° $-CH_2-$ 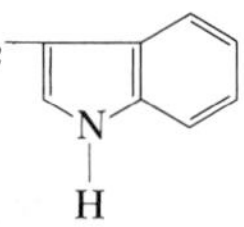
18. Histidine $-CH_2-$

20. Hydroxyproline $HOHC$——CH_2 / H_2C N $CHCO_2H$ / H

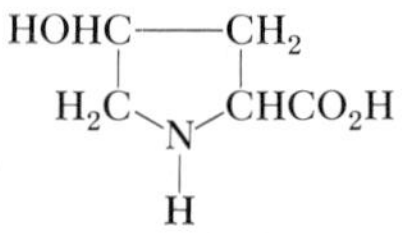

° Essential amino acids

$$R-\overset{\overset{\displaystyle H}{|}}{\underset{\underset{\displaystyle NH_2}{|}}{C}}-\overset{\displaystyle O}{\underset{\displaystyle OH}{C}}$$

where R is a characteristic group for each amino acid. The simplest amino acid is *glycine*, in which R is a hydrogen atom:

$$H-\overset{\overset{\displaystyle H}{|}}{\underset{\underset{\displaystyle NH_2}{|}}{C}}-\overset{\displaystyle O}{\underset{\displaystyle OH}{C}}$$

Glycine

The human body is capable of synthesizing some amino acids needed for protein structures, but it is unable to provide others at a rate necessary for normal growth and development. The latter are designated *essential* amino acids and must be ingested in the food supply. The *nonessential* amino acids are just as necessary as the essential amino acids for life but can be made from any protein or nonprotein food supply. The essential amino acids are indicated in Table 9–1.

THE PEPTIDE LINKAGE

Amino acid units are linked together in protein structures by the peptide linkage. This same linkage was illustrated in nylon 66, in which an acid and an amine were condensed to form the polymer and the peptide bond. As it applies to proteins, this type of chemical bond can be understood in terms of a hypothetical reaction, which is a gross oversimplification of natural protein synthesis. If the acid group of one glycine molecule reacts with the basic amine group of another, the two are joined through the peptide linkage, and one molecule of water is eliminated for each link formed.

$$H-\overset{\overset{\displaystyle H}{|}}{\underset{\underset{\displaystyle H}{|}}{N}}-\overset{\overset{\displaystyle O}{\|}}{\underset{\underset{\displaystyle H}{|}}{C}}-C-OH + H-\overset{\overset{\displaystyle H}{|}}{\underset{\underset{\displaystyle H}{|}}{N}}-\overset{\overset{\displaystyle O}{\|}}{\underset{\underset{\displaystyle H}{|}}{C}}-C-OH \longrightarrow$$

Glycine   *Glycine*

Peptide (amide) linkage

$$H-\overset{\overset{\displaystyle H}{|}}{\underset{\underset{\displaystyle H}{|}}{N}}-\overset{\overset{\displaystyle O}{\|}}{\underset{\underset{\displaystyle H}{|}}{C}}-C-N-\overset{\overset{\displaystyle H}{|}}{\underset{\underset{\displaystyle H}{|}}{C}}-\overset{\overset{\displaystyle O}{\|}}{C}-OH + HOH$$

Glycylglycine

If this hypothetical reaction is carried out with two different amino acids, glycine and alanine, two different *dipeptides* are possible.

189

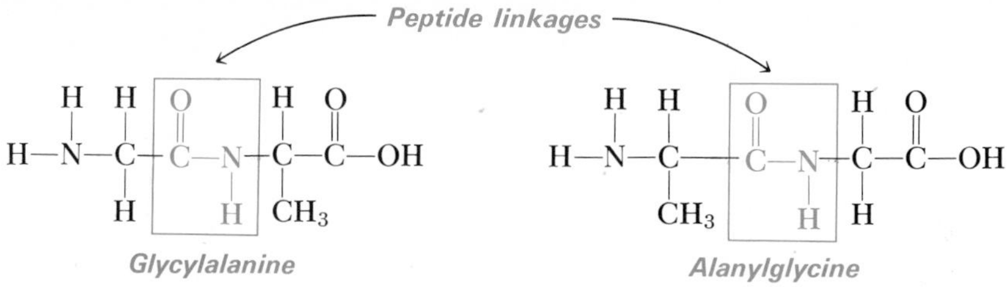

Peptide linkages

$$H-N-C-C-N-C-C-OH$$

Glycylalanine

$$H-N-C-C-N-C-C-OH$$

Alanylglycine

Twenty-four *tetra*peptides are possible if four amino acids (for example, glycine, Gly; alanine, Ala; serine, Ser; and cystine, Cy) are linked in all possible combinations. They are:

Gly-Ala-Ser-Cy	Ala-Gly-Ser-Cy	Ser-Ala-Gly-Cy	Cy-Ala-Gly-Ser
Gly-Ala-Cy-Ser	Ala-Gly-Cy-Ser	Ser-Ala-Cy-Gly	Cy-Ala-Ser-Gly
Gly-Ser-Ala-Cy	Ala-Ser-Gly-Cy	Ser-Gly-Ala-Cy	Cy-Gly-Ala-Ser
Gly-Ser-Cy-Ala	Ala-Ser-Cy-Gly	Ser-Gly-Cy-Ala	Cy-Gly-Ser-Ala
Gly-Cy-Ser-Ala	Ala-Cy-Gly-Ser	Ser-Cy-Ala-Gly	Cy-Ser-Ala-Gly
Gly-Cy-Ala-Ser	Ala-Cy-Ser-Gly	Ser-Cy-Gly-Ala	Cy-Ser-Gly-Ala

If 17 different amino acids are used, the sequences alone would make 3.56×10^{14} uniquely different 17-unit molecules.[*] Although there are numerous protein structures in nature, these represent an extremely small fraction of the possible structures. Of all the many different proteins which could possibly be made from a set of amino acids, a living cell will make only a relatively small, select number.

PROTEIN STRUCTURES

The *primary structure* of a protein indicates only the order of the sequence of amino acid units in the *polypeptide* chain. Since the single bonds in the chain

[*] If the amino acids are all different, the number of arrangements is n! (read n factorially). For five different amino acids, the number of different arrangements is 5! (or $5 \times 4 \times 3 \times 2 \times 1 = 120$).

Linus Pauling (1901–   ): A scientist of great versatility and accomplishment. His interests have included the determination of the molecular structures of crystals by x-ray diffraction and theories of the chemical bond. His work on the structure of spiral polypeptide chains led to the Nobel Prize in 1954.

allow free rotation around the bond, it is reasonable to assume that there is an almost infinite number of possible conformations. Because of interactions, such as hydrogen bonding, between atoms in the same chain, certain conformations, called *secondary structures*, are favored. In 1954, Linus Pauling received the Nobel Prize for his work on secondary protein structure. Along with R. B. Corey, he suggested the two secondary structures discussed in the following paragraphs.

Polyglycine is the synthetic protein made entirely of the amino acid glycine. In polyglycine the hydrogen attached to the nitrogen atom and the oxygen bonded to the carbon are both well suited to engage in hydrogen bonding. In the two stable conformations of polyglycine, maximum advantage is taken of the hydrogen bonding available. In one conformation, the hydrogen bonding is between adjacent chains of the polypeptide; in the other, hydrogen bonding occurs between atoms within the same chain.

Figure 9–8 illustrates a sheetlike structure in which row after row of the polypeptide are joined by hydrogen bonding. Note that all the oxygen and nitrogen atoms are involved in hydrogen bonding. Most of the properties of silk can be explained in terms of this type of structure for fibroin, the protein of silk, which contains 44 per cent glycine.

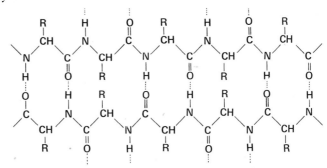

A

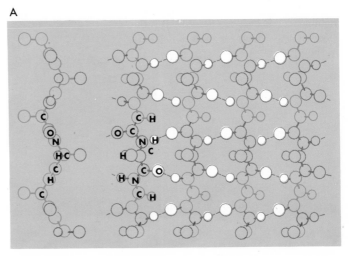

B

FIGURE 9–8   Sheet structure for polypeptide. In A, the two-dimensional drawing emphasizes that all of the oxygen and nitrogen atoms are involved in hydrogen bonding for the most stable structure. B illustrates the bonding in perspective showing that the sheet is not flat; rather, it is sometimes called a pleated sheet structure.

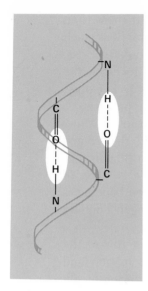

FIGURE 9-9   Helix structure for a polypeptide in which each oxygen atom can be hydrogen bonded to a nitrogen atom in the third amino acid unit down the chain.

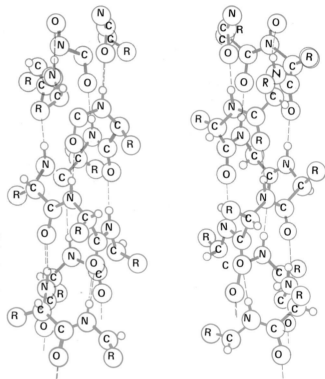

FIGURE 9-10   Drawing of the helix structure for polypeptides suggested by Pauling and Corey. Both left-handed and right-handed helices are possible. R is the characteristic group for each amino acid as shown in Table 9–1.

FIGURE 9-11   The structure of collagen.

Hydrogen bonding is possible within a single polypeptide chain if the secondary structure is helical (Figure 9–9). Bond angles and bond lengths are such that the nitrogen atom forms a hydrogen bond with the oxygen atom in the third amino acid unit down the chain (Figure 9–10).

Collagen is the principal fibrous protein in mammalian tissue. It has remarkable tensile strength which makes it useful in structuring bones, tendons, teeth, and cartilage. Three polypeptide chains, each of which is twisted into a left-handed helix, are twisted into a right-handed super helix to form an extremely strong fibril, as shown in Figure 9–11. A bundle of such fibrils forms the macroscopic protein.

The structure of collagen illustrates a third level of protein structure, *tertiary* structure. The primary structure is the sequence of amino acids in the protein; the secondary structure is the helical form of the protein chain; and the tertiary structure is the twisted or folded form of the helix. Another tertiary structure is illustrated by globular proteins. In these structures, the helix chain (secondary structure) is folded and twisted into a definite geometric pattern. This pattern may be held in place by one or more of several different kinds of chemical bonds, depending on the particular functional groups in the amino acids involved (Table 9–1). Figure 9–12 illustrates the folded structure of a typical globular protein. Abnormal hemoglobin structures are unable to transfer oxygen in the blood if the wrong amino acid is in a given position in the polypeptide structure (Figure 9–13).

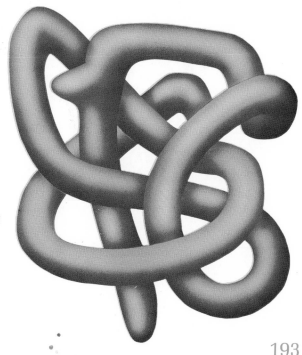

FIGURE 9–12   Typical folded structure of the helix in a globular protein.

The structure of heme (chemical structure diagram showing CH₂, CH, CH₃ groups around a porphyrin ring with central Fe and four N atoms, with CH=CH₂, HC, CH, CH₃ substituents, and two CH₂—CH₂—COOH chains at the bottom)

A

FIGURE 9–13   *A*, The structure of heme; *B*, a model, from two views, of the hemoglobin structure. The heme structures are indicated by disks. (Courtesy of M. F. Perutz and *Science,* 140:863, 1963.)

B

FATS AND OILS

Fats and oils are esters of glycerol (glycerin) and one or more fatty acids. Glycerol is a tri-alcohol with the molecular structure:

$$
\begin{array}{c}
H \\
| \\
H - C - OH \\
| \\
H - C - OH \\
| \\
H - C - OH \\
| \\
H
\end{array}
$$

A molecule of a fatty acid is composed of a carboxyl group, —COOH, attached to a long hydrocarbon chain. An example is stearic acid,

$$H-\underset{\underset{H}{|}}{\overset{\overset{H}{|}}{C}}-\underset{\underset{H}{|}}{\overset{\overset{H}{|}}{C}}-\underset{\underset{H}{|}}{\overset{\overset{H}{|}}{C}}-\underset{\underset{H}{|}}{\overset{\overset{H}{|}}{C}}-\underset{\underset{H}{|}}{\overset{\overset{H}{|}}{C}}-\underset{\underset{H}{|}}{\overset{\overset{H}{|}}{C}}-\underset{\underset{H}{|}}{\overset{\overset{H}{|}}{C}}-\underset{\underset{H}{|}}{\overset{\overset{H}{|}}{C}}-\underset{\underset{H}{|}}{\overset{\overset{H}{|}}{C}}-\underset{\underset{H}{|}}{\overset{\overset{H}{|}}{C}}-\underset{\underset{H}{|}}{\overset{\overset{H}{|}}{C}}-\underset{\underset{H}{|}}{\overset{\overset{H}{|}}{C}}-\underset{\underset{H}{|}}{\overset{\overset{H}{|}}{C}}-\underset{\underset{H}{|}}{\overset{\overset{H}{|}}{C}}-\underset{\underset{H}{|}}{\overset{\overset{H}{|}}{C}}-\underset{\underset{H}{|}}{\overset{\overset{H}{|}}{C}}-\underset{\underset{H}{|}}{\overset{\overset{H}{|}}{C}}-C\overset{\nearrow O}{\underset{\searrow OH}{}}$$

or $CH_3(CH_2)_{16}COOH$

The term *fat* is usually reserved for solids (butter, lard, tallow) and oil for liquids (castor, olive, linseed, tung, and so forth). Since fats and oils are products of living organisms, the biochemist is naturally interested in them, and uses the term *lipid* to apply to fats, oils, and other fat-soluble compounds.

$$
\begin{array}{ccccccc}
CH_2-OH & & HO-\overset{\overset{O}{\|}}{C}-R & & CH_2-O-\overset{\overset{O}{\|}}{C}-R & & \\
| & & HO-\overset{\overset{O}{\|}}{C}-R' & \rightleftharpoons & | & & \\
CH-OH & + & HO-\overset{\overset{O}{\|}}{C}-R' & & CH-O-\overset{\overset{O}{\|}}{C}-R' & + & 3H_2O \\
| & & HO-\overset{\overset{O}{\|}}{C}-R'' & & CH_2-O-\overset{\overset{O}{\|}}{C}-R'' & & \\
CH_2-OH & & & & & & \\
\end{array}
$$

Glycerol   *Fatty acid*   *A fat or oil*   *3 molecules*

(one molecule)   *(3 molecules, which may*   *molecule*   *of water*

or may not be the same)

R, R′, and R″ symbolize three different hydrocarbon chains.

Saturation (all single bonds with maximum hydrogen content) in the carbon chain of the fatty acids is usually found in solid or semi-solid fats, whereas unsaturated fatty acids (one or more double bonds) are usually found in oils. Hydrogen can be catalytically added to the double bonds of an oil to convert it into a semi-solid fat. For example, liquid soybean and other vegetable oils are hydrogenated to produce cooking fats and margarine.

Consumers in Europe and North America have historically valued butter as a source of fat. As the population increased the advantages of a substitute for butter became apparent, and efforts to prepare such a product began about a hundred years ago. One problem which arose was the fact that common fats are almost all *animal* products with very pronounced tastes of their own. Analogous compounds from vegetable oils, which have mixed flavors, were generally *unsaturated* and consequently *oils*. A solid fat could be made from the much cheaper vegetable oils if an inexpensive way could be discovered to add hydrogen across their double bonds. After extensive experiments, many catalysts were found, of which finely divided nickel is among the most effective. The nature of the process can be illustrated by the reaction:

$$
\begin{array}{l}
H_2C-O-\overset{\overset{O}{\|}}{C}-(CH_2)_7-CH=CH(CH_2)_7CH_3 \\
| \\
HC-O-\overset{\overset{O}{\|}}{C}-(CH_2)_7-CH=CH(CH_2)_7CH_3 \\
| \\
H_2C-O-\overset{\overset{O}{\|}}{C}-(CH_2)_7-CH=CH(CH_2)_7CH_3 \\
\end{array}
\xrightarrow[\sim 200°C]{H_2, Ni}
\begin{array}{l}
H_2C-O-\overset{\overset{O}{\|}}{C}-(CH_2)_7-CH_2-CH_2-(CH_2)_7CH_3 \\
| \\
HC-O-\overset{\overset{O}{\|}}{C}-(CH_2)_7-CH_2-CH_2-(CH_2)_7CH_3 \\
| \\
H_2C-O-\overset{\overset{O}{\|}}{C}-(CH_2)_7-CH_2-CH_2-(CH_2)_7CH_3 \\
\end{array}
$$

Triolein, a liquid fat (oil)   *Tristearin, a solid fat*

195

Oils commonly subjected to this process include those obtained by pressing cottonseed, peanuts, corn germ, soya bean, coconut, and safflower seeds. In recent years, as it became apparent that hydrogenation destroys essential fatty acids, soft, partially hydrogenated products have been placed on the market. For the three essential fatty acids, the best sources are safflower oil for linoleic, soya bean oil for linolenic, and peanut oil for arachidonic, and these have been incorporated in many margarines, as their labels testify.

There is considerable interest in the relationship between saturated fats in the diet and heart disease. It is known that an increase in solid fats increases the concentration of the complicated biochemical cholesterol (see Chapter 14) in the blood stream. Since it is thought by some physicians that a rise in cholesterol is associated with hardening of the arteries, there has been much interest in replacing solid fats with liquid oils in the diet.

WAXES

In its broadest sense, a wax is anything with a waxy feel and a melting point below body temperature. However, the chemical definition is more restrictive: a wax is an ester of a long chain alcohol and a long chain fatty acid. Its structure is

$$R-\overset{\overset{\textstyle O}{\|}}{C}-O-R'$$

TABLE 9-2   COMMON WAXES, COMPOSITION AND PROPERTIES

WAX (MELTING POINT)	SOURCE	ACIDS (IN ESTER FORM)	ALCOHOLS (IN ESTER FORM)	OTHER MOLECULES PRESENT	COMMENTS
Spermaceti 42–47°C	Whale Oil	Chiefly C_{16}	Chiefly C_{16}	Chiefly C_{18}	Used in cosmetics
Beeswax 60–82°C	Lining of cell of honeycomb	C_{26} and C_{28}	C_{30} and C_{32}	10–14 percent Hydrocarbons plus numerous hydroxy acids	C_{31} is the primary hydrocarbon present
Carnauba	Coating on leaves of a Brazilian palm	C_{24} and C_{28}	C_{32} and C_{34} alcohols	Hydroxy acids	A valuable natural wax because of its high melting point. Hard and impervious, used for floor and auto coverings
Wool Grease (Lanolin)	Scouring of wool	Wide variety	Wide variety	Alcohols, fatty acids, hydrocarbons	Emulsifies with water (up to 80 percent water); widely used in cosmetics. Lanolin is purified wool grease.

wherein R and R′ are long chain hydrocarbon groups. In a natural or even a semi-purified commercial product, a wax will consist of many different molecules with different R and R′. In addition, the natural product will always contain hydrocarbons and more complex esters, such as fats.

Spermaceti is an unusual wax in that it is mostly composed of a single kind of molecule. It is crystallized from whale oil and is primarily cetyl palmitate,

$$C_{15}H_{31}-\overset{\overset{\displaystyle O}{\|}}{C}-O-C_{16}H_{33}.$$

Cetyl alcohol, a 16-carbon alcohol, has the structure $CH_3(CH_2)_{14}CH_2OH$; and palmitic acid (16 carbons) has two carbon atoms less than does stearic acid(18): $CH_3(CH_2)_{14}COOH$.

Other common waxes are described in Table 9–2.

QUESTIONS

1. Show the structure of the product which would be obtained if two alanine molecules (Table 9–1) react to form a dipeptide.

2. What is an essential amino acid?

3. The ketone structure of D-fructose has three asymmetric carbon atoms per molecule. How many isomers result from these asymmetric centers?

4. What polysaccharide yields only D-glucose upon complete hydrolysis? What disaccharide yields the same hydrolysis product?

5. What is the difference between the starch, amylopectin, and the "animal starch," glycogen?

6. What is the chief function of glycogen in animal tissue?

7. Explain the basic difference between starch, amylose, and cellulose.

8. What functional groups are always present in each molecule of an amino acid?

9. Give the name and formula for the simplest amino acid in structure. What natural product has a high percentage of this amino acid?

10. If three different acids form all of the possible tripeptides, how many would there be?

11. What is the meaning of the terms: primary, secondary, and tertiary structures of proteins?

12. From a molecular point of view, what is the difference between a saturated fat and an unsaturated fat?

13. Would you classify sugars as low-energy or high-energy substances? Give a reason for your answer.

14. Why is it necessary to refer to two molecular structures in describing D-glucose?

15. How many monosaccharides are known? How many occur naturally?

16. Give an example of a monosaccharide that contains an aldehyde functional group.

17. How many simple sugars could be easily obtained from a mixture of sucrose, maltose, and lactose? What are they?

18. Refer to Figure 9–7 to explain why cotton has such a great attraction for water.

19. Show the characteristic bonding in a peptide linkage.

20. Why is it appropriate to refer to a protein as a polypeptide?

21. What is collagen and where is it found?

22. What holds the spiral configuration in place in the protein structure?

23. From a molecular-structure point of view, how are fats and waxes kin?

SUGGESTIONS FOR FURTHER READING

Battista, O. A., "Sugar—The Chemical with a Thousand Uses," *Chemistry*, Vol. 38, No. 4, p. 12 (1965).

"Bio-Organic Chemistry," Readings from *Scientific American*, W. H. Freeman and Co., San Francisco, 1968.

Mazur, A. and Harrow, B., *Biochemistry: A Brief Course*, W. B. Saunders Company, Philadelphia, 1968.

Pauling, L., Corey, R., and Hayword, R., "The Structure of Protein Molecules," *Scientific American*, Vol. 191, No. 1, p. 51. July (1954).

10

ENZYMES, VITAMINS, AND HORMONES

The tremendous importance of enzymes, vitamins, and hormones is surprising when we consider the amounts of these substances that we have in our bodies. Much less than one per cent of the total weight of your body is composed of enzymes, vitamins, and hormones, yet they control the chemical reactions that sustain life. In this chapter we shall attempt to see how such minute amounts of these substances can have such overwhelming control over the chemistry of our bodies.

ENZYMES

An enzyme is a biochemical catalyst. Like other catalysts, a given enzyme increases the rate of a reaction without requiring an increase in temperature. As an example of a simple type of catalysis, consider the oxidation of glucose, a sugar which burns in air with some difficulty and is hard to light with a match. If cigarette ashes, or other catalysts, are placed on its surface, combustion can be initiated easily with a match. When the sugar burns, it liberates a large amount of energy.

$$C_6H_{12}O_6 + 6O_2 \longrightarrow 6CO_2 + 6H_2O \quad \Delta G = -688 \text{ Kcal}$$

Sugar   Oxygen   Carbon   Water
dioxide

Recall that a negative value of ΔG, the free energy change, indicates a spontaneous reaction capable of producing energy.

The energy required to get the reaction started is the activation energy; catalysts, in general, work by lowering the activation energy. If an enzyme can lower the activation energy to a point where the average kinetic energy of the molecules in a living cell (or in a laboratory system) is sufficient for reaction, then the reaction can proceed rapidly. Glucose is oxidized rapidly and efficiently at ordinary temperatures in the presence of biochemical catalysts. To be sure,

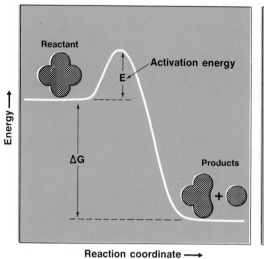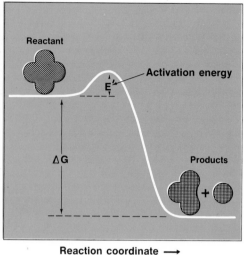

A. No enzyme present   **B. Enzyme present**

FIGURE 10-1   Enzyme Effect on Activation Energy. The vertical coordinate represents increasing energy and the horizontal one the course of the reaction in going from reactants to products. For energy-producing reactions the reactant molecule is at a higher energy than the product molecule, as illustrated in A. The difference between these energies is the net free energy change of the reaction. However, it is necessary for the reactant molecule to "get over" the energy barrier (acquire the activation energy, E) in going from reactant to product. Note that the activation energy is given back along with the free energy. The enzyme lowers the activation energy, E', as illustrated in B, while the free energy change remains the same. The net effect is to obtain the free energy of the reaction with a smaller expenditure of activation energy.

the oxidation of glucose in a living cell requires many enzymes and many steps, but enzymatic catalysis produces the same final result as combustion at elevated temperature, namely carbon dioxide, water, and 688 kilocalories of usable energy per mole of sugar. Figure 10–1 graphically illustrates the concepts of activation energy, the energy available from an energy-producing reaction, and the reduction of the activation energy by an enzyme.

Enzymes are remarkable catalysts in that they are highly specific for a given reaction. Maltase is an enzyme that catalyzes the hydrolysis of maltose into two molecules of D-glucose. This is the only known function of maltase, and no other enzyme can substitute for it. The explanation for the specific activity of enzymes can be found in their molecular geometry. Enzymes are globular proteins with very definite tertiary protein structures (Figure 9–12). The highly specific action of maltase can be explained if its globular structure accurately accommodates a maltose molecule at the point where the reaction occurs, the reactive site. When the two units come together, strain is placed on the bonds holding the two simple sugar units together. As a result, water is allowed to enter and hydrolysis occurs. Sucrose cannot be hydrolyzed by maltase because of the different geometry involved. Another enzyme, sucrase, hydrolyzes sucrose effectively. Some enzymes, however, are less specific. The digestive enzyme, trypsin, for example, acts predominantly on peptide bonds in proteins, but it will also catalyze the hydrolysis of some esters because of somewhat similar geometry and polarity at the active site.

The Human Side

James B. Sumner

It has been estimated that the human body contains nearly 150,000 different kinds of working enzymes. No one knows the total number of enzymes produced by all kinds of living things. The possibilities seem infinite. Over 1500 enzymes have been purified since 1926 when James B. Sumner (1887–1955), an American biochemist at Cornell, separated, crystallized, identified, and characterized the first enzyme, urease. Interestingly enough, Sumner had been advised by his teachers not to enter chemistry because he had only one arm and they thought this would handicap his laboratory work. He won a Nobel Prize in 1946 for his work in chemistry.

How does an enzyme work? How can it lower the activation energy and be so specific for a given reaction? While a definitive answer cannot be given at this time (the matter is presently under intensive research), a "lock-and-key" analogy has been a fruitful approach to the problem. Just as a key can separate a padlock into two parts and subsequently remain unchanged, ready to unlock other identical locks, so the enzyme makes possible a molecular change (Figure 10–2). With enough energy the lock could be separated without the key, and with enough energy the molecular alteration could occur without the enzyme. An enzyme cannot make a nonspontaneous chemical reaction occur. An enzyme just releases the brakes.

The tremendous speed of enzyme-catalyzed reactions requires more than just random collisions to fit "the key in the lock." For example, a molecule β-amylase catalyzes the breaking of four thousand bonds in amylose per second. Speed like this requires something to attract the key into the lock, such as electrically polar regions, partially charged groupings, or ionic sections on the enzyme and the substrate (the reactant molecule). These regions attract as well

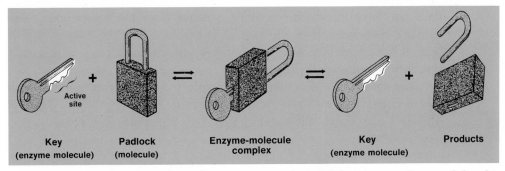

| Key (enzyme molecule) | Padlock (molecule) | Enzyme-molecule complex | Key (enzyme molecule) | Products |

FIGURE 10-2   Lock-and-Key theory for enzymatic catalysis. While it is generally agreed that this analogy is an oversimplification, it does make one very important point: the enzyme makes a difficult job easy by reducing the energy required to get the job started. It also suggests that the enzyme has a particular structure at an active site which will allow it to work only for certain molecules, similar to a key that fits the shape of a particular keyhole.

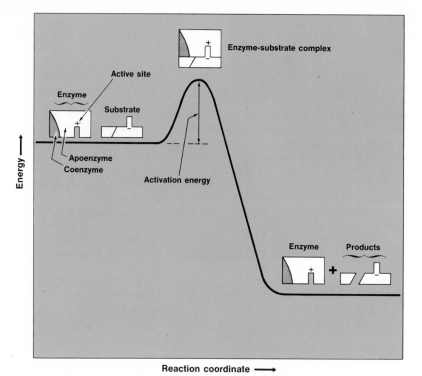

FIGURE 10-3   A reaction utilizing an enzyme which requires a coenzyme. Here the essential feature is that the combination of enzyme plus coenzyme allows the reaction to proceed with a lower activation energy. Apoenzyme is the name given to the enzyme structure that combines with the coenzyme.

as guide the substrate to the proper position on the enzyme and thereby speed up the reaction. The electrically charged portions of the enzyme are believed to be the chemically *active sites* in the enzyme.

Sometimes an enzyme requires another species, called a *coenzyme,* in order to function as a catalyst. The coenzyme is generally a nonprotein and may be an ion, such as Co^{3+}, Fe^{3+}, or Mg^{2+}, in one of the essential minerals, or it may have a vitamin as part of its structure (as we shall see in the next section). The protein part with the coenzyme is called the *apoenzyme.* The coenzyme alone does not have enzymatic activity; neither does the apoenzyme. Before the enzyme becomes active, the two must combine like the two keys required to open a bank lock-box. Neither your key nor the bank's key will open the box, but both together will (Figure 10–3).

Let's use a simple chemical example to demonstrate the relationship between coenzyme, substrate, and apoenzyme. The peptide glycylglycine is hydrolyzed very slowly by water to give two molecules of glycine.

202

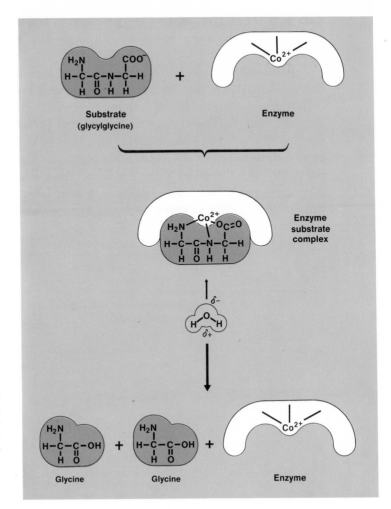

FIGURE 10-4   Operation of an enzyme. The substrate molecule is bonded to the enzyme (glycylglycine dipeptidase) through chemical bonding, the negative oxygen and the nitrogen atoms binding to the positive cobalt ion in the enzyme. The bonding of the substrate makes it more susceptible to attack by water. Hydrolysis occurs and the glycine molecules are released by the enzyme which is then ready to play its catalytic role again.

The glycylglycine substrate forms an intermediate compound with an enzyme, and as a result the substrate is activated for further reaction. Activation can result from extensive hydrogen bonding, interaction with a metal ion in the enzyme, or a number of other processes. In this case the coordination of the glycylglycine to the positive charge of a cobalt ion (Co^{2+}), the coenzyme, makes it more susceptible to attack by the negative end of a water molecule (Figure 10-4).

Most of the names of enzymes end in -ase. There are a few exceptions, such as pepsin and trypsin, which are both digestive enzymes. Hydrolases break down foodstuffs and other substances by hydrolysis. For example, carbohydrases (such as maltase, lactase, sucrase, ptyalin and amylase) affect hydrolysis of carbohydrates. Proteases, such as pepsin, trypsin, and chymotrypsin, hydrolyze large protein molecules into smaller groups of proteins. Lipases hydrolyze esters such as fats and oils. The oxidizing enzymes are called oxidases, such as catalase, which converts hydrogen peroxide to water and oxygen, and dehydrogenases, such as NAD, NADP, and FAD (described in the next section). There are other categories of enzymes, but these illustrate the wide variety of biochemical catalysts.

Besides being biochemical middlemen, speeding up and directing all the chemical reactions that go into the continuous breakdown and buildup of our cells (three million red blood cells are renewed in the human body every second), enzymes may be the answer to future food problems. Scientists have already developed a way to produce sugar (on a limited experimental basis) by bubbling carbon dioxide into water containing enzymes. Trash fish can be converted into palatable animal feed by using enzymes. Work is underway to convert oil spills into edible products for sea organisms. A little later in this text we will encounter the use of enzymes in detergents and as meat tenderizers.

TABLE 10-1   PURPOSES OF VITAMINS°

VITAMIN	SELECTED BIOCHEMICAL FUNCTION	DEFICIENCY EFFECTS
A	Regeneration of rhodopsin (visual purple)	Excessive light sensitivity; night blindness; increased susceptibility to infection
B_1 (Thiamine)	Nerve activity, carbohydrate metabolism	Beriberi; serious nervous disorders; muscular atrophy; serious circulatory changes
B_2 (Riboflavin)	Coenzyme; affects sight	Sores on the lips; bloodshot and burning eyes; excessive light sensitivity
B_3 (Pantothenic acid)	Growth factor; component coenzyme A	Retarded growth
B_4 (Choline)	Source of transferrable methyl groups	Hepatosis; may allow alcoholic cirrhosis of the liver; dermatosis; anemia
B_5 (Niacin; Nicotinic acid)	Metabolism of ATP	Stunted growth; pellagra
B_6 (Pyridoxine)	Coenzyme; metabolism of fatty acids	Retarded growth; anemia; leukocytosis; insomnia, lesions about eyes, nose, mouth
B_7 (Biotin)	Growth factor; affects scaly and greasy skin; CO_2 fixation; coenzyme	Scaly and greasy skin
B_9 (Folic acid)	Coenzyme; tyrosine metabolism	Inhibition of cell division (mitosis)
B_{12}	Growth factor; involved in synthesis of DNA	Degeneration of the spinal cord; pernicious anemia
C	Coenzyme; reducing agent; cholesterol metabolism	Hemorrhages; lesions in the mouth; muscular degeneration; sterility; scurvy
D	Ca and P metabolism	Abnormal development of bones and teeth; rickets
E	Antioxidant; cofactor between cytochromes b and c	Sterility
K	Synthesis of prothrombin, coenzyme Q	Hemorrhage; slow clotting of blood

° These vitamins participate in biochemical changes concerned with the utilization of foodstuffs.

VITAMINS

We are constantly reminded of our need for vitamins. Cereal boxes and the advertising media have proclaimed the need for vitamins for years. From childhood, we have been told daily, "Don't forget your vitamins." We need vitamins daily because the body cannot synthesize most vitamins, nor store them. Fortunately, a large variety of foods generally contains vitamins, either naturally or added, and a good diet usually supplies our needs without taking vitamin pills.

Although most vitamins cannot be stored by the body and must be supplied each day, Vitamin A is an exception. It is stored in small amounts in the liver. Some animals are much more efficient in this storage than man. For example, the polar bear stores so much Vitamin A in his liver that polar bear liver is poisonous to man. The codfish and halibut are also very efficient in storing vitamins in their livers. Perhaps you have unpleasant childhood memories of taking daily doses of codliver oil (a source of Vitamin A).

It might seem strange for the polar bear liver to be poisonous because it has so much Vitamin A. If vitamins are needed, how can too much be bad? Too much of a given vitamin can get in the way of the intricate chemical processes of the body and cause harm. Since some vitamins (D, for example) can cause adverse effects when taken in excess, *the unlimited use of vitamins is not advised.* However, most vitamins are much like the B vitamins in that they are retained by the body like water is in a glass. If too much is put in, the excess runs over and out of the system.

Most of our problems with vitamins are not with an excess, but with a deficiency. You are aware of most of the vitamin-deficiency diseases by name at least—beriberi, scurvy, pellagra, pernicious anemia, nightblindness, rickets, and so on. The specific vitamins and their related vitamin-deficient diseases are listed in Table 10–1.

The Human Side

Christiaan Eijkman

The discovery of another cause of disease besides germs came at a time when skilled physicians were proving Pasteur's germ theory to be true. Partly by "accident," the Dutch physician, Christiaan Eijkman (ike´mahn) (1858–1930; Nobel prize in 1929), discovered the cause of beriberi by studying some sick chickens in Batavia (the modern Djakarta) in 1896. When some of the chickens being used for laboratory experiments contracted a disease resembling beriberi, Eijkman tackled the problem with fervor. He tried in vain to transfer the disease from a sick chicken to a healthy one. Then as suddenly as the disease appeared, it disappeared. Investigation revealed that the chickens had been fed polished rice meant for hospital patients. When the cook who had been feeding the chickens was transferred, his replacement did not think it right to use rice intended for patients on chickens. He went back to commercial chicken feed and the disease disappeared. Eijkman found that he could induce the disease by feeding polished rice (hulls removed) and could cure the disease by feeding unpolished (or whole) rice. Although Eijkman was the first to pinpoint what we now call a "dietary-deficiency disease," he did not realize the significance of his findings at first. He thought there was some sort of toxin in rice that was neutralized by something in the hulls. It remained for several others to suggest the correct explanation. The most prominent was an Englishman, Sir Frederick Gowland Hopkins (1861–1947), with whom Eijkman shared the Nobel prize in 1929.

We need vitamins because they are intimately involved in our chemistry. The detailed chemical action of all vitamins is not yet certain. However, a considerable number of facts are known about some. For example, the structures of all the known vitamins have been worked out, and interestingly, they are not much alike in structure. Where all proteins are similar in that they are built from 20 or so amino acids, vitamins have no common relationship. The extremes in structure are represented by niacin (vitamin B_5) and vitamin B_{12}. It took a group of experts seven years to work out the structure of vitamin B_{12}.

Nicotinic acid

Nicotinamide

Niacin, Vitamin B_5 (A mixture of nicotinic acid and nicotinamide)

$A = -CH_2\overset{\overset{O}{\|}}{C}NH_2$

$B = -CH_2CH_2\overset{\overset{O}{\|}}{C}NH_2$

$M = -CH_3$

Vitamin B_{12}

Many vitamins function as coenzymes. Vitamins participate as long as their apoenzyme participates. They not only trigger specific enzymes, they pitch in and help out. The B vitamins are found in every cell as coenzymes in various oxidative processes. For example, niacin (vitamin B_5) becomes part of an enzyme that prevents pellagra, the most common vitamin-deficiency disease in the United States.

Doctors and biochemists now know that the body suffers from pellagra when it lacks sufficient pyridine rings.

Pyridine

Some foods such as yeast, liver, meats, fish, eggs, whole wheat, brown rice, and peanuts contain pyridine rings in the form of niacin. The body needs only one two-thousandth of an ounce of niacin each day in order to prevent pellagra. This isn't much but it is vital. The coenzyme of which niacin is a part is vital to the energy production in the body. If energy is not provided, the whole process of renewing cells and building needed compounds slows down and eventually stops. The coenzyme involved is quite large, as is its name, nicotinamide adenine dinucleotide (NAD).

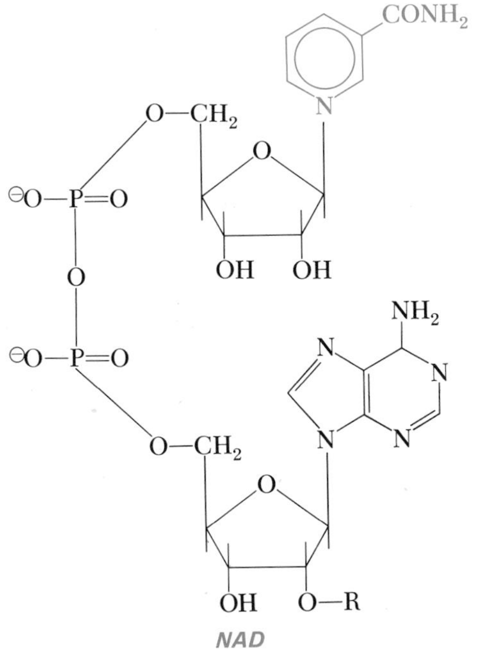

NAD

Niacin is also an integral part of another oxidative enzyme, nicotinamide adenine dinucleotide phosphate (NADP). In NAD, R = H; in NADP, R = H_2PO_3. The part of the structure printed in color comes from niacin and is the active part in oxidation-reduction. It accepts hydrogen during biological oxidation processes.

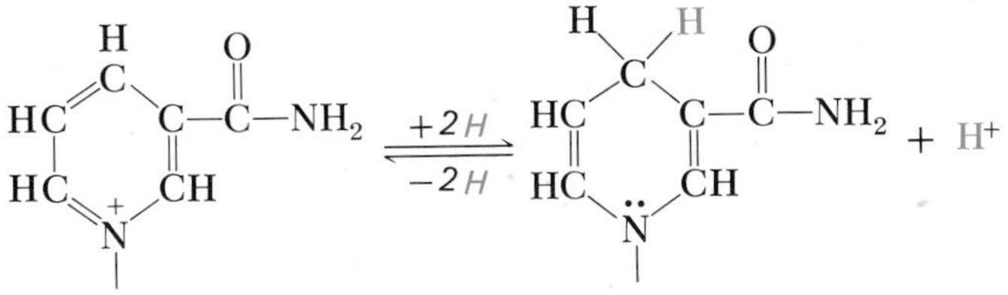

This reaction appears to be necessary to supply the energy you need to turn this page. How energy is provided and stored will be discussed in the next chapter. The important thing here is that niacin is a vital and necessary part of the structure of a particular coenzyme that is required if the enzyme is to bring off its oxidation-reduction action.

Riboflavin (Vitamin B_2) is a necessary part of another important hydrogen acceptor coenzyme, flavin adenine dinucleotide (FAD).

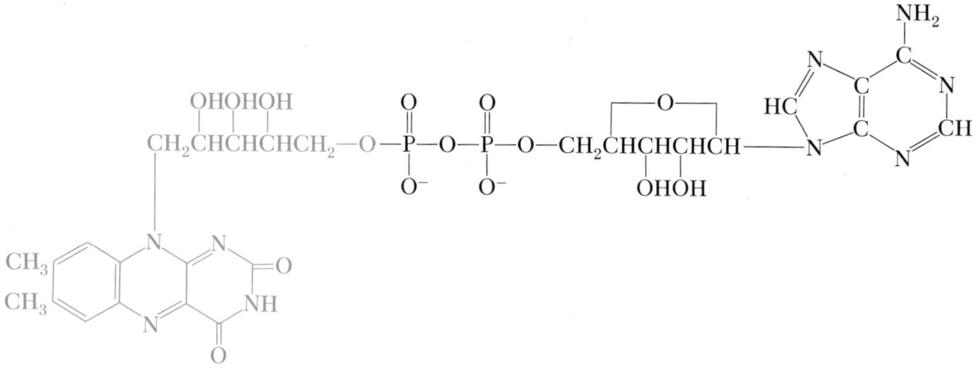

Flavin adenine dinucleotide (FAD)
(The portion in color is riboflavin.)

FAD accepts hydrogens in the following manner:

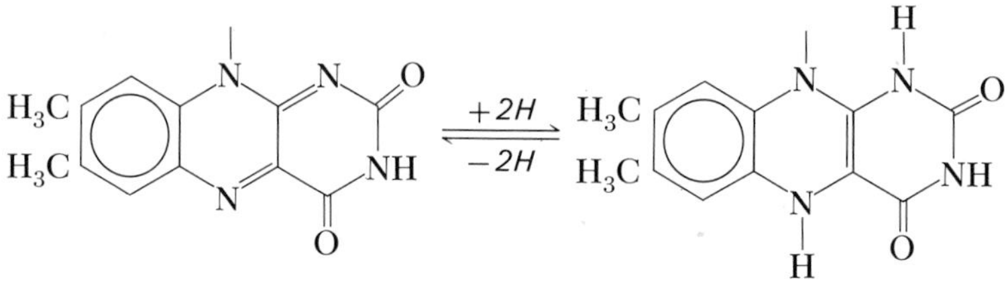

Glucose and glycogen are the principal sources of energy in the body. How NAD, NADP, and FAD fit into the oxidation of glucose and glycogen will be shown in the next chapter.

The vitamin drawing the most attention in the early seventies was vitamin C (ascorbic acid).

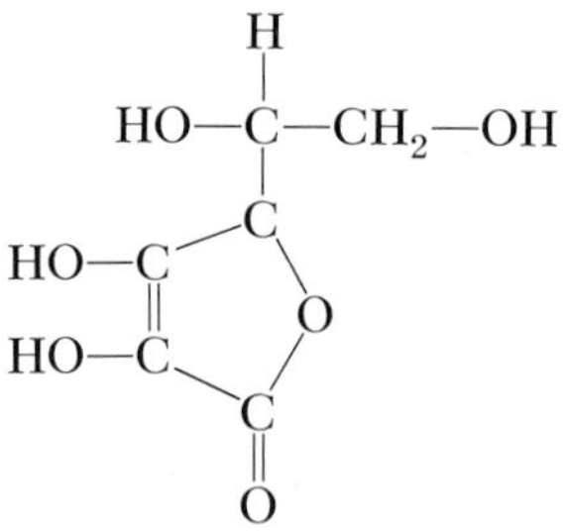

Vitamin C

Vitamin C received considerable attention as a possible cure and preventative for the common cold. Professor Linus Pauling was a major proponent of this theory. In 1971, Dr. Pauling recommended that one should take 0.25 to 5 grams of C a day to prevent colds, and up to 10 grams a day to fight existing colds.

He and other reputable scientists extolled the virtues of the vitamin when used in these, or even greater amounts. Still others found vitamin C to have no effect in preventing or fighting colds. All seem to agree, however, that no toxic effects are produced by large doses of the vitamin. In view of the possibly conflicting evidence, we might draw the conclusion that taking vitamin C is a personal matter—it will help some and not others.

Vitamin C is found in citrus and other fruits. The demand generated by the recent publicity has called for the commercial preparation of vitamin C to swing into high gear. It is prepared commercially by converting glucose into five intermediate compounds, one after the other, and finally into ascorbic acid. One step requires the enzymes of some microorganisms, named *A. suboxydans*, to facilitate the removal of two hydrogens. There is no chemical difference between the synthetic vitamin C and the "natural" kind.

HORMONES

Hormones, like vitamins and enzymes, are organic compounds needed in very small amounts in the body. They are messenger chemicals, rather than

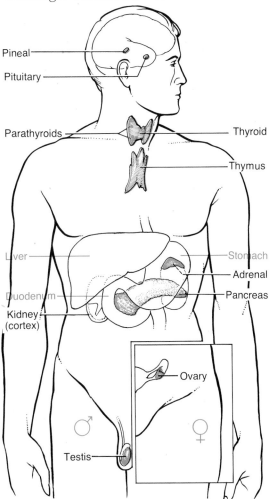

FIGURE 10-5   The approximate locations of the endocrine glands in man. Although the pineal body, thymus and stomach are shown, they are not definitely known to secrete hormones.

TABLE 10-2   PRINCIPAL HORMONES OF MAN AND THEIR PHYSIOLOGICAL EFFECTS°

HORMONE	EFFECT
Protein Hormones	
Anterior Pituitary Gland	
Human growth hormone (HGH)	Controls bond growth and general body growth; affects protein, fat and carbohydrate metabolism
Thyrotropin (TSH)	Stimulates growth of thyroid and production of thyroxin
Adrenocorticotropin (ACTH)	Stimulates adrenal cortex to grow and produce cortical hormones
Follicle-stimulating hormone (FSH)	Stimulates growth of graafian follicles in ovary and of seminiferous tubules in testis
Luteinizing hormone (LH)	Controls production and release of estrogens and progesterone by ovary and of testosterone by testis
Prolactin	Maintains secretion of estrogens and progesterone by ovary; stimulates milk production by breast; controls "maternal instinct"
Hypothalamus, via Posterior Pituitary Gland	
Oxytocin	Stimulates contraction of uterine muscles and secretion of milk
Vasopressin	Stimulates contraction of smooth muscles; antidiuretic action on kidney tubules
Parathyroid Glands	
Parathormone	Regulates calcium and phosphorous metabolism
Pancreas	
Insulin	Increases glucose utilization, decreases blood sugar concentration, increases glycogen storage and metabolism of glucose
Glucagon	Stimulates conversion of liver glycogen to blood glucose
Amino-Acid-Derivative Hormones	
Thyroid Gland	
Thyroxin	Increases metabolism rate
Adrenal Medulla	
Epinephrine	Reinforces action of sympathetic nerves; stimulates breakdown of liver and muscle glycogen
Norepinephrine	Constricts blood vessels
Sterol Hormones	
Adrenal Cortex	
Cortisol	Stimulates conversion of proteins to carbohydrates
Aldosterone	Regulates metabolism of sodium and potassium
Dehydropiandrosterone	Stimulates development of male sex characteristics
Testis	
Testosterone	Stimulates development and maintenance of male sex characteristics
Ovary (Follicle)	
Estradiol	Stimulates development and maintenance of female sex characteristics
Ovary (Corpus Luteum)	
Progesterone	Acts with estradiol to regulate estrous and menstrual cycles

° From, Villee, C. A., *Biology*, 4th. ed., W. B. Saunders Co., Philadelphia, p. 390 (1962).

catalysts or building blocks. Although the methods are not clearly understood, the release of a hormone in one part of the body causes chemical reactions to occur in other parts of the body. Since each hormone affects only a certain organ or type of cell, the hormones as a group exhibit specificity similar to the lock-and-key effect of enzymes. In fact some hormones, such as insulin, are proteins like the enzymes. Unlike vitamins, hormones are synthesized in the body by the endocrine (Greek *endo*, within; *krinein*, to separate, i.e., to secrete) glands. The endocrines secrete their hormones directly into the bloodstream without the aid of ducts or tubes. The locations of the endocrine glands are shown in Figure 10–5, and the hormones they secrete are given in Table 10–2.

The general classification of hormones in Table 10–2 is based on general features of structure. The structures of a protein hormone (insulin) and an amino acid derivative (epinephrine) are given later in this section. The steroid structure shows up in many different plant and animal compounds.

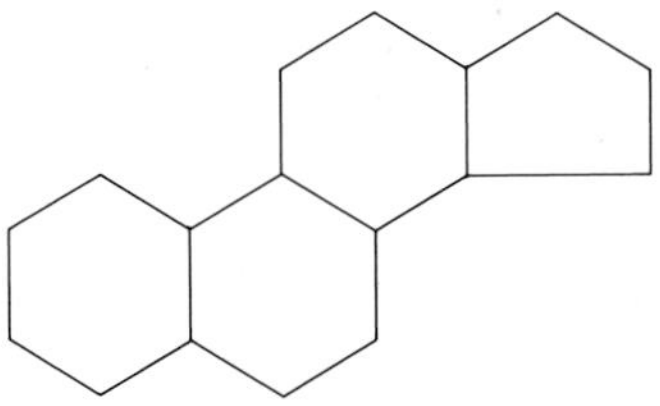

The general steroid structure

In addition to the hormones listed in Table 10–2, other compounds with no hormone activity (such as cholesterol) incorporate the steroid structure. Cholesterol shows up in several common places. For example, gallstones are, in part, hardened cholesterol, and hardening of the arteries results from deposits of cholesterol, that diminish the flexibility and pumping action of the blood vessels. About one-tenth of the brain is cholesterol, but its function there is unknown.

Hormones and vitamins generally differ in their action. Most hormones seem to initiate chemical activity and are then inactive while the action takes place. In fact the meaning of the word hormone (Greek *hormon*, arousing, exciting) is consistent with its chemical role. Most vitamins, on the other hand, initiate chemical activity in their role as coenzymes and continue to participate as part of the enzyme as long as the reaction occurs. Hormones are the "straw bosses" of the chemical hierarchy of the body; vitamins and enzymes are "traffic managers" that depend on each other to direct and facilitate chemical activity.

Hormones appear to function in one of three ways: gene activation, changing the permeability of cell walls, or enzyme activation. An example of each will illustrate these points.

Some recent experiments support the possibility that the female hormones (estrone and estradiol, for example) initiate the menstrual cycle by first combining with enzymes in cells lining the uterus. This complex then activates the release of templates (prepared from the genes) which are used over and over in the production of proteins. The hormone can wait while the templates prepare the lining of the uterus for receiving an ovum. The preparation requires about two weeks. When an ovum is released from the ovary, production of the hormone, progesterone, begins. If the ovum is fertilized, the production of progesterone

211

continues, the lining of the uterus remains intact, and the menstrual cycle is shut off during pregnancy. If the ovum is not fertilized, progesterone diminishes, the lining breaks down, and menstrual discharge occurs. Progesterone deactivates the pituitary hormone which produces the estrogen hormones. The active ingredient of "the pill" is a compound structurally similar to progesterone, as described in Chapter 14. The important point to be made here is that the estrogen hormones seem to act chemically by activating genes to produce proteins.

More is known about how hormones exert their effect in changing the permeability of cell walls. This is especially true with the hormone insulin. Insulin appears to be somewhat of a doorkeeper for glucose. If insulin is present, glucose can enter the cell; if insulin is not present, glucose is shut out, its level rises in the bloodstream, and if this continues diabetes ensues. Diabetes is acquired in a variety of ways such as through inheritance, infection of the pancreas, stress (as during pregnancy), or continually eating too much carbohydrates. In nonhereditary cases the disease can be subdued by watching the diet, by oral diabetic drugs such as tolbutamine and phenformin, although there is some doubt about the safety and usefulness of these drugs (see the Journal of the American Medical Association, August 9, 1971, page 817). In extreme cases, the disease is treated with insulin injections. When you see the size of the insulin molecule (Figure 10–6), it is clear that a molecule this size would have great difficulty migrating through the walls of the digestive tract into the bloodstream. Furthermore, the protein digesting enzymes of the gastrointestinal tract would destroy insulin before it had a chance to be absorbed.

The actual mechanism of insulin action is clouded with mystery at present. One current theory is that insulin attaches itself to "pores" in the cell membrane (protein-to-protein) and there serves as a specific escort for glucose through the membrane and into the cytoplasm of the cell.

The human growth hormone (HGH) is thought to alter the permeability

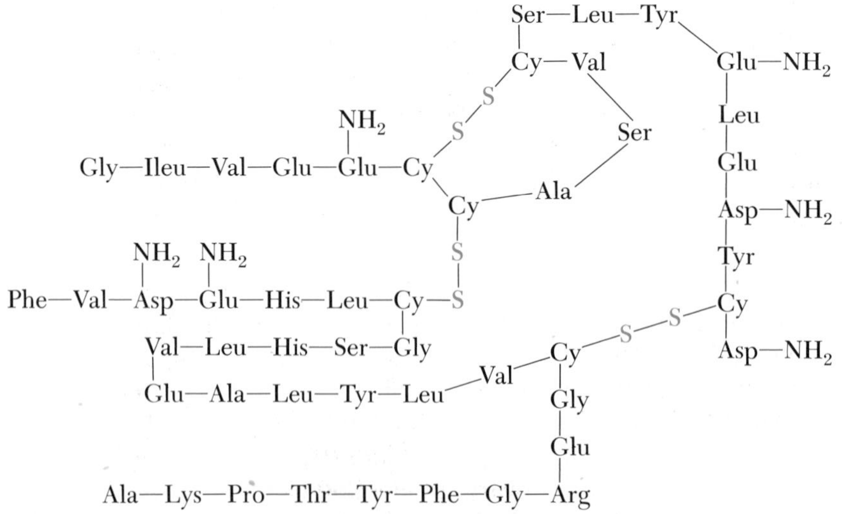

FIGURE 10-6   Beel insulin.

of the cell wall to amino acids in a manner that may be similar to the insulin-glucose mechanism. The 188-unit HGH protein molecule has been synthesized recently from individual amino acids by a group led by Professor Li at the University of California (see Chapter 12).

We know more about the action of hormones that activate enzymes. Epinephrine (or adrenalin), the "emergency hormone," is generally believed to be an enzyme activator. In times of stress, small amounts of epinephrine are discharged from the adrenal glands into the bloodstream. This triggers the release of a flood of glucose molecules for quick energy (described in the next chapter). One epinephrine molecule is thought to cause the release of about 30,000 molecules of glucose. This quick release of energy allows "superhuman" feats such as lifting a car off a person after an accident. Epinephrine is one of a pair of optical isomers (C^* is the asymmetric carbon atom). Only the isomer that rotates polarized light to the left is effective in starting a heart that has stopped beating or in giving a person unusual strength during times of great emotional stress.

Epinephrine
(Adrenalin)

The probable action of epinephrine is summarized in Figure 10–7. The cascading effect is begun by "knocking on the cell membrane." Its chemical action with a cell-membrane-bound enzyme, adenyl cyclase, converts the enzyme into an active form to catalyze the production of an intracellular messenger, cyclic AMP (adenosine monophosphate). Then, depending upon the cell, the cyclic AMP sets in motion such actions as stopping glucose from forming glycogen, stimulating the breakdown of glycogen into glucose, or activating cellular functions, such as the formation of insulin. All this is very rapid since the hormone does not have to be transported into the cell. Epinephrine would be the Western Union boy if cyclic AMP is the inhouse bellhop. Epinephrine is a bloodstream messenger and cyclic AMP is an inside-the-cell messenger.

Cyclic AMP is prepared from ATP (adenosine triphosphate), the energy bank of the body. We might think of ATP as the energy "currency" of the body. If a process needs energy, ATP is sent to provide the energy and, in a sense, to pay for the process. ATP is made during the oxidation of glucose to carbon dioxide and water. This interesting and important process will be described in the next chapter. The work on cyclic AMP as a cellular messenger was signally honored in 1971 when the Nobel prize in medicine was awarded to Dr. Earl Sutherland of Vanderbilt University for his contribution to this field.

213

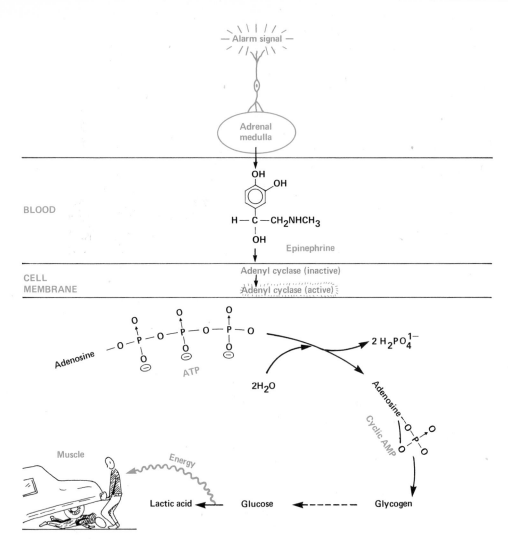

FIGURE 10-7   The cascading action of the hormone epinephrine. The adrenal medulla is stimulated to release epinephrine into the bloodstream by signals from the central nervous system. Epinephrine enters the muscles and activates adenyl cyclase, which catalyzes the formation of cyclic AMP. Cyclic AMP triggers, among other things, several fast chemical reactions, which quickly lead to the breakdown of glycogen into glucose and then into lactic acid. The quick energy derived from this process provides superhuman strength. Only a minute amount of epinephrine is required for a large effect.

ENZYMES ARE NOT ETERNAL

The work of enzymes goes on eternally it seems, but individual enzymes have a relatively short lifetime. The body purposely destroys enzymes in order to economize materials and space and also to prepare other enzymes needed at the time. There is limited space and raw materials (amino acids, glucose, and so on) in the body. The wisest use of materials and space is to make enzymes as they are needed and to destroy those not needed at the moment.

Enzymes are easily destroyed by heat, acid, and regulatory enzymes. Even shaking can occasionally destroy, that is, "denature" an enzyme so it will no longer be active.

The body carries out its processes of denaturing and hydrolyzing proteins and enzymes without our being aware of it. However, there are chemical agents that can destroy or deactivate enzymes to such an extent that these agents are considered to be toxic or poisonous. Heavy metals, such as mercury and lead, and the metal-like element, arsenic, are toxic because they render enzymes ineffective.

Metals owe their toxicity primarily to their ability to react with and inhibit sulfhydryl (—S—H) enzyme systems such as those involved in the production of cellular energy. For example, in glutathione, a tripeptide of glutamic acid, cysteine, and glycine that occurs in most tissues, the metal ions, such as Hg^{2+} or Pb^{2+}, replace the hydrogen on the sulfhydryl group.

$$2 \text{ Glutathione} + \text{Metal ion } (M^{2+}) \longrightarrow M \text{ (Glutathione)}_2 + 2H^+$$

Glutathione

British Anti-Lewisite (BAL) is a standard therapeutic item in a hospital's poison emergency center and is used routinely to treat mercury poisoning. The BAL bonds to the metal at several sites and forms a chelate (Greek: *chela,* claw), a term applied to a reacting agent that envelopes a species of metal ion. BAL is one of a large number of compounds that can act as chelating agents for metals.

BAL   *Heavy metal ion*   *Chelated metal ion*

With the heavy metal ion tied up, the sulfhydryl groups in vital enzymes are freed and can resume their normal functions.

A vivid description of the psychic changes produced in an individual by mercury poisoning can be seen in the Mad Hatter, a character in Lewis Caroll's "Alice in Wonderland." An old practice in the fur felt hat industry involved the use of mercuric nitrate, $Hg(NO_3)_2$, to stiffen the felt. This not only accounted for the Mad Hatter's odd behavior, but also gave the workers in the hat factories chronic mercury poisoning, with physical (nervous) symptoms known as "hatter's shakes."

Some insecticides and nerve gases are enzyme deactivators. A case in point is DFP, a nerve gas used in World War II. The DFP deactivates the enzyme cholinesterase which clears the way for new messages to be transmitted across the opening (the synapse) between the ends of two nerve cells.

A nerve impulse is transmitted along a nerve fiber by electrical impulses. When the impulse reaches the synapse, a small quantity of acetylcholine is liberated. The acetylcholine moves away from the incoming nerve ending and activates a receptor on an adjacent nerve. To enable the receptor to receive further impulses, the enzyme cholinesterase breaks down acetylcholine into acetic acid and choline.

$$CH_3\overset{\overset{\displaystyle O}{\|}}{C}-OCH_2-CH_2-\overset{\overset{\displaystyle CH_3}{|}}{\underset{\underset{\displaystyle CH_3}{|}}{N}}\overset{+}{}-CH_3 \quad OH^-$$

Acetylcholine

In the presence of potassium and magnesium ions, other enzymes, such as acetylase, resynthesize new acetylcholine from the acetic acid and the choline within the sending nerve ending:

$$\text{Acetic Acid} + \text{Choline} \xrightarrow{\textit{Acetylase}} \text{Acetylcholine} + H_2O$$

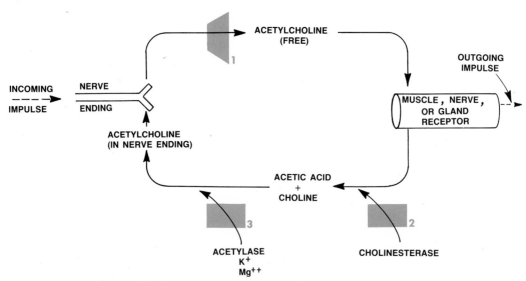

FIGURE 10–8 The acetylcholine cycle, a fundamental mechanism in nerve impulse transmission, is often affected by many poisons. An impulse reaching a nerve ending in the normal cycle liberates acetylcholine, which then stimulates a receptor. To enable the receptor to receive further impulses, the enzyme *cholinesterase* breaks down acetylcholine into acetic acid and choline; other enzymes resynthesize these into more acetylcholine. Botulinus and dinoflagellate toxins inhibit the synthesis, or the release, of acetylcholine (1). The "anticholinesterase" poisons inactivate cholinesterase, and therefore prevent the breakdown of acetylcholine (2). Curare and atropine desensitize the receptor to the chemical stimulus. (3).

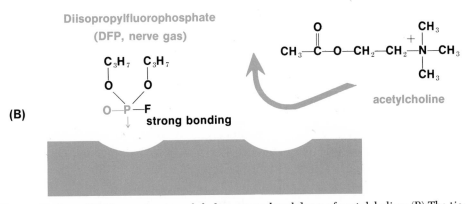

FIGURE 10-9 (A) The mechanism of cholinesterase breakdown of acetylcholine; (B) The tie-up of cholinesterase by an anticholinesterase poison like the nerve gas DFP blocks the normal hydrolysis of acetylcholine since the molecule cannot bind to the enzyme.

The new acetylcholine is available for transmitting another impulse across the gap.

Neurotoxins can affect the transmission of nerve impulses at nerve endings in a variety of ways (Figure 10–8). The anticholinesterase poisons prevent the breakdown of acetylcholine by deactivating cholinesterase. These poisons are usually structurally analogous to acetylcholine, so they bond to the enzyme cholinesterase and deactivate it (Figure 10–9). The cholinesterase molecules bound by the poison are held so effectively that restoration of proper nerve

217

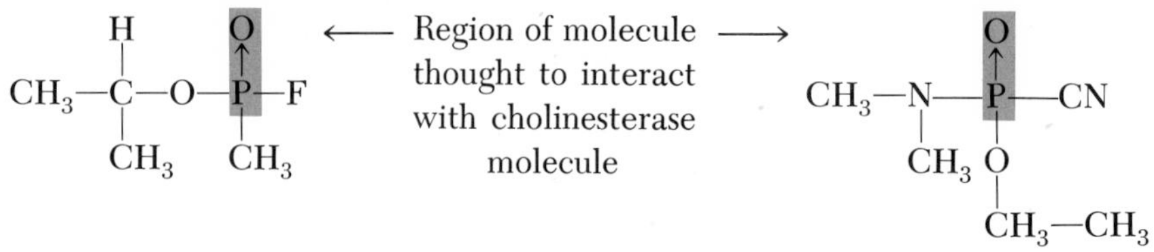

Sarin
(A nerve gas developed during World War II)

Tabun
(A nerve gas)

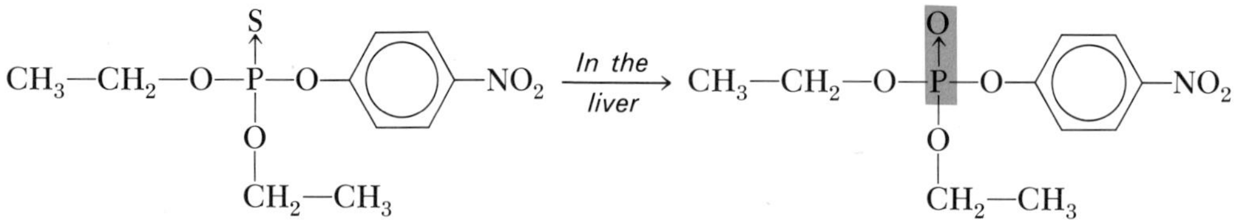

Parathion
(very toxic)

FIGURE 10–10 Some anticholinesterase poisons. The potent insecticide Parathion is converted in the liver to a molecule much like the nerve gases (colored area).

function must await the manufacture of new cholinesterase. In the meantime, the excess acetylcholine overstimulates nerves, glands, and muscles, producing irregular heart rhythms, convulsions, and death. Many of the organic phosphates which are widely used as insecticides are metabolized in the body to produce anticholinesterase poisons. For this reason they should be treated with extreme care. Some toxic mushrooms also contain an anticholinesterase poison. The structures of some of these substances are given in Figure 10–10.

Other poisons affect acetylcholine by either blocking its formation or its reception. Toxic proteins secreted by the botulinus bacillus, a microorganism that multiplies in spoiled food and causes botulism, are the most powerful poisons yet discovered. They act to block the synthesis of acetylcholine. The lack of acetylcholine means signals are not sent, paralysis occurs, and death is caused by respiratory failure. Nicotine, atropine (used to dilate the eye), morphine, codeine, cocaine, and certain local anesthetics (nupercaine, procaine, tetracaine) (see Chapter 14) combine with the receptor protein to block its reaction with acetylcholine. Hence the message is not received and cannot be sent on. The paralytic poisons used by primitive peoples on their poisoned arrows act similarly. Although the effects have been known for centuries, only in recent years have we begun to understand how and why such poisons act.

QUESTIONS

1. What is the meaning of the term enzyme specificity?

2. What is a substrate? Give an example of a specific substrate and its enzyme.

3. Give two chemical characteristics of all enzymes.

4. What do you think would be the result if the body was deprived of its supply of niacin? To what general class of compounds does niacin belong?

5. Using the energy diagrams, explain the concept of activation energy for a chemical reaction and show the effect of an enzyme on the activation energy.

6. Point out three important similarities of the lock-and-key analogy to enzymatic activity.

7. What is a coenzyme and why are they sometimes necessary? Give two categories of substances that are often coenzymes.

8. Give two significant differences between enzymes, vitamins, and hormones.

9. What is a function of

   (a) a lipase?
   (b) NAD?
   (c) a hydrolase?
   (d) a protease?
   (e) cholinesterase?
   (f) insulin?

10. Explain how epinephrine gets its "message" across so quickly.

11. How does Parathion poison a human being?

12. What specific chemical effect does mercury have on the chemicals of the body?

13. Health experts estimate that up to 225,000 children become ill from lead poisoning each year. Much of the lead comes from eating chips of old lead-pigmented paint. What chemical effect does lead have on enzymes?

14. Suggest two possible chemical conditions that would provide active sites on an enzyme.

15. Match the following:

   ( ) Riboflavin   a. pyridine rings in too little supply
   ( ) Vitamin A   b. affects permeability of cell walls
   ( ) pellagra   c. integral part of the coenzyme FAD
   ( ) NAD   d. internal messenger
   ( ) insulin   e. enzyme deactivator
   ( ) cyclic AMP   f. stored in liver
   ( ) mercury   g. poison antidote
   ( ) BAL   h. oxidation-reduction coenzyme

16. True or False.

   (a) Enzymes cause reactions that would not otherwise occur.
   (b) Insulin is a dehydrogenase enzyme.
   (c) Mercury is a cumulative poison.
   (d) An enzyme is very effective because it lasts a long time before wearing out.

17. Suppose a person injected too much insulin. What effect would this have on the level of glucose in the blood? How is this counteracted?

18. Is an enzyme a good way to shift an equilibrium toward more products? Explain.

SUGGESTIONS FOR FURTHER READING ▬▬▬▬▬▬▬

References

Lehnigner, A. L., "Biochemistry," Worth Publishers, Inc., New York (1970).
Locke, D. M., "Enzymes—The Agents of Life," Crown Publishers, Inc., New York (1970).

Paperbacks

Asimov, I., "The Chemicals of Life," A Signet Science Library Book T3147 (1954).
Frieden, E., and Lysner, H., "Biochemical Endocrinology of the Vertebrates," Prentice-Hall, Inc., Englewood Cliffs, New Jersey (1971).
Routh, J. I., "Introduction to Biochemistry," W. B. Saunders Co., Philadelphia (1971).

Articles

Adams, E., "Poisons," *Scientific American,* Vol. 201, No. 11, p. 76 (1959).
Chisolm, J. J., Jr., "Lead Poisoning," *Scientific American,* Vol. 224, No. 2, p. 15 (1971).
Goldwater, L. J., "Mercury in the Environment," *Scientific American,* Vol. 224, No. 5, p. 15 (1971).
Raw, I., "Enzymes, How They Operate," *Chemistry,* Vol. 40, No. 6, p. 8 (1967).

11

THE H. A. KREBS
ENERGY FACTORY

The study of the human body leaves a person in awe of how well things work out and fit together. In this chapter, we shall examine a bit of chemistry of the human body that leads to a deeper understanding and appreciation of our chemical and physical selves.

The body is organized into billions and billions of small chemical communities—the cells. Each cell is somewhat independent in that it generates many of the services it needs, but like any modern community it must call on outside help to survive. The cells need blood to bring chemicals for energy, growth, and maintenance and to take away waste products. As we have already seen in the last chapter, cells need other cells to supply direction (hormones and nerves) and other materials for support. Thus, a human cell is a semi-dependent community of life and chemicals.

Among the many activities of each cell are several different kinds of chemical factories. Almost every cell has a nucleus (red blood cells do not), which is a chemical factory for making new cells. Most cells have ribosomes, which are factories for making structural materials for the body—proteins. These two kinds of factories will be studied in the next chapter. In this chapter, we will tour the factories where energy is produced and stored—the power plants— which every community **MUST** have. These are called the mitochondria. Or with some license, we might call them the H. A. Krebs Energy Factories. You see, Sir Hans Adolf Krebs, a German-British biochemist (1900–   ) discovered the principles of their operation around 1940, and from Sir Adolf comes the name of our energy factories.

Before we tour one of the factories, it will be necessary to begin outside of the factory itself and trace the flow of power and raw materials.

MEET SIR HANS ADOLF KREBS

As we begin our tour of the energy factory, it would be bad manners not to indulge the tour guide a few moments to praise the discoverer of the metabolic energy process we shall soon see. The guide begins, "Sir Hans was born in

Hanover, Germany, August 25, 1900. He attended several German Universities and obtained his degree in medicine in 1925 from the University of Hamburg. In his research, he soon grew interested in the breakdown of amino acids to obtain energy. The first step is the removal of the element nitrogen, and Sir Hans was the first to observe this process. In 1932, Dr. Krebs worked out the manner in which urea is formed when the amino acids are used for energy. The understanding has been refined, but the overall process remains pretty much as he proposed it more than 40 years ago.

"When Hitler came to power, Sir Hans could no longer remain in Germany. He went to England, studied at Cambridge, and soon joined the faculty at Oxford University. It was at Oxford that he discovered the cycle of chemical reactions by which much of the energy of the body is produced. He took up where Otto Fritz Meyerhof and Carl and Gerty Cori had left off in unravelling the process of converting glucose to carbon dioxide and water to obtain usable energy. The details will be pointed out as you tour the factory so we will not take the time to recount them here.

"In 1953, Dr. Krebs shared the Nobel Prize in physiology and medicine with Dr. Fritz Albert Lipmann, the discoverer of Coenzyme A. In 1958, Hans Adolf Krebs was knighted by the queen of his adopted land."

THE RAW MATERIALS

Every factory needs raw materials for conversion into marketable products. The raw materials that provide energy in the body are the foodstuff fuels (carbohydrates, fats, and proteins) and molecular oxygen. Like the burning of coal or gas, the energy-producing process is an oxidation of a fuel, but an oxidation at body temperature under controlled conditions. The principal source of energy and the most basic foodstuff is glucose. The ultimate source of the energy in glucose is the sun, via the process of photosynthesis in growing plants.

PHOTOSYNTHESIS

Recall that a synthetic chemical reaction involves the build-up of a chemical structure from starting materials. Photosynthesis, then, is an appropriate name for a synthetic reaction which employs light energy as a driving force in the reaction. While many reactions might be included under this definition, the name photosynthesis has been reserved for the natural synthesis of carbohydrates and sugars from carbon dioxide and water, using solar energy to power the process.

Photosynthesis is a very complex process which belies the relatively simple overall reaction:

$$6CO_2 + 6H_2O + 688 \text{ Kcal} \longrightarrow C_6H_{12}O_6 + 6O_2$$

| Carbon dioxide | Water | Energy (Sunlight) | Glucose | Oxygen |

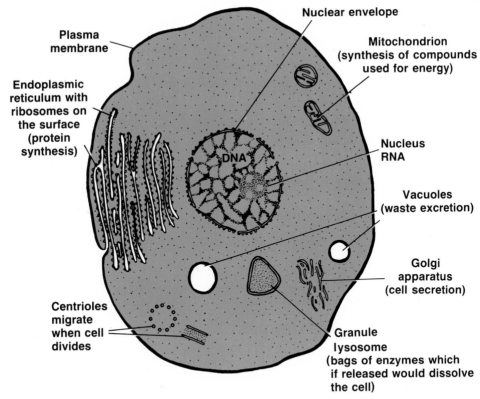

Plasma membrane

Endoplasmic reticulum with ribosomes on the surface (protein synthesis)

Nuclear envelope

Mitochondrion (synthesis of compounds used for energy)

DNA

Nucleus RNA

Vacuoles (waste excretion)

Golgi apparatus (cell secretion)

Centrioles migrate when cell divides

Granule
lysosome (bags of enzymes which if released would dissolve the cell)

FIGURE 11-1   Diagrammatic generalized cell to show the relationships between the various components of the cell.

In photosynthesis carbon dioxide is reduced to sugar (reduction is the gain of electrons; see Chapter 5):

$$6CO_2 + 24H^+ + 24e^- \longrightarrow C_6H_{12}O_6 + 6H_2O$$

and water is oxidized (oxidation is the loss of electrons):

$$12H_2O \longrightarrow 6O_2 + 24H^+ + 24e^-$$

Note that these two half reactions, the one reduction and the other oxidation, give the overall reaction when added together.

All the details of photosynthesis are not fully understood. However, some aspects are presented here as part of our examination of the energy flow through biochemical systems.

Photosynthesis is generally considered in terms of the *light reaction,* which can occur only in the presence of light energy, and the *dark reaction,* which can occur in the dark, but feeds on the high energy structures produced in the light reaction. Actually, both the light and dark reactions are a series of reactions, all occurring simultaneously in the green plant cell. As you might suspect, the light reaction is unique to green plants while the dark reaction is characteristic of both plant and some animal cells.

223

THE LIGHT REACTION

Photosynthesis is initiated by a quantum of light energy. The green plant contains certain pigments that readily absorb light in the visible region of the spectrum. The most important of these are the chlorophylls, *chlorophyll a* and *chlorophyll b*.

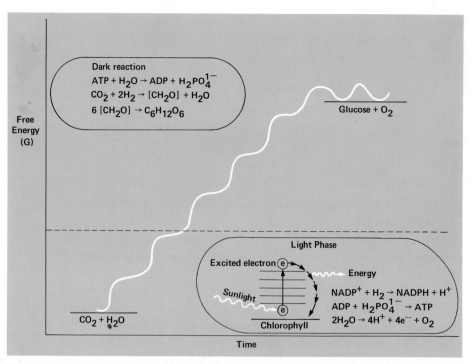

FIGURE 11-2   Free energy within the chemical system is increased as carbon dioxide and water are converted into glucose and oxygen by photosynthesis. This results in stored, useful energy in glucose. The very simplified mechanism for photosynthesis is discussed in the text.

Note that both are coordination compounds of magnesium and both have complex ring systems. Such ring systems usually absorb light in the visible region of the spectrum, consequently, they are colored. For example, chlorophyll is green because it absorbs light in the violet region (about 400 nanometers) and the red region (about 650 nanometers) and allows the green light in between those wavelengths to be reflected, or transmitted.

When chlorophyll absorbs photons of light, electrons are raised to higher energy levels. As these electrons move back down to the ground state, the very efficient chloroplasts of the green plant grab this energy, and through a series of steps which are not all completely known, store the energy as chemical potential energy. As shown in Figure 11–2, one of the chemicals is the niacin coenzyme, nicotinamide adenine dinucleotide phosphate ($NADP^+$), of Chapter 10. Energy is absorbed by $NADP^+$ in the process of being reduced, by gaining electrons, to $NADPH + H^+$.

$$NADP^+ + H_2 + energy \longrightarrow NADPH + H^+$$

Through a series of reactions, as yet unknown, NADPH eventually transfers its precious energy to the universal chemical storehouse of biochemical energy, adenosine triphosphate, ATP.

ATP

The energy is stored in the P—O—P bonds of ATP. Each of the two end bonds stores about 8000 calories/mole of free energy. ATP is the principal energy bank of living things. It is in ATP that energy is stored and doled out for the ongoing chemical processes in the cell. As you might expect, ATP structures occupy a position of honor and busy activity among biochemicals. Enzymes, vitamins, and hormones get the work done principally by administration, but ATP provides the "cash," or the energy. The breaking off of one mole of phosphate units provides about 8000 calories of energy.

$$ATP + H_2O \longrightarrow ADP + H_2PO_4^- \; [\Delta G = -8000 \text{ calories}]$$

The product ADP is adenosine diphosphate.

Thus far little mention has been made of the electrochemical charge balance and the source of oxygen in photosynthesis, both of which are important parts

225

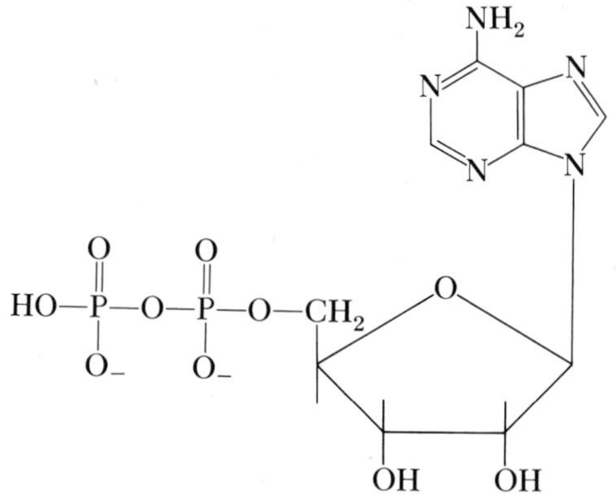

ADP

of the light reaction. Water, in the presence of chlorophyll and light, is decomposed into oxygen, hydrogen ions, and electrons:

$$2H_2O \longrightarrow 4H^+ + 4e^- + O_2$$

The hydrogen ions and electrons are available for maintaining the balance of charge, and oxygen is liberated from the plant cell.

At this point we no longer need the energy of the sun. We have the energy necessary to run biochemical systems in the ATP structure. If the plant cell is given the minerals along with carbon dioxide and water, it, or subsequent living cells, can use the energy in ATP in the dark to help carry out the complex biochemical reactions that take place.

THE DARK REACTION

The dark reaction which is responsible for the ultimate conversion of gaseous carbon dioxide to glucose was discovered by Melvin Calvin (1911–   ; Nobel Prize in 1961). Calvin studied the uptake of radioactive carbon in carbon dioxide by plant cell chloroplasts. He illuminated the plants for definite short periods of time and then analyzed the plant cells to determine which compounds contained the most radioactive carbon. As the time periods were reduced, more of the radioactive carbon was found in those compounds into which it had been initially incorporated and less in compounds which had been formed in subsequent reactions. For example, after only five seconds illumination, radioactive carbon is found in the compound 3-phosphoglyceric acid:

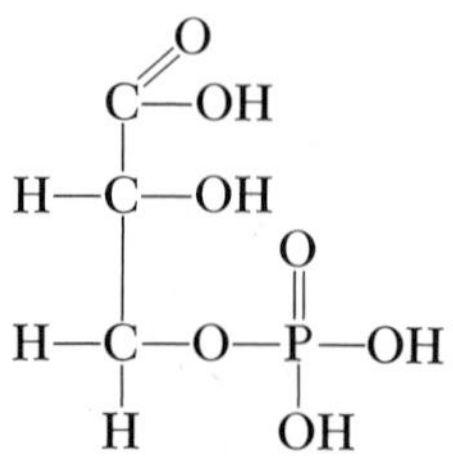

3-Phosphoglyceric acid

226

This compound is apparently formed in the initial reaction in which CO_2 from the air reacts with some molecule present in the plant. Calvin discovered that the key was the reaction of atmospheric CO_2 with ribulose 1,5-diphosphate to give two molecules of 3-phosphoglycerate:

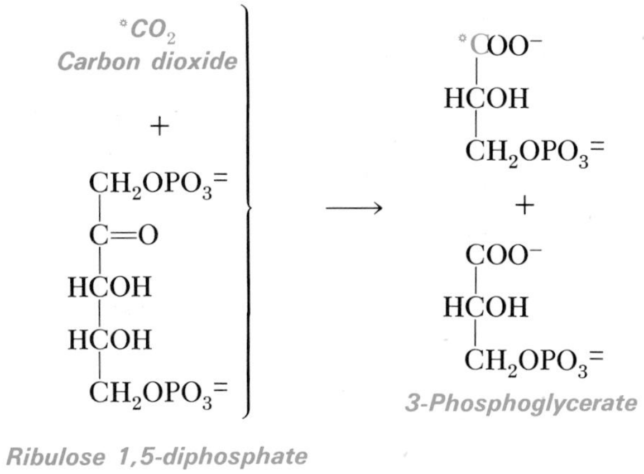

$$
\begin{array}{l}
°CO_2 \quad \text{Carbon dioxide} \\
+ \\
CH_2OPO_3^= \\
C=O \\
HCOH \\
HCOH \\
CH_2OPO_3^= \\
\textit{Ribulose 1,5-diphosphate}
\end{array}
\longrightarrow
\begin{array}{l}
°COO^- \\
HCOH \\
CH_2OPO_3^= \\
+ \\
COO^- \\
HCOH \\
CH_2OPO_3^= \\
\textit{3-Phosphoglycerate}
\end{array}
$$

° C indicates the fate of radioactive carbon as determined by Calvin's experiments.

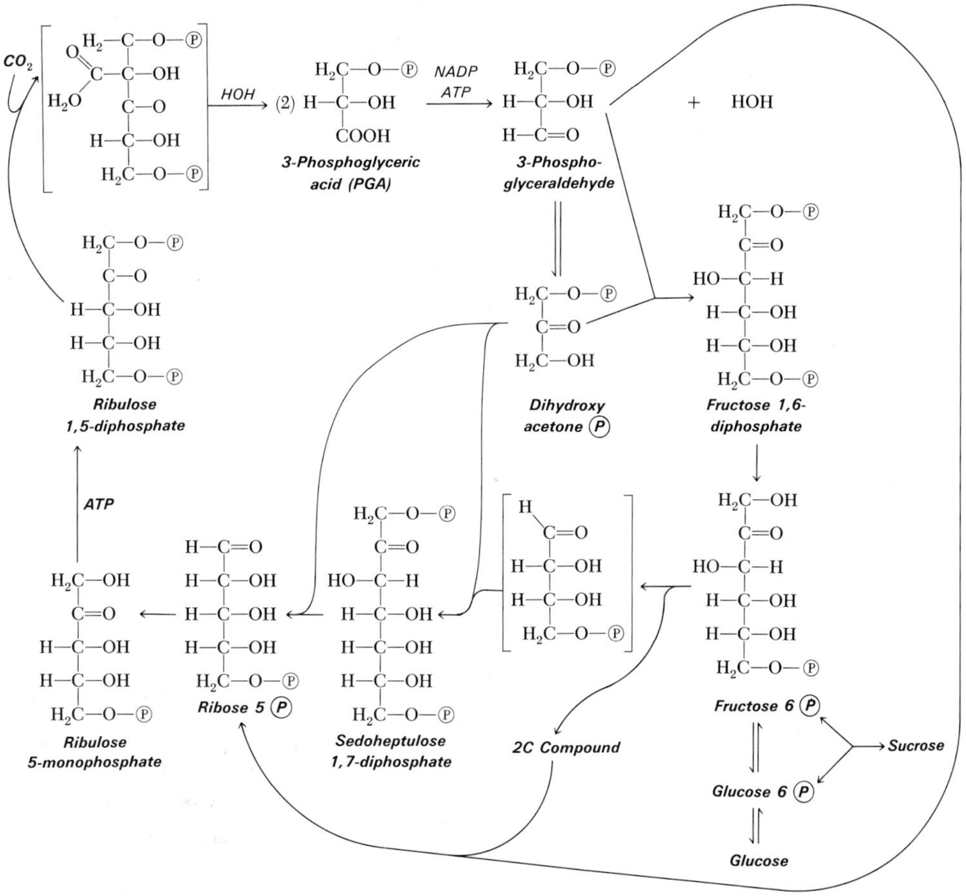

FIGURE 11-3 Abbreviated version of the dark reaction of photosynthesis. The energy needed to carry out the dark reactions is furnished from the high energy compounds produced in the light reaction.

The 3-phosphoglycerate is then transformed into other carbohydrates in reactions which produce glucose and regenerate ribulose 1,5-diphosphate (for further uptake of more atmospheric CO_2). The energy needed to carry out these reactions is furnished by the NADPH$^+$ and ATP generated from the light reaction.

Figure 11–3 presents the cyclic character of the dark reaction of photosynthesis. It is fantastic that this much detail has been worked out for such an intricate process. Note that carbon dioxide enters the cycle at the upper left and that sugars are removed at the lower right. Since just one carbon atom enters the cycle per round, only one fructose 6-phosphate molecule out of seven is converted to glucose.

When the light and dark aspects are included, the net equation for photosynthesis can be expanded over the equation given at the beginning of this chapter.

$$6CO_2 + 6H_2O + 18ATP + 12NADPH^+ \longrightarrow C_6H_{12}O_6 + 6O_2 + 18ADP + 18H_3PO_4 + 12NADP^+$$

DIGESTION

Before the glucose produced by photosynthesis is ready for the H. A. Krebs Energy Factory, practically all of it is put into storage. It may be converted first into either a disaccharide, polysaccharide, fat, or protein. The condensation of glucose into either amylose (plant starch) or glycogen (animal starch) was described in Chapter 9. The conversion of glucose into a fat or a protein is a bit of involved chemistry that will not be discussed here.

Once the foodstuff has been eaten, it is on its way to providing the body with energy and structural materials. The first step is digestion. From the chemical point of view, digestion is the breakdown of ingested foods through hydrolysis (reaction with water), catalyzed by specific enzymes. The products of these hydrolytic reactions are relatively small molecules that can be absorbed through the walls of the alimentary canal into the body fluids where they are used for metabolic processes. The hydrolysis of carbohydrates ultimately yields simple sugars, proteins yield amino acids, and fats yield fatty acids.

The digestion of starch begins in the mouth where salivary amylase, or ptyalin, catalyzes the formation of limited amounts of the disaccharide, maltose. Ptyalin is inactivated by the high acidity in the stomach, so its activity stops there. Carbohydrate digestion picks up again in the small intestine where pancreatic amylase, sucrase, maltase, and lactase convert starches and disaccharides into simple sugars such as glucose, fructose, and galactose. These simple sugars are then absorbed into the bloodstream where the control of blood sugar is regulated by the hormone insulin as described in Chapter 10. If the sugar level is too high, glycogen is produced and stored in the liver; if the blood sugar is too low, the stored glycogen is hydrolyzed into glucose.

Fats and oils are not digested, that is, hydrolyzed, in either the mouth or stomach. The enzyme lipase, which is secreted by the pancreas, catalyzes hydrolysis of the ester linkages in the fat or oil in the small intestine. For a typical fat, such as palmitooleostearin, the hydrolysis products are fatty acids and glycerol as shown in the following reaction.

MINERALS

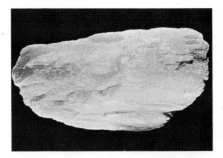

Wollastonite — $CaSiO_3$, a native calcium silicate from metamorphic rock.

Pyrite — FeS_2, iron disulfide.

Talc — $3MgO \cdot 4SiO_2 \cdot H_2O$, hydrated magnesium silicate.

Limonite — $2Fe_2O_3 \cdot 3H_2O$, hydrous ferric oxide.

Calcite — $CaCO_3$, calcium carbonate.

Selenite — $CaSO_4 \cdot 2H_2O$, waferlike crystal cluster.

Barytes — $BaSO_4$, blue and yellow crystals, on purple hematite.

Pyrolucite — MnO_2, dark dendrites on sandstone.

Plate I

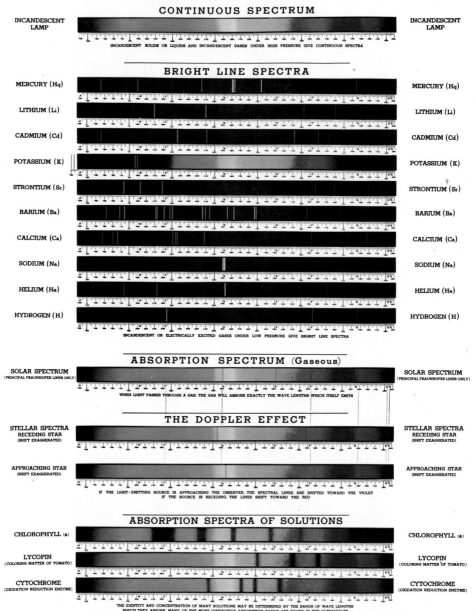

SPECTRUM CHART

Some typical spectra in the visible region. An emission spectrum (Continuous and Bright Line) represents all of the light emitted by the source of light while an absorption spectrum (Solar Spectrum) represents the light from the source minus the light absorbed by an intervening absorbing medium. Spectral lines from sources moving toward or away from the observer are shifted to shorter or longer wave lengths respectively, the Doppler effect. Wave lengths for spectral lines are given in angstrom units (one-hundredth million of a centimeter). (Courtesy of Sargent-Welch Scientific Company, Skokie, Illinois)

Plate II

$$H_2C-O-\overset{\overset{O}{\|}}{C}-(CH_2)_{14}CH_3$$

$$HC-O-\overset{\overset{O}{\|}}{C}-(CH_2)_7CH=CH(CH_2)_7CH_3 \; + \; 3HOH$$

$$H_2C-O-\overset{\overset{O}{\|}}{C}-(CH_2)_{16}CH_3$$

Palmitooleostearin
(a triglyceride)

Water

$CH_3(CH_2)_{14}COOH$
Palmitic Acid

$CH_3(CH_2)_7CH=CH(CH_2)_7COOH \; + \; C_3H_5(OH)_3$
Oleic acid   *Glycerol*

$CH_3(CH_2)_{16}COOH$
Stearic acid

Without the aid of bile salts secreted by the liver, the digestion of fats and oils would be very difficult. The enzyme is water-soluble and the fats and oils are insoluble in water. Bile salts emulsify the oil, that is, they break up the oil into very tiny drops and prevent the drops from combining readily. The tiny drops provide more surface area for the enzyme to attack so digestion can occur. The bile salts form an interface between the nonpolar oil and the polar water and make it possible for the oil to "dissolve" in water. For a molecule to form a buffer zone between polar and nonpolar molecules, the intermediate molecules must have characteristics of both. One of our principal bile salts is derived from glycocholic acid.

Sodium salt of glycocholic acid

Notice: (a) the bulky hydrocarbon groups (the rings form a steroid structure) which are compatible with oil or fat and (b) the —OH and ionic groups which anchor to water molecules. The bile salts emulsify oil in a manner similar to the action of a soap or a detergent during the cleaning process (Chapter 15).

The hydrolysis of proteins begins in the stomach and continues in the small intestine. Several different types of enzymes are known to be involved, such as pepsin, trypsin, and peptidases. These enzymatic systems must be controlled very carefully, for they have the potential to digest the walls of the stomach and intestines. A number of these enzymes are secreted in an inactive form. For example, pepsin, which is secreted in the stomach, is first present in a form called prepepsin. The molecular weight of prepepsin is 42,600. Prepepsin, in the presence of the acid in the stomach, is broken down by still another enzyme to pepsin. The molecular weight of pepsin is 34,500. It is reasonable to believe

229

that both prepepsin (the enzyme that breaks down) and the dilute hydrochloric acid normally present in the stomach have no effect on the stomach wall. After mixing pepsin is formed, and this enzyme would have considerable action on the stomach protein had it been formed under the mucous lining, a lining which is constantly sloughing off like our outer skin.

Pepsin facilitates the breakdown of only about 10 per cent of the bonds in a typical protein, leaving polypeptides with molecular weights from 600 up to 3000. In the small intestine peptidases complete the hydrolysis to small peptides which are absorbed through the intestinal wall.

Some protein enzymes are sold commercially. Meat tenderizers are protein materials that speed up cooking by partial digestion of meat. Enzymes were used as stain removers in detergents, although their effect on the skin is open to some question. Enzymes are also used to free the lens of the eye prior to cataract surgery.

THE KREBS ENERGY FACTORY

Like many of our modern factories, the Krebs Energy Factory makes full use of the assembly line technique (Figure 11–4) and is supplied by other factories. The assembly line is located in the mitochondria of the cells (Figure 11–1). The process is summarized in Figure 11–5. Details are given in Figure 11–7. At each location along the assembly line is an enzyme. The separate enzymes

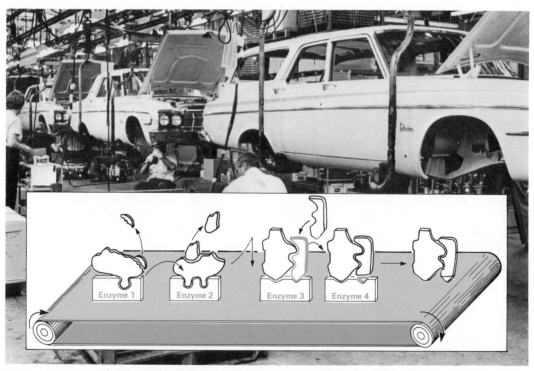

FIGURE 11–4   Molecules produce carbon dioxide, water, and energy in the Krebs Factory by moving from enzyme to enzyme on an assembly-line basis in the mitochondria.

230

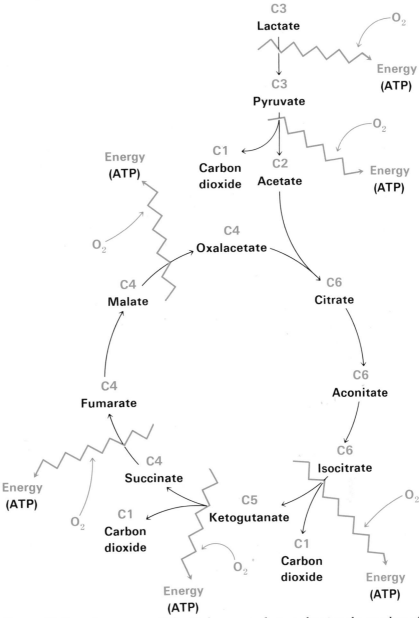

FIGURE 11-5   A summary of the Krebs energy factory showing the number of carbon atoms per molecule for the principal substances in the cycle.

catalyze, for example, the removal of hydrogen, the transfer of a phosphate (PO_4^{3-}) group or a carboxyl group (—COOH), and release the molecule each time to be passed on to the next enzyme. The products coming off the assembly line are carbon dioxide and water, and the energy is stored in ATP molecules. The raw materials (glucose, fatty acids, and amino acids) enter the assembly line at different places. Most of our energy comes from glucose as it appears to be best able to enter the assembly line. Fats and amino acids must be altered

231

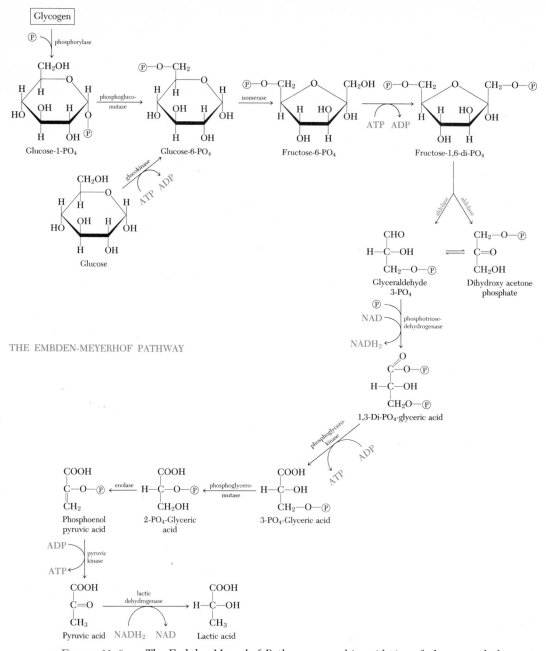

THE EMBDEN-MEYERHOF PATHWAY

FIGURE 11-6   The Embden-Meyerhof Pathway-anaerobic oxidation of glucose and glycogen. (Adapted from Routh: *Introduction to Biochemistry*, W. B. Saunders Co., Philadelphia, page 100, 1971).

FIGURE 11-7   The Krebs Cycle. Derivatives of fats enter the cycle as acetyl coenzyme A. Derivatives of proteins enter at various points depending upon the specific amino acid. (Adapted from Routh: *Introduction to Biochemistry*, W. B. Saunders Co., Philadelphia, page 102, 1971).

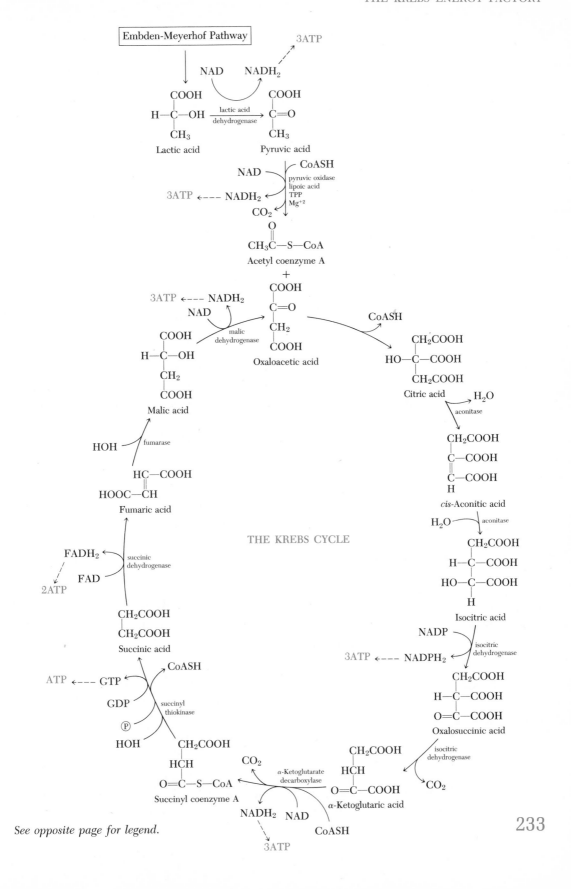

THE KREBS CYCLE

See opposite page for legend.

233

before they can be used as energy sources, but they can be used to provide energy when glucose is in short supply.

When you bend your arm, you "turn on" the energy factory in the mito-chondria of muscle cells in your arm. While the arm muscle is working, glucose is converted by a series of steps known as the Embden-Meyerhof pathway to lactic acid (Figure 11–6). We are concerned here only with the starting material, the energy flow, and the final products. This process occurs without oxygen and is called an anaerobic (without air) process. The overall reaction can be repre-sented by the equation:

$$C_6H_{12}O_6 + 2ADP + 2H_2PO_4^- \longrightarrow 2CH_3-\overset{\displaystyle H}{\underset{\displaystyle OH}{C}}-\overset{\displaystyle O}{C}-OH + 2ATP$$

Glucose   *Lactic acid*

If you continue flexing your arm long enough, the lactic acid buildup will produce tiredness and a painful sensation in the arm. Oxygen is needed to convert the lactic acid to carbon dioxide and water, which are excreted. This is accom-plished by the slower, aerobic (with air) process known as the Krebs Cycle. As you look at Figure 11–7 imagine how a molecule of citric acid sits down on the enzyme aconitase. The citric acid has —OH and —H exchange sites and the exchange produces isocitric acid. The isocitric acid is bumped off of aconitase and goes to the next stop on the assembly line, isocitric dehydrogenase, where two hydrogen atoms are removed by NADP (remember the coenzyme NADP from the last chapter?) and so on around the cyclic assembly line. Hydrogen atoms are removed, carboxyl groups are destroyed as carbon dioxide is formed, and hydrolysis occurs. Since two carbon atoms are fed into the cycle at oxalo-acetic acid by acetyl coenzyme A, two carbons in the form of two molecules of carbon dioxide (CO_2) must be eliminated before the cycle returns to oxalo-acetic acid. This is exactly what happens. See whether you can find the two molecules of CO_2 formed in the Krebs cycle. Figure 11–5 might help.

If the Krebs cycle is aerobic, where does the oxygen enter the cycle? Oxygen is required to regenerate the subassembly coenzyme NADP from $NADPH_2$. It is also required to remove the H_2's from $NADH_2$ and $FADH_2$ as well. There are about seven known steps involved in the removal of hydrogen from $NADH_2$ and $NADPH_2$ and six steps for $FADH_2$. The steps can be sum-marized by the following equations:

$$NADPH_2 + 3ADP + 3H_2PO_4^- + \tfrac{1}{2}O_2 \longrightarrow NADP + 3ATP + H_2O$$

$$NADH_2 + 3ADP + 3H_2PO_4^- + \tfrac{1}{2}O_2 \longrightarrow NAD + 3ATP + H_2O$$

$$FADH_2 + 2ADP + 2H_2PO_4^- + \tfrac{1}{2}O_2 \longrightarrow FAD + 2ATP + H_2O$$

The NADP, NAD, and FAD are now ready to be fed back into the Krebs cycle. They keep the assembly line running by extracting other pairs of hydrogen atoms.

When these equations are combined with the principal equation for the Krebs cycle, we have an equation for the conversion of lactic acid into carbon dioxide and water aerobically.

$$C_3H_6O_3 + 18ADP + 18H_2PO_4^- + 3O_2 \longrightarrow 3CO_2 + 3H_2O + 18ATP$$
Lactic acid

The key product is ATP. Energy derived from glucose (and lactic acid) is stored and shipped in the phosphorus-oxygen bonds of ATP. This is the usable product of the Krebs Energy Factory. Waste products are carbon dioxide and water. Since two lactic acid molecules are formed from one glucose molecule, 36 ATP's are formed per glucose molecule oxidized via the Krebs cycle. Two more ATP's are formed in the Embden-Meyerhof pathway, making a total of 38 ATP molecules formed in the complete oxidation of a glucose molecule. If we burn glucose to carbon dioxide and water outside the body, 688,000 calories of heat (enthalpy) are emitted. This is the total heat energy available from which usable energy is extracted. Earlier in this chapter, the point was made that breaking a P—O—P bond and forming $H_2PO_4^-$ gives up about 8,000 calories of usable, directed, free energy (ΔG). Thirty-eight ATP molecules would provide

$$38 \text{ moles ATP} \times 8,000 \frac{\text{calories}}{\text{mole ATP}} = 304,000 \text{ calories}$$

This is an efficiency of 44 per cent for deriving usable energy from the total energy available.

$$\frac{304,000 \text{ calories} \times 100\%}{688,000 \text{ calories}} = 44\%$$

This is remarkable when you consider that the efficiency of the automobile engine is only about 20 per cent, and real heat engines of any size seldom go above 35 per cent.

When ATP converts back to ADP by losing a phosphate group, the energy is used to move muscles, such as the pumping action of the heart, twitching the nose, moving the diaphragm so we can breathe, and hundreds of other movements required for the everyday ongoing of life. ATP also provides energy for cell reproduction (described in the next chapter) and for many different kinds of endothermic chemical reactions required in the chemistry of our bodies.

The energy is stored in ATP and released from ATP by means of *coupled reactions*. In Figures 11–6 and 11–7, the curved arrows denote reactions in which sufficient free energy is provided by one reaction to drive another reaction. These are examples of coupled reactions. In fact, the very first step in the oxidation of glucose is a specific example of a coupled reaction. The substitution of a phosphate group for a hydrogen on glucose is an unfavorable reaction, as far as energy is concerned.

Glucose + $H_2PO_4^-$ $\rightleftharpoons$ Glucose-6-PO_4 + H_2O ΔG = about + 3000 cal
(Unfavorable—requires energy)

If this reaction is coupled with the favorable hydrolysis of ATP,

ATP + H_2O $\rightleftharpoons$ ADP + $H_2PO_4^-$ ΔG = about − 8000 cal
(Favorable—releases energy)

235

the net reaction is favorable:

$$\text{Glucose} + \text{ATP} \rightleftharpoons \text{Glucose-6-PO}_4 + \text{ADP} \qquad \Delta G_{net} = -8000 + 3000$$

$$= -5000 \text{ cal (favorable)}$$

Even with this favorable free energy change, the enzyme glucokinase is required to hasten the process.

Fats and proteins (as amino acids) can enter the Krebs cycle also. Amino acids can enter at several places after first being converted by a series of reactions into one of the compounds in the cycle. Each amino acid has its particular point of entry. For example, aspartic acid is converted into α-ketoglutaric acid and enters at that point in the Krebs cycle.

Fatty acids go through a series of at least five reactions in which the hydrocarbon chain of the fatty acid is decreased by a two-carbon fragment. The fragment completes a molecule of acetyl coenzyme A. Palmitic acid, a 16-carbon fatty acid (see Chapter 7), would do seven turns around its cycle to form eight acetyl CoA molecules.

In conclusion, you now have seen some of the detailed chemistry involved in simply raising your arm, and you are now aware of what happens to some of the sugars and starches that disappear down the hatch. Of course, there is much more known than presented here and there appears to be no end to what is left to be discovered.

A WRENCH IN THE MACHINERY

If you wanted to sabotage the Krebs Energy Factory, the obvious place to destroy or deactivate it is at one of the crucial enzyme action points. This would jam up the whole process, and if continued long enough it would bring death to the individual. The fluoroacetate ion is toxic to the body because it blocks the oxidation of glucose. In this respect, it is a wrench in the machinery of the Krebs Energy Factory.

Fluoroacetic acid

Nature has used the synthesis of this molecule as part of a defense mechanism for certain plants. Native to South Africa, the gifblaar plant contains lethal quantities of fluoroacetic acid. Cattle that eat these leaves usually sicken and die.

Sodium fluoroacetate, the sodium salt of this acid (also called compound 1080), is a potent rodenticide. Because it is odorless and tasteless, it is especially dangerous, and its sale in this country is strictly regulated by law.

Fluoroacetate is toxic because the body uses it to synthesize fluorocitric acid, and fluorocitric acid blocks the Krebs cycle (Figure 11–8). The C—F linkage

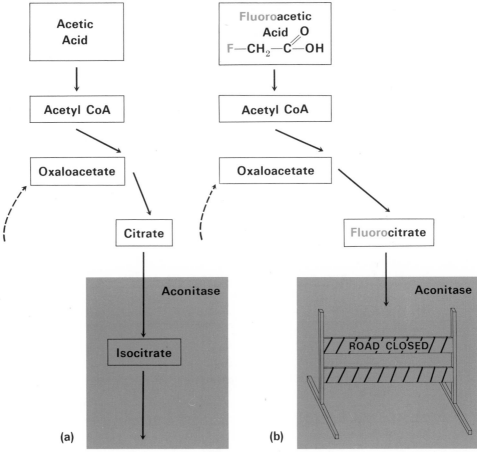

FIGURE 11-8   Fluoroacetic acid is synthesized into fluorocitrate, which then forms a stable bond with enzyme aconitase. This blocks the normal Krebs cycle, a portion of which is shown.

ties up the enzyme aconitase, thus preventing it from converting citrate to isocitrate.

In this instance, the poison is sufficiently similar to the substance (the substrate) which normally reacts with the enzyme so that it provides effective competition for the active sites on the enzyme. If the poison has sufficient affinity for the active groups of the enzyme, it occupies the active sites on the enzyme in much the same manner that the normal substrate occupies the sites, but even more firmly. The blocking of the Krebs cycle by fluorocitrate is a typical example of this. The fluorocitrate competes with citrate for the active sites on aconitase, thereby inhibiting the enzyme and blocking the Krebs cycle. If fluoroacetate is not present in excessive amounts, its action can be reversed simply by increasing the concentration of available citrate.

QUESTIONS

1. Write a basic equation for the digestion of: (a) starch to a disaccharide; (b) a disaccharide to a simple sugar; (c) a protein to amino acids; (d) a triglyceride (fat) to fatty acids.

2. What is the metal in chlorophyll?

3. If you were to "feed" radioactive carbon dioxide to a green plant, what would be the first radioactive carbon compound formed? Who made this discovery?

4. Give the structure of ATP and point out the region of the molecule that contains the bonds where hydrolysis occurs in coupled reactions.

5. What is meant by coupled reactions? Give an example. Why do the energetics of biochemical systems make coupled reactions necessary?

6. What type of compound first absorbs light energy in photosynthesis? Give an example.

7. In photosynthesis, why is it partially correct to say that light is an oxidizing agent?

8. What are the two major divisions in photosynthesis? Express in words what is accomplished in each.

9. What is the source of oxygen in photosynthesis?

10. Since chlorophyll loses electrons because of light, it must subsequently gain electrons from somewhere. Where do they come from?

11. If protein digestion is facilitated by enzymes, and these enzymes are produced in body organs made of proteins, explain why the enzymes do not cause rapid digestion of the organs themselves?

12. If a certain muscle requires 10 calories for contraction and obtains this ultimately by the hydrolysis of ATP to ADP, what is the minimum number of molecules of ATP that are needed to furnish the energy for such a contraction process?

13. What other fatty acids would you expect to find as intermediates in the oxidation of caproic acid?

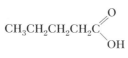

Caproic acid

14. What is the basic nature of the digestion processes for large molecules?

15. The chemical changes in the Krebs cycle can be classified as dehydrogenations (removal of hydrogens, a type of oxidation), dehydration (removal of water), hydrolysis (reaction with water in which water loses its molecular identity), decarboxylation (removal of COOH group and formation of CO_2), and phosphorylations (adding a phosphate group, such as $H_2PO_4^-$, represented as Pi in the diagram). Beginning with citric acid and progressing around the cycle to oxaloacetic acid, determine the total number of each kind of chemical change.

16. What compound produces soreness after a period of exercise?

17. Which compounds in the CO_2 fixation scheme in photosynthesis could enter directly the reactions of the Embden-Meyerhof pathway or the Krebs cycle?

18. Why is fluoroacetic acid toxic to the body?

19. Do you think that taking large doses of citrate would help in treating a case of fluoroacetic acid poisoning? Why?

SUGGESTIONS FOR FURTHER READINGS

General

"Bio-Organic Chemistry," Readings From *Scientific American*, W. H. Freeman and Company, San Francisco, 1968.

Lehninger, A. L., "Bioenergetics," W. A. Benjamin, Inc., New York, 1965.

Mazur, A., and Harrow, B., "Biochemistry: A Brief Course," W. B. Saunders Co., Philadelphia, 1968.

Routh, J. I., "Introduction to Biochemistry," W. B. Saunders Co., Philadelphia, 1971.

Carbohydrate Metabolism

Horecker, R. L., "Pathways of Carbohydrate Metabolism and Their Biological Significance," *J. Chem. Ed.* 42, p. 244 (1965).

Oesper, P., "Error and Trial: The Story of the Oxidative Reaction of Glycolysis," *J. Chem. Ed.* 45, p. 607 (1968).

Photosynthesis

Levine, "The Mechanism of Photosynthesis," *Scientific American*, Vol. 221, No. 6, p. 58 (1969).

Hydrogen Bonding

McClellan, A. L., "Significance of Hydrogen Bonds in Biological Structures," *J. Chem. Ed.* 44, p. 547 (1967).

Toxic Substances

Maxwell, K. E., "Chemicals and Life," Dickerson Publishing Co., Inc., Belmont, Calif., 1970.

12
LIFE BEGETS LIFE

Until the seventeenth century, people believed in spontaneous generation theory (or abiogenesis), which stated that living things, especially lower types, arise spontaneously from nonliving substances. For example, it was believed that field mice arise spontaneously from the mud of the Nile River or from pieces of cheese placed in bundles of rags. Flies were thought to originate from dirt, manure, and decaying meats. Francisco Redi (1626–1698) and Louis Pasteur (1822–1895) were instrumental in dispelling the theory of spontaneous generation. Redi, an Italian scientist, demonstrated by a simple experiment that by protecting decaying meat from contamination by flies, maggots did not arise spontaneously from the spoiled meat. Instead, they developed from living eggs deposited by flies. Redi's results showed that "life arises from living organisms." Pasteur devised several experiments that proved the air contained microorganisms in a somewhat dormant condition that were ready to become active when suitable environmental conditions of moisture, temperature, and food were encountered.

With the idea of spontaneous generation behind us, the details of the chemistry of propagation have progressed to the point where custom-made life has become a real possibility. The eradication of inherited diseases appears to be possible, based on our present understanding of how a new living organism springs from another living organism and how the organism maintains itself during its life cycle. This intricate and fascinating bit of chemistry requires an understanding of nucleic acids. Life begets life through nucleic acids.

NUCLEIC ACIDS

Like the polysaccharides and the polypeptides, the *nucleic acids* are high molecular weight substances, with molecular weights up to several million. Nucleic acids are found in all living cells, with the exception of the red blood cells of mammals, and the structures of these compounds are believed to be directly related to the characteristics not only of the individual cell but of the gross organism itself. The almost infinite variety of possible structures for nucleic acids allow information in coded form to be recorded in giant molecules in somewhat similar fashion to the complex of language symbols used to convey ideas in this book. Such stored information controls the inherited characteristics of the next generation, as well as many of the ongoing life processes of the organism.

Hydrolysis of nucleic acids yields one of two simple sugars, phosphoric acid (H_3PO_4), and a group of nitrogen compounds that have basic (alkaline) properties. Based on these hydrolysis products, the nucleic acids can be classified as either those that contain the sugar D-*deoxyribose*, or those that contain D-*ribose*. The former are called *deoxyribonucleic acids* (DNA) and the latter *ribonucleic acids* (RNA). DNA is found primarily in the nucleus of the cell, whereas RNA is found mainly in the cytoplasm outside the nucleus.

RIBOSE AND DEOXYRIBOSE

The structures for the two sugars in nucleic acids are shown in Figure 12–1. The names and formulas for the basic nitrogen compounds in nucleic acids are given in Figure 12–2.

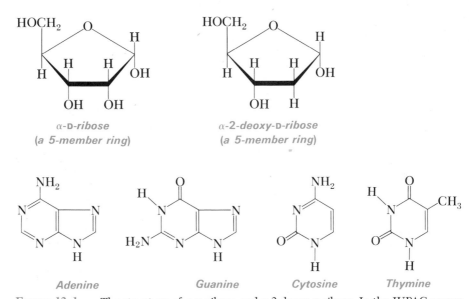

FIGURE 12–1 The structure of α-D-ribose and α-2-deoxy-D-ribose. In the IUPAC names given, α indicates the one of two ring-forms possible; D distinguishes the isomers that rotate plane polarized light in opposite directions and the 2 indicates the carbon to which no oxygen is attached in the second sugar.

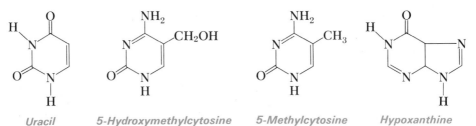

FIGURE 12–2 Nitrogenous bases obtained from the hydrolysis of nucleic acids.

241

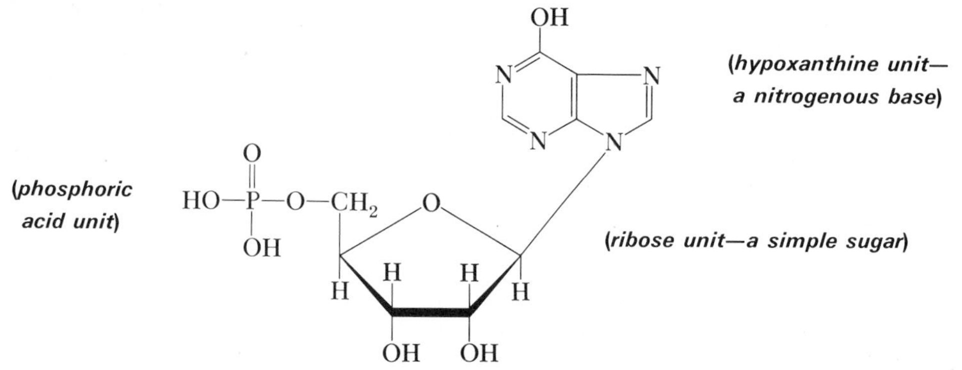

OH

(*hypoxanthine unit—
a nitrogenous base*)

(*phosphoric
acid unit*)

(*ribose unit—a simple sugar*)

FIGURE 12–3   Inosinic acid (a nucleotide). If other bases are substituted for hypoxanthine, a number of nucleotides are possible for each of the two sugars. There is ample evidence that the nucleotides found in both DNA and RNA have the general structure indicated for inosinic acid.

NUCLEOTIDES

Incomplete hydrolysis of DNA or RNA yields *nucleotides*. These substances contain a simple sugar unit, one of the nitrogenous base units, and one or two units of phosphoric acid. An example of the nucleotide structure is illustrated by inosinic acid, Figure 12–3.

POLYNUCLEOTIDES

In addition to the mononucleotides, partial hydrolysis of DNA or RNA yields oligonucleotides which have a few nucleotide units in their molecular structure. The structure for a trinucleotide is illustrated in Figure 12–4. Obviously, a large number of oligonucleotides are possible when one considers the choice of base structures and the different sequence possibilities for the chain of nucleotides.

DNA and RNA are polynucleotides. The number of possible structures for these molecules, which have molecular weights as high as a few million, appears to be almost limitless. Since DNA is a major part of the chromosome material in the nucleus of a cell, it seems reasonable to assume that the organism's characteristics are coded in the DNA structure. It has been estimated that there are over two million different species of organisms. If each individual requires a different DNA structure, there are ample combinations of nucleotides for each individual to be unique. It is now believed that some kinds of RNA transfer the information coded in the DNA structure to control the chemistry in the cytoplasmic region of the cell.

Three major types of RNA have been identified. They are messenger RNA (mRNA), transfer RNA (tRNA), and ribosomal RNA (rRNA). Each has a characteristic molecular weight and base composition. Messenger RNA's are generally the largest, with molecular weights between 25,000 and one million. They contain from 75 to 3000 mononucleotide units. Transfer RNA's have molecular weights in the range of 23,000 to 30,000 and contain 75 to 90 mononucleotide units. Ribosomal RNA's make up as much as 80 per cent of the total cell RNA and have molecular weights between those of mRNA's and tRNA's.

242

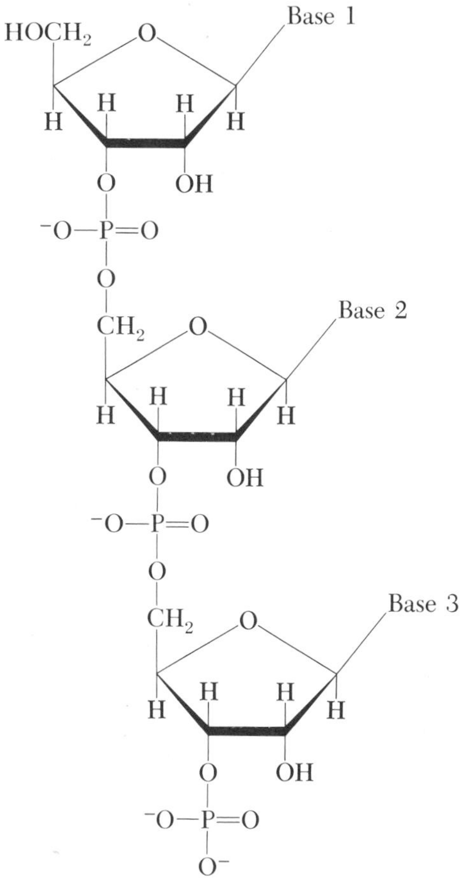

FIGURE 12-4 Bonding structure of a trinucleotide. Bases 1, 2, and 3 represent any of the nitrogenous bases obtained in the hydrolysis of DNA and RNA. The primary structure of both DNA and RNA is an extension of this structure to produce molecular weights as high as a few million.

FIGURE 12-5 F. H. C. Crick (1916–   ) (right) and J. D. Watson (1928–   ) (left), working in the Cavendish Laboratory at Cambridge, built scale models of the double helical structure of DNA based on the x-ray data of M. H. F. Wilkins. Knowing distances and angles between atoms, they compared the task to the working of a three-dimensional jigsaw puzzle. Watson, Crick, and Wilkins received the Nobel Prize in 1962 for their work relating to the structure of DNA.

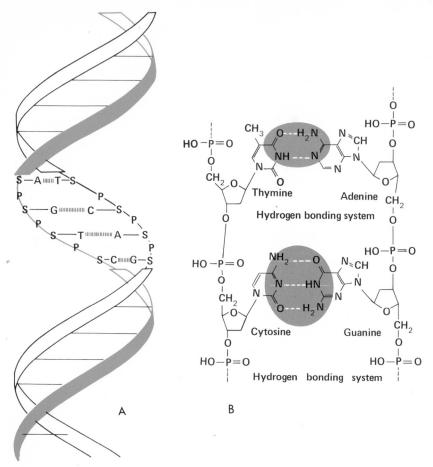

FIGURE 12–6   A, Double helix structure proposed by Watson and Crick for DNA.
S-sugar, P-phosphate, A-adenine, T-thymine, G-guanine, C-cytosine. B, Hydrogen
bonds in the thymine—adenine and cytosine—guanine pairs stabilize the double
helix.

All RNA's consist of a single polyribonucleotide strand. Most RNA is found
in the cytoplasm and ribosomes of the cell (Figure 11–1), but in liver cells as
much as 11 per cent (largely mRNA) of the total cell RNA is found in the nucleus.
Besides having different molecular weights, the three types of RNA appear to
differ in function. One difference in function is described later in this chapter
in the discussion of natural protein synthesis.

In 1953, two scientists, J. D. Watson and F. H. C. Crick, proposed a secondary structure for DNA that has since gained wide acceptance. Figure 12–6
illustrates the proposed structure in which two polynucleotides are arranged in
a double helix stabilized by hydrogen bonding between the adjoining base groups.
Note that only certain pairs of bases "fit" for hydrogen bonding.

VIRUS STRUCTURE

A virus is a parasitic chemical complex that can reproduce only when it
has invaded a host cell. It has the ability to disrupt the life processes of the

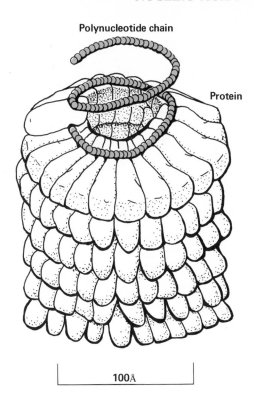

Polynucleotide chain

Protein

100Å

FIGURE 12-7   The structure of the tobacco mosaic virus. The polynucleotide chain is coiled, and there is one protein unit for each three nucleotide units. Part of the polynucleotide chain is exposed for clarity. This structure is based on X-ray studies.

host cell and order the cell contents therein to reproduce the virus structure. The isolated virus unit has neither the enzymes nor the smaller molecules necessary to reproduce itself alone.

A virus is a polynucleotide surrounded by a layer of protein. One virus that has been studied in detail is the tobacco mosaic virus, illustrated in Figure 12–7.

REPLICATION OF DNA

Almost all of the cells of one organism contain the same chromosome structure in their nuclei. This structure remains constant regardless of whether the cell is starving or has an ample supply of food materials. Each organism begins life as a single cell with this same chromosome structure; in sexual reproduction one-half of this structure comes from each parent. These well known biological facts, along with recent discoveries concerning polynucleotide structures, lead to the conclusion that the DNA structure is faithfully copied during normal cell division (mitosis—both strands) and is only partly copied in cell division producing reproductive cells (meiosis—only one strand).

A prominent theory of DNA replication, based on verifiable experimental facts, suggests that the double helix of the DNA structure unwinds and each half of the structure serves as a template or pattern to reproduce the other half from the molecular environment (Figure 12–8).

245

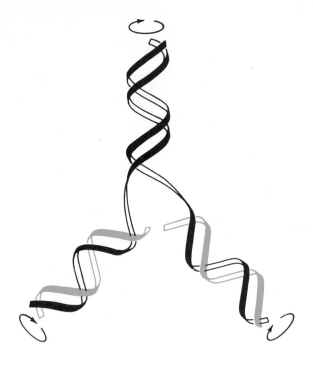

FIGURE 12-8   Replication of DNA structure. When the double helix of DNA (black) unwinds, each half serves as a template on which to assemble subunits (color) from the environment.

NATURAL PROTEIN SYNTHESIS

The proteins of the body are being replaced and resynthesized continuously from the amino acids available to the body, many of which are obtained from the diet. The amino acids and proteins in the body can be considered as constituents of a "nitrogen pool"; additions to and losses from the pool are shown in Figure 12-9.

The use of isotopically labeled amino acids has made possible studies on the average lifetimes of amino acids as constituents in proteins—that is, the time it takes the body to replace a protein in a tissue. For a process that must be extremely complex, it is very rapid. Only minutes after radioactive amino acids are injected into animals, radioactive protein can be found. Although all the proteins in the body are continually being replaced, the rates of replacement have been found to vary. Half of the proteins in the liver and plasma are replaced in *six days*. The time is much longer for muscle proteins, about 180 days, and replacement of protein in other tissues, such as bone collagen, takes even longer.

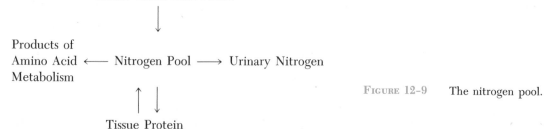

FIGURE 12-9   The nitrogen pool.

Recall that each organism has its own kind of protein. The number of possible arrangements of 20 amino acid units is more than the number of atoms in the known universe, yet proteins characteristic of a given organism can be synthesized in a matter of a few minutes. It should come as no surprise, then, that a vast amount of research has been devoted to this problem in recent years. Although it is thought that the general scheme of protein synthesis is now understood, you should realize that many of the details are still to be worked out.

The DNA in the cell nucleus holds the code for protein synthesis. Messenger RNA, like all forms of RNA, is synthesized in the cell nucleus. The sequence of bases in one strand of the chromosomal DNA serves as the template for

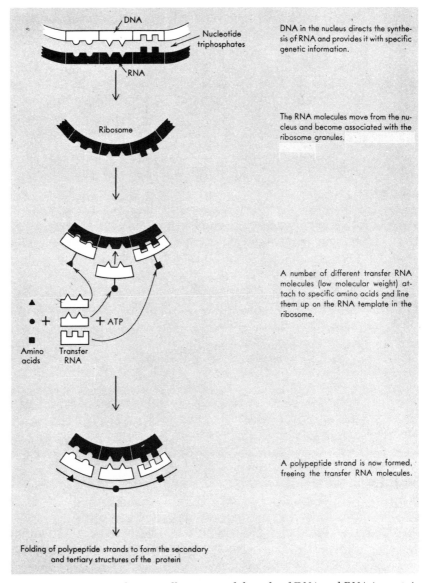

DNA in the nucleus directs the synthesis of RNA and provides it with specific genetic information.

The RNA molecules move from the nucleus and become associated with the ribosome granules.

A number of different transfer RNA molecules (low molecular weight) attach to specific amino acids and line them up on the RNA template in the ribosome.

A polypeptide strand is now formed, freeing the transfer RNA molecules.

FIGURE 12-10   A schematic illustration of the role of DNA and RNA in protein synthesis.

247

monoribonucleotides to order themselves into a single strand of mRNA (Figure 12–10). The bases of the mRNA strand complement those of the DNA strand. A pair of complementary bases is so structured that each one fits the other and forms one or more hydrogen bonds. Messenger RNA contains only the four bases adenine (A), guanine (G), cytosine (C), and uracil (U). DNA contains principally the four bases adenine (A), guanine (G), cytosine (C), and thymine (T). The base pairs are as follows:

DNA	mRNA
A	U
G	C
C	G
T	A

This means that every place a DNA has an adenine base (A), the mRNA will transcribe a uracil base (U), and so on, provided the necessary enzymes are present.

After transcription, mRNA passes from the nucleus of the cell to a ribosome, where it serves as the template for the sequential ordering of amino acids during protein synthesis. As its name implies, messenger RNA contains the sequence message for ordering amino acids into proteins. Each of the thousands of different proteins synthesized by cells is coded by a specific mRNA or segment of a mRNA molecule.

Transfer RNA's carry the specific amino acids to the messenger RNA. Each of the 20 amino acids found in proteins has at least one corresponding tRNA, and some have multiple tRNA's. For example, there are five distinctly different tRNA molecules specifically for the transfer of the amino acid leucine in cells of the bacterium *Escherichia coli*. At one end of a tRNA molecule is a trinucleotide base sequence that fits a trinucleotide base sequence on mRNA. At the other end of a tRNA molecule is a specific base sequence of three terminal nucleotides—CCA—with a hydroxyl group exposed on the terminal adenine nucleotide group. This hydroxyl group reacts with a specific amino acid by an esterification reaction and with the aid of enzymes.

$$(\textit{Mononucleotides})_{75-90}\text{CCA—OH} + \text{HOCCH(NH}_2)\text{R} \longrightarrow$$
tRNA   Amino acid

$$(\textit{Mononucleotides})_{75-90}\text{CCA—OCCH(NH}_2)\text{R} + \text{H}_2\text{O}$$
tRNA-amino acid

ATP provides the energy for the reaction between an amino acid and its transfer RNA. A molecule of ATP first activates an amino acid.

$$\text{ATP} + \text{Amino acid} \longrightarrow \text{ATP—amino acid}$$
activated species

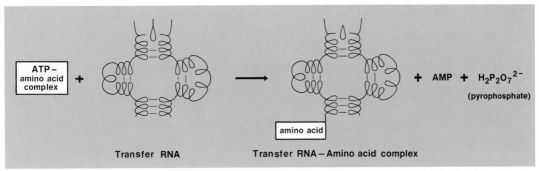

FIGURE 12-11   Bonding of activated amino acid to transfer RNA.

The activated complex then reacts with a specific transfer RNA and forms the products shown in Figure 12–11.

The transfer RNA and its amino acid migrate to the ribosome where the amino acid is given up in the formation of a polypeptide. The transfer RNA is then free to migrate back and repeat the process.

Messenger RNA is used only once, or at most a few times, before it is depolymerized. While this may seem to be a terrible waste, it allows the cell to produce different proteins on very short notice. As conditions change, a different type of messenger RNA comes from the nucleus, a different protein is made, and the cell adequately responds to a changing environment.

SYNTHESIS OF LIVING SYSTEMS

In his quest for a molecular understanding of living systems, man must finally put his theories to the ultimate test of synthesis. If he can synthesize some of the complex molecules described above and see them successfully participate in the life processes, he can be reassured that he is on the right track. It should be emphasized that the biochemist is presently working at the molecular level, and as yet only relatively few of the giant molecules have been characterized. The syntheses in a living cell are far too complex for our present methods to duplicate. However, in spite of the enormity of this undertaking, remarkable strides have been made recently, and interest in current research in this area is intense.

SYNTHETIC PROTEIN

A mixture of amino acids will react to form polypeptide structures outside a living cell. Only relatively simple catalytic agents are required, and the complex enzymatic environment of cytoplasm is not necessary. However, a mixture of only a few amino acids will result in a multitude of different protein structures.

Methods too complex to be detailed here have been devised to construct protein structures with a desired ordered sequence of amino acids. In order to

249

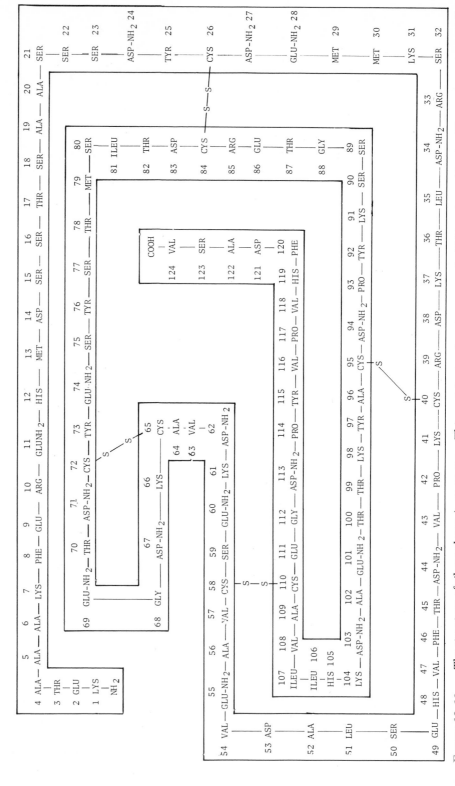

FIGURE 12-12   The structure of ribonuclease A, an enzyme. This protein structure is composed of 124 amino acid units. The structure is partially explained by sulfur-sulfur bonds between some of the acid units—for example, between 40 and 95.

obtain the desired bond between an amino acid and the peptide already con-structed, it is necessary to block all the functional groups except those which are to undergo the peptide reaction. With the chemical blocking groups in place, the particular amino acid is added and the peptide bond is made. This is followed by removal of the blocking groups. To build a polypeptide with 20 amino acid units, this complicated process would have to be repeated 19 times. As the peptide chain grows, it becomes increasingly difficult to carry out the chemical operations without disturbing the bonds previously formed.

In spite of the difficulties, the customized synthesis of a prescribed protein has progressed steadily. In 1953, Vincent du Vigneaud synthesized a hormone from 9 amino acids. For his work he received a Nobel Prize in 1955. Other, more lengthy molecules have been synthesized similarly by starting with the

FIGURE 12-13   Merrifield solid phase method of synthesizing a pro-tein is carried out stepwise from the carboxyl end toward the amino end of the peptide. An aromatic ring of the polystyrene (1) is activated by attaching a chloromethyl group (2). The first amino acid (black), protected by a butyloxycarbonyl (Boc) group (black box), is coupled to the site (3) by a benzyl ester bond and is then deprotected (4). Subse-quent amino acid units are supplied in one of two activated forms; a second unit is shown in one of these forms, the nitrophenyl ester of the amino acid (5). The ester (colored box) is eliminated as the second unit couples to the first. Then the second unit is deprotected, leaving a dipeptide (6). These processes are repeated to lengthen the peptide chain.

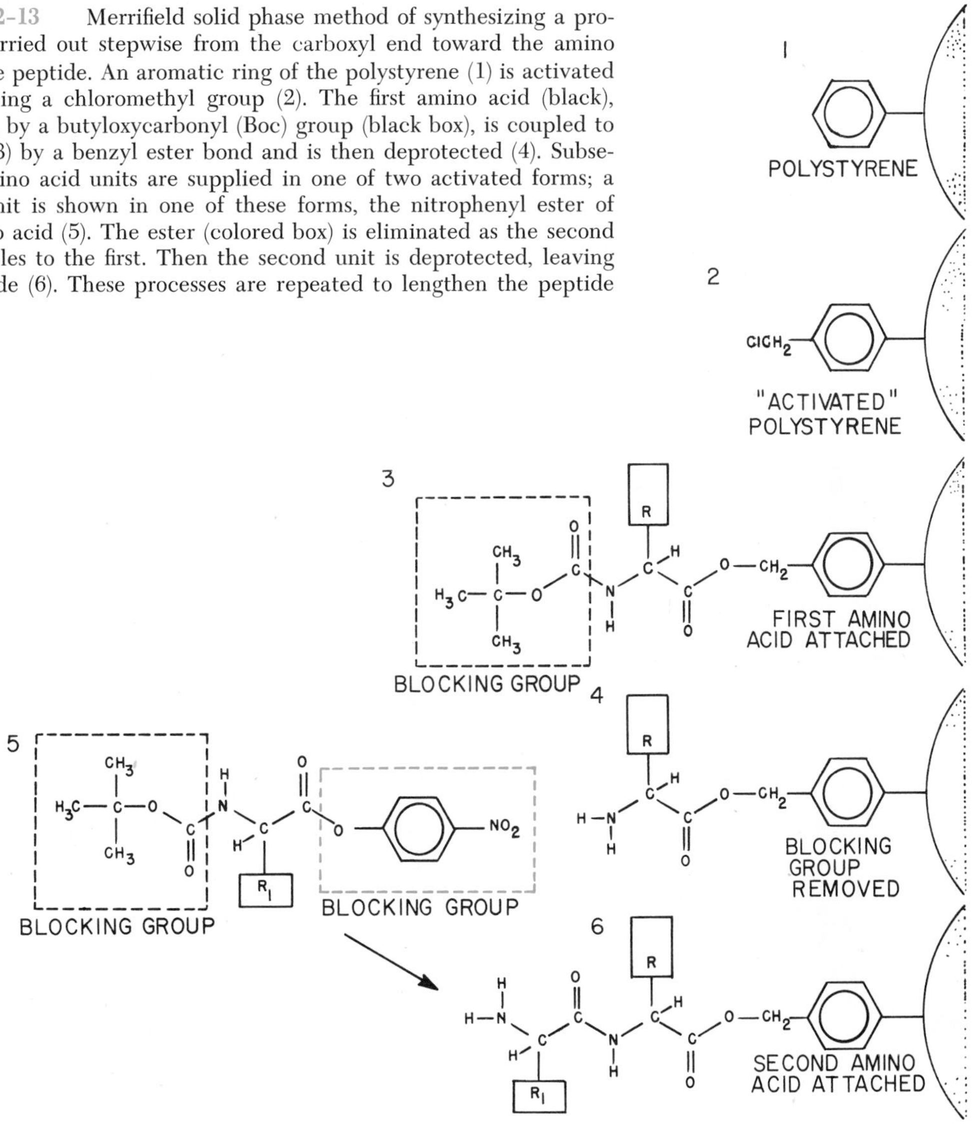

individual amino acids and adding them in proper sequence to make the desired protein. Notable hormone syntheses were those of β-corticotropin with 39 amino acid units, and insulin, with 58 units. In 1969, two teams of researchers synthesized the first enzyme, ribonuclease (Figure 12–12), to be assembled outside the living cell from individual amino acids. Ribonuclease contains 124 amino acid units. One team was led by Rockefeller University's Robert B. Merrifield and Bernard Gutte, and the other team, at Merck Sharp & Dohme research laboratories, was led by Robert G. Denkewalter and Ralph F. Hirschmann.

Merrifield's method has had wider application. In this automated technique, an insoluble solid support, polystyrene, acts as an anchor for the peptide chain during the synthesis (Figure 12–13). The first amino acid is firmly bonded to a small polystyrene bead, and each of the other amino acids is then added one at a time in a stepwise manner. The synthesis of ribonuclease required 369 chemical reactions and 11,931 steps in a continuous operation on a machine developed for this purpose.

In 1970, a group led by Choah Hao Li at the University of California used Merrifield's method to synthesize human growth hormone (HGH) (Figure 12–14). HGH has 188 amino acid units and a molecular weight of about 21,500. This hormone is produced naturally by the front lobe of the pituitary gland, a pea-sized body located at the base of the brain. In humans, it has a dual function—it controls milk formation and it regulates many aspects of growth. In childhood, excess secretions of HGH can cause giantism; too little HGH causes dwarfism.

These synthetic proteins are identical in every respect to the natural products. They perform the same functions and give the same results on analysis. The limits of the present methods are unknown; in time even longer molecules are expected to be duplicated.

FIGURE 12-14   Dr. Li (center) checks experiment with Dr. Richard Noble (left) and Dr. Donald Yanashiro (right). This group was the first to synthesize the HGH polypeptide. (From *Chemical and Engineering News*, January 11, 1971.)

SYNTHETIC NUCLEIC ACIDS

Progress in the synthesis of polynucleotides has been difficult, principally because of the difficulties involved in determining the proper blocking groups. Progress, although slow, is being made.

In 1959, Arthur Kornberg synthesized a DNA-type polynucleotide, for which he received a Nobel Prize. He used natural enzymes as templates to arrange the nucleotides in the order of a desired polynucleotide. His product was not biologically active. In 1965, Sol Spiegelman synthesized the polynucleotide portion of an RNA virus. This polynucleotide was biologically active and reproduced itself readily when introduced into living cells. In 1967, Mehran Goulian and Kornberg synthesized a fully infectious virus of the more complicated DNA type.

In 1970, Gobind Khorana synthesized a complete, double-stranded, 77-nucleotide gene. He, too, used natural enzymes to join previously synthesized, short, single-stranded polynucleotides into the double-stranded gene.

Work is under way to synthesize other genes such as the tyrosine-suppressor transfer-RNA found in the bacterium *Escherichia coli*. This work, as have all previous syntheses of polynucleotides, relies on natural enzymes and extracts from natural systems to order the nucleotides. The real breakthrough of using strictly chemical means to align and bond the nucleotides in a desired sequence is on the horizon. This goal may have already been reached by the time you read this material since research in this area is moving very rapidly.

If man can construct DNA, can he then control the genetic code? Genes are the submicroscopic, theoretical bodies proposed by early geneticists to explain the transmission of characteristics from parents to progeny. It was thought that genes composed the chromosomes, which are large enough to be observed through the microscope as the central figures in cell division. It is now generally believed that DNA structures carry the message of the genes; hence, DNA contains the *genetic code*. If man can construct DNA, he could very well alter its structure and thereby control the genetic code.

A *mutation* occurs when an individual characteristic appears, that has not been inherited by the parent, and is passed along as an inherited factor to the next generation. A mutation can readily be accounted for in terms of the DNA genetic code; that is, some force alters the nucleotide structure in a reproductive cell. Some sources of energy, such as gamma radiation, are known to produce mutations. This is entirely reasonable because certain kinds of energy can disrupt some bonds, which will re-form in another sequence.

One very interesting theory of aging, developed by Professor Henry Eyring at the University of Utah, involves the mutation of DNA to the point where it cannot reproduce perfectly. It has been proposed that, as more and more chemicals interact with DNA, it is changed and the resulting RNA either does not form proteins or forms proteins incapable of surviving and functioning in their intended environments. Hence, there are no cells to replace those that die, and the organism diminishes, its functions are not performed, and we die from "old age." Mutations via chemical agents are discussed as a separate topic in Chapter 20. Free radicals (species with unpaired valence electrons) are especially

253

devastating on DNA and chromosomes. The caffeine and sodium nitrite used to color meat may be especially active in this respect. Eyring and others recommend Vitamin E as a health promoter, especially for slowing the aging process. Vitamin E is an antioxidant and a free-radical scavenger. Part of its role may be to tie up free radicals, thereby reducing mutations, and allowing normal cells and protein to be produced by non-mutated DNA.

If man can control the genetic code, can he control hereditary diseases such as sickle cell anemia, gout, some forms of diabetes, or mental retardation? If the understanding of detailed DNA structure and the enzymatic activity in building these structures continues to grow, it is reasonable to believe that some detailed relationships between structure and gross properties will emerge. If this happens, it may be possible to build compounds which, when introduced into living cells, can combat or block inherited characteristics.

QUESTIONS

1. What three molecular units are a part of nucleotides?

2. List the different kinds of RNA that are employed in protein synthesis and briefly describe the role played by each.

3. Based on the structure in Figure 12–1, explain the meaning of the prefix deoxy- in deoxyribonucleic acid.

4. How many trinucleotides with the structure indicated in Figure 12–4 could be made with the nitrogenous bases listed in Figure 12–2?

5. What are the basic differences between DNA and RNA structures?

6. What stabilizing forces hold the double helix together in the secondary DNA structure proposed by Watson and Crick?

7. What two molecular structures are present in viruses?

8. Does a strand of DNA actually duplicate itself base for base in the formation of a strand of messenger RNA? Explain.

9. Distinguish between the three types of RNA as to molecular weight, number of amino acid units per molecule, and function.

10. (a) Describe the general method of synthesizing DNA in vitro (in a test tube) at present.
    (b) Why would the synthesis of a polynucleotide from the individual phosphoric acid, sugar, and nitrogenous bases be a breakthrough in controlling the genetic code?

11. Discuss the feasibility of solving sociological and psychological problems via religion, DNA alteration, chemical suppressants, political pressures, or persuasive dialogue.

12. Check the recent issues of *Science* or other scientific news publications to update the work done on synthesis of proteins and polynucleotides.

13. Describe the general method of natural protein synthesis.

14. Describe a general method of synthesizing proteins in vitro. What general problems arise in the process described?

15. Since RNA is so complicated, it looks as if our bodies would retain the synthesized RNA instead of resynthesizing it each time it needs a protein. This is not generally the case. Justify this.

SUGGESTIONS FOR FURTHER READING

"A Step Toward Synthetic Life," *Chemistry*, Vol. 41, No. 2, p. 27 (1968).

"Anatomy of a Virus," *Chemistry*, Vol. 39, No. 4, p. 20 (1966).

"Bonding Habits of DNA Bases," *Chemistry*, Vol. 41, No. 8, p. 34 (1968).

Davies, D. R., "X-ray Diffraction and Nucleic Acids," *Chemistry*, Vol. 40, No. 2, p. 8 (1967).

Fraenkel-Conrat, H., "The Genetic Code of a Virus," *Scientific American,* p. 46, October (1964).

Hofmann, K., Khorana, H., and Spiegelman, S., "The Synthesis of Living Systems," *Chemical and Engineering News*, Vol. 45, August 7, p. 144 (1967).

"Interlocking Rings of DNA," *Chemistry*, Vol. 41, No. 2, p. 26 (1968).

"Is DNA the Master Molecule," *Chemistry*, Vol. 41, No. 6, p. 23 (1968).

Merrifield, R. B., "The Automated Synthesis of Proteins," *Scientific American*, Vol. 218, No. 3, p. 56 (1968).

Pauling, L., Corey, R., and Hayward, R., "The Structure of Protein Molecules," *Scientific American,* p. 51, July (1954).

"Portrait of a Gene," *Chemistry*, Vol. 42, No. 8, p. 20 (1969).

"Trouble on the DNA Front," *Chemistry*, Vol. 43, No. 9, p. 24 (1970).

"Viruses as Invaders of Living Cells," *Chemistry*, Vol. 41, No. 5, p. 29 (1968).

Watson, J. D., *The Double Helix,* Atheneum, New York, 1968.

Zimmerman, J., "First Synthesis of an Enzyme, Ribonuclease," *Chemistry*, Vol. 42, No. 4, p. 21 (1969).

Part 4
Chemistry for Better Living— Consumer Products

13

FOOD ADDITIVES—OR —MODIFIED NATURAL FOODS

No material thing is more important to us than food. Early in our history foods were eaten as they were found—"natural" foods. Notions, superstition and religion, along with our immediate sensory response, controlled our list of acceptable foods. Changing food to preserve it is older than recorded history; salt was added or water was removed to keep meat in an eatable condition. We have since learned to add a wide variety of chemicals to preserve food, to enhance its taste, or to increase its nutritional value.

In recent years, the use of food additives has grown faster than the general public's awareness of them. Only in very recent months have the following facts become a part of our general information:

(a) At least 2500 chemicals are added to our food.
(b) Approximately 1500 chemicals enter foods through packaging and processing.
(c) 1000 to 1500 new food products are introduced annually; these depend heavily on food additives.
(d) Food additives fall into at least 40 different categories based on function.

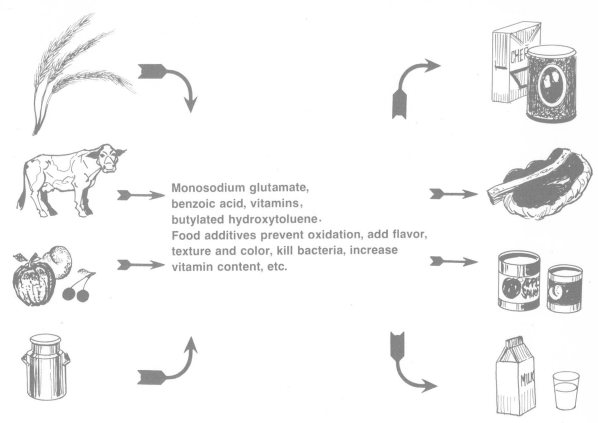

Monosodium glutamate, benzoic acid, vitamins, butylated hydroxytoluene. Food additives prevent oxidation, add flavor, texture and color, kill bacteria, increase vitamin content, etc.

FIGURE 13-1   Between the harvested and the consumer-ready food is the addition of a large variety of food additives, which attempt to enhance or preserve the nutritional value of the food.

(e) 2500 health food stores, specializing in "natural, health and organic" foods, are now operating in the USA.

(f) The cyclamates have been removed from the GRAS list (a list of food additives that are generally recognized as safe by the Federal Food and Drug Administration).

The immense use of food additives is not free from controversy. Extreme positions range from "Food Pollution!" and "Poisoned Food" to "No Danger in Food Additives." The experts in the field admit some dangers but believe we are in a generally safe situation. Dr. Fredrick J. Stare, head of the Department of Nutrition at Harvard, said in *Life,* "As a physician and a student of nutrition for the past 30 years, I am convinced that food additives are far safer in actual use than the basic foods themselves . . ."

In this chapter we shall examine some of the chemicals used as food additives.

SYNERGISM AND POTENTIATION

We should enter this discussion of food modification with the notion that most food additives are mixtures and that mixtures of chemicals can have effects   257 that are not necessarily the sum of individual effects. In some formulations, the individual properties of certain chemicals are reinforced and enhanced by other chemicals in the mixture. This is called *synergism*. For example, citric acid has a small antioxidant effect on foods; butylated hydroxyanisole (BHA) has considerable effect in preventing oxidation. When these two substances are used together in foods, the antioxidative powers of the BHA are increased several fold. The citric acid is said to have synergistic action on BHA. The reasons behind this will be discussed later in this chapter.

A closely related term is *potentiation*. Potentiators do not have the particular effect themselves, as in the case of citric acid, but merely exaggerate an effect of another chemical. The 5′-nucleotides, for example, have no taste, but they increase the flavor of meat or the effectiveness of salt.

THE GRAS LIST

To protect the food consumer, the U.S. Government has enacted many pieces of legislation called Food, Drug, and Cosmetic Acts. The Federal Food and Drug Administration (FDA) administers these laws and, in this sense, is the guardian of our health. For example, the FDA is required by law (Delaney Amendments of 1958) to withdraw sanction on or ban any food additive which "is found to induce cancer when ingested by man or animal." To recognize a food additive as safe, the substance must not be harmful in the amounts used in the food.

The most difficult point to check is the synergistic action of a food additive. An additive may produce no harmful effects by itself. In combination with other substances in the foodstuff, a sleeping giant may be awakened and effects could be amplified much beyond those of the individual chemicals. Synergism and potentiation are the most difficult checks to make because of the scope of the testing. This is being done, however. Our increasing understanding of molecular structure and interaction facilitates the predictability of synergism.

At this time the FDA lists about 600 chemical substances "generally recognized as safe" (GRAS list) for their intended use. A small portion of this list is given in Table 13–1. It must be emphasized that an additive on the GRAS list is safe **only if it is used in the amounts and foods specified.** The GRAS list was published in several installments in 1959 and 1960. It was compiled by asking experts in nutrition, toxicology and related fields to give their opinion about the safety of using various materials in foods. Since its publication, no substance has been added to the GRAS list and a few, such as the cyclamates, have been removed.

It is evident, in view of the 2500 known food additives, that many more chemicals than those that appear on the GRAS list are approved (or at least, not banned) for use as food additives by the FDA. It is quite expensive to introduce a new food additive with the approval of the FDA. Allied Chemical Corporation has been working since 1964 on a new synthetic food color, Allura Red AC. It has recently been approved by FDA and went on the market in 1972. The cost for introducing this product was $500,000 and about one-half of this amount was spent on safety testing.

258

Each food additive is included in a consumer foodstuff for a specific purpose. Table 13–1 is organized according to some of the functions of the additives. Some additives have more than one function; only the major function is given in this

TABLE 13-1   A PARTIAL LIST OF THE GRAS LIST OF FOOD ADDITIVES°

FOOD COLORS	PRESERVATIVES	SEQUESTRANTS
Annatto (yellow)	Benzoic acid	Citrate esters:
Carbon black	Na benzoate	isopropyl, stearyl
Carotene (yellow-orange)	Methylparaben	Citric acid
Cochineal (red)	Oxytetracycline	EDTA, Ca and Na salts
Food dyes and colors	Propylparaben	Pyrophosphate, Na
Red No. 2	Propionic acid	Sorbitol
Red No. 3	Ca propionate	Tartaric acid
Blue No. 1	Na propionate	Na tartrate
Titanium dioxide (white)	Sorbic acid	
	Ca sorbate	STABILIZERS AND
ACIDS, ALKALIES,	K sorbate	THICKENERS
AND BUFFERS	Na sorbate	Agar-agar
Acetates: Ca, K, Na	Sulfites, Na, K	Algins: NH_4^+, Ca, K, Na
Acetic acid		Carrageen
Calcium lactate	ANTIOXIDANTS	Gum acacia
Citrates: Ca, K, Na	Ascorbic acid	Gum tragacanth
Citric acid	Ca ascorbate	Sodium carboxymethyl cellulose
Fumaric acid	Na ascorbate	
Lactic acid	Butylated hydroxyanisole	FLAVORINGS
Phosphates, CaH, Ca, Na,	(BHA)	Acetanisole (slight haylike)
Na, NaAl	Butylated hydroxytoluene	Allyl caproate (pineapple)
Potassium acid	(BHT)	Amyl acetate (banana)
tartrate	Lecithin	Amyl butyrate (pear-like)
Sorbic acid	Propyl gallate	Bornyl acetate (piney, camphor)
Tartaric acid	Sulfur dioxide and sulfites	Carvone (spearmint)
	Nordihydroguaiaretic acid	Cinnamaldehyde (cinnamon)
SURFACE ACTIVE AGENTS	(NDGA)	Citral (lemon)
Cholic acid	Trishydroxybutyrophenone	Ethyl cinnamate (spicy)
Glycerides: mono- and	(THBP)	Ethyl formate (rum)
diglycerides of fatty acids		Ethyl propionate (fruity)
Polyoxyethylene (20)		Ethyl vanillin (vanilla)
sorbitan mono-		Eucalyptus oil (bittersweet)
palmitate		Eugenol (spice, clove)
Sorbitan mono-		Geraniol (rose)
stearate		Geranyl acetate (geranium)
		Ginger oil (ginger)
POLYHYDRIC ALCOHOLS		Linalool (light floral)
Glycerol		Menthol (peppermint)
Sorbitol		Methyl anthranilate (grape)
Mannitol		Methyl salicylate (wintergreen)
Propylene glycol		Orange oil (orange)
		Peppermint oil (peppermint)
NONNUTRITIVE SWEETENERS		(menthol)
Saccharin: NH_4^+, Ca, Na		Pimenta leaf oil (allspice)
		(eugenol, cineole)
FLAVOR ENHANCERS		Vanillin (vanilla)
Monosodium glutamate (MSG)		Wintergreen oil (wintergreen)
5'-nucleotides		(methyl salicylate)
Maltol		

° For precise and authoritative information on levels of use permitted in specific applications, the regulations of the U.S. Food and Drug Administration and the Meat Inspection Division of the U.S. Department of Agriculture should be consulted.

table. An understanding of some of the chemistry involved can make the morning's reading of the cereal box more enlightening.

PRESERVATION OF FOODS

The major causes of food deterioration are oxygen and microorganisms. Both are aided by water and room temperatures. Any process that prevents the growth of or presence of microorganisms and excludes oxygen or water, or both, is generally an effective preservative process for food. Perhaps the oldest technique is the drying of grains, fruits, fish, and meat. Water is necessary for the growth and metabolism of microorganisms. Without water they dehydrate and die. Water is also important in food oxidation and in this capacity will be discussed in the next section. Dryness thwarts both oxidation and microorganisms.

Salted meat and fruit preserved in a concentrated sugar solution are protected from microorganisms in a similar fashion. The abundance of sodium chloride and sucrose in the immediate environment of the microorganisms forms a hypertonic condition in which water flows by osmosis from the microorganism to its environment (Figure 13–2). The salt and sucrose have the same effect on the microorganism as does dryness. All dehydrate the microorganism and thus destroy its ability to grow and reproduce and exude distasteful and dangerous products into the food.

The canning process was developed by 1810 and involves first the heating of food to kill all bacteria and then sealing it in bottles or cans to prevent the access of other microorganisms and oxygen. This technique is widely used for fish, meat, and vegetables, and some canned meat has been successfully preserved for over a century.

With the increasing growth of urban population and the need for preserving food for longer periods of time, the necessity of developing improved techniques for food preservation has become urgent. As a result of this need, newer techniques have developed such as dehydration, freezing, pasteurization, cold storage,

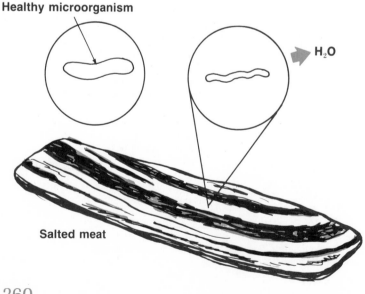

Healthy microorganism

H_2O

Salted meat

FIGURE 13–2 A salt solution dehydrates microorganisms by osmosis. Without water, microorganisms cannot carry on the chemical reactions required for growth and reproduction.

irradiation, and chemical preservation resulting from food additives. Before these methods came into general use, a very large percentage of many kinds of foods perished in the journey from harvester to consumer. Today we can expect food to be fresh for at least several days after it has been processed.

ANTIMICROBIAL PRESERVATIVES

Antimicrobial preservatives are widely used in a large variety of foods. For example, in the United States sodium benzoate is permitted in nonalcoholic beverages and in some fruit juices, fountain syrups, margarines, pickles, relishes, olives, salads, pie fillings, jams, jellies, and preserves. Propionates are legal in bread, chocolate products, cheese, pie crust, and fillings. Depending on the food, the weight of the additive permitted ranges up to a maximum of 0.1 per cent for sodium benzoate and 0.3 per cent for the propionates.

Food spoilage caused by microorganisms is a result of their large population, excretion of toxins, and decay. A preservative is effective if it prevents multiplication of the microbes during the shelf-life of the product. Of course, this can be attained by complete sterilization (usually by heat or radiation) or by inactivation of the microorganisms (freezing). It is seldom achieved by safe chemical agents.

Despite the fact that chemical preservatives of various kinds have been used for several decades, their mode of action remains largely unexplained. Postulated mechanisms could be grouped into three possible categories: (1) interference with the permeability of cell membranes, (2) interference with genetic mechanism, and (3) interference with intracellular enzyme activity.

Interference with the cell membrane and cell wall permeability can have vast effects on the flow of cell nutrients into and waste products out of the cell. For example, propionic acid is known to have an antimicrobial action on certain types of bacteria. It is thought that the propionic acid coats the cell surface with a substance that reduces its permeability.

The coat effectively blocks passage of materials into and out of the cell and the cell ceases to function. On the other hand, oxytetracycline, an antibiotic added to meats, may destroy bacteria by creating a leakage in the cell membrane. Some antibiotics, such as penicillin, are known to prevent cell wall synthesis. Benzoic acid and salicylic acid are known to accumulate on the cell membrane and are thought to inhibit the microorganisms by rendering the cell wall less permeable.

There is no evidence that the generally accepted food preservatives interfere with genetic mechanisms of microbes. However, certain antibiotics such as streptomycin and chloramphenicol are known to exert their effects by chemically combining with the ribosome and inhibiting protein synthesis (Chapter 12). The comparatively simple preservatives with less complex structures are expected to be less reactive with cellular sites of attachment, and even less specific.

There is some evidence that food preservatives interfere with cellular enzymes in microbes. They could interrupt the metabolism of enzymes, and thus destroy them. For example, sorbic acid is known to interfere with cellular dehydrogenases, which normally dehydrogenate fatty acids as the first step in metabolism of molds that grow on cheese. This interrupts their use of food and renders the mold inactive. Several food preservatives are known to inhibit enzyme systems isolated from the cell, but such observations cannot necessarily be extrapolated to cellular conditions. The preservative either may not penetrate the cell wall or may not attain sufficient concentration at the enzyme site.

ATMOSPHERIC OXIDATION

While spoilage by microbial activity is one of the most important factors to be considered in preserving carbohydrate and protein portions of food products, *atmospheric oxidation* is the chief factor in destroying fats and fatty portions of foods. Chemically, oxygen reacts with the fat to form a hydroperoxide (R—OOH).

Portion of an unsaturated   Hydroperoxide
fat molecule

The mechanism involves the formation of very reactive free radicals (species with one or more unpaired electrons) in a chain reaction process. For example:

$$RH + O_2 \longrightarrow R \cdot + HO_2 \cdot$$
(Fat)   (Free radicals)

$$R \cdot + O_2 \longrightarrow ROO \cdot$$
$$ROO \cdot + RH \longrightarrow ROOH + R \cdot$$
(Hydroperoxide)

Foods kept wrapped airtight, cold, and dry are relatively free of oxidation. An antioxidant added to the food can also hinder oxidation. Antioxidants most commonly used in edible products contain various combinations of butylated hydroxyanisole (BHA), butylated hydroxytoluene (BHT), or propyl gallate:

BHA

BHT   Propyl gallate

The word *butylated* is not widely used in organic chemical nomenclature. It is applied here because the usual names for BHA and BHT include the words *cresol* or *phenol,* and these have a history of being toxic. To avoid consumer rejection of these "safe" compounds—added within the prescribed limits (maximum of 0.02 per cent or less, depending on the food)—the names were made "safe" too.

To prevent the oxidation of fats, the antioxidant can donate its phenolic hydrogen atoms (—OH) to the free radicals and stop the chain reaction. The bulky aromatic radicals formed are relatively stable and unreactive; they add the free electrons to their supply of delocalized electrons:

$$\text{R} \cdot + (CH_3)_3C \underset{CH_3}{\overset{OH}{\bigcirc}} C(CH_3)_3 \longrightarrow \text{RH} + (CH_3)_3C \underset{CH_3}{\overset{\dot{O}}{\bigcirc}} C(CH_3)_3$$

If antioxidants are not present, the hydroperoxy group will attack a double bond where present. This reaction leads to a complex mixture of volatile aldehydes, ketones, and acids, which cause the odor and taste of rancid fat.

SEQUESTRANTS

As mentioned earlier in this chapter, citric acid is a synergist for antioxidants. It deters oxidation by bonding with traces of metal ions in foods. Metals get into food from the soil and from machinery during harvesting and processing. Copper, iron, and nickel, and their ions catalyze the oxidation of fats. A molecule of citric acid wraps itself around the metal ion, thereby rendering it ineffective

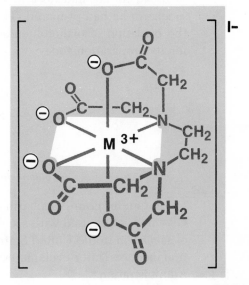

FIGURE 13-3   The structural formula for the metal chelate of ethylene diaminetetraacetic acid (EDTA).

263

as a catalyst. With the antagonist, competitor metal ions tied up, antioxidants such as BHA and BHT can accomplish their task much more effectively.

Citric acid belongs to a class of food additives known as *sequestrants*. (To sequester means "to withdraw from public use.") For the most part sequestrants react with trace metals in foods, tying them up in complexes so the metals will not catalyze the decomposition of food. They are used in shortenings, mayonnaise, lard, soup, salad dressings, margarine, cheese, vegetable oils, pudding mixes, vinegar, confectionery, and other foods. Sodium and calcium salts of EDTA (ethylenediamine tetraacetic acid) are sequestrants permitted in beverages, cooked crab meat, salad dressings, margarine, and vinegar from 0.0025 to 0.15 per cent. The bonding pattern of EDTA is shown in Figure 13–3. Note the five-membered rings (five- or six-membered rings are usually required for stable bonding angles), and the octahedral arrangement of the donor atoms about the central metal ion, M.

SWEETNESS WITHOUT CALORIES

Sweetness is characteristic of a wide range of compounds, many of which are completely unrelated to sugars. Lead acetate, $Pb(CH_3COO)_2$, is sweet but poisonous. A number of *artificial sweeteners* are allowed in foods. These are primarily used for special diets such as those of diabetics. Artificial sweeteners have no known metabolic use in the body and therefore are not involved in the insulin balance.

SACCHARIN

The most common artificial sweetener is saccharin. Its sweet taste was discovered accidentally when the chemist who synthesized it noted that a piece of bread he ate had an obvious sweet taste. He traced the taste back to the saccharin he had prepared. The material was found in the course of studies on the oxidation of orthotoluenesulfonamide, which is the starting material used for its preparation today:

o-Toluenesulfonamide   *Saccharin*

Gram for gram, saccharin is at least 200 times sweeter than ordinary sugar. Glycine is often used with saccharin to cut the bitter aftertaste. The production of saccharin in the United States exceeds 5,000,000 pounds per year. Its utilization increased after cyclamates were taken off the market by the Food and Drug Administration.

Cyclamate

The sweetness of cyclamate salts was also discovered accidentally. Sodium cyclamate is about 30 times sweeter than sugar. Because a cyclamate does not have the aftertaste of saccharin, and because it could be used in cooked or baked products, it rapidly replaced saccharin in a wide variety of dietary products. It was removed from the market as a result of suspicions that large doses could be carcinogenic, although no such results have been reported in man.

Sodium cyclamate

Because of the great potential usefulness of synthetic sweeteners, other suitable compounds will probably be developed which meet the high FDA requirements for safety.

The sweetness factor is determined by panels of tasters who sample different dilutions of solutions of the sweeteners in water. If a 0.01 M artificial sweetener solution provides the same sweetness sensation as a 1 M solution of sucrose, the artificial sweetener is said to be 100 times sweeter than sucrose.

FLAVORS

Flavors constitute the largest class of food additives; it is estimated that the number of natural and synthetic flavors range up to about 1400.

NATURAL FLAVORS

Flavors result from a complex mixture of volatile chemicals. Since we have only four tastes (sweet, sour, salt, bitter), most of the sensation of food is smell. For example, the flavor of coffee is determined largely by its aroma, and this in turn is due to a very complex mixture of over 100 compounds. The reproduction of a particular flavor is usually attained by mixing coffee from different sources rather than trying to match it chemically. Because roasted coffee loses its volatile compounds in air (and, consequently, its flavor), it must be kept sealed to prevent staleness. The caffeine content of tea is considerably higher than that of coffee, but coffee contains a variety of other ingredients and generally produces more side effects than tea. Roasted chicory, which is often mixed in with coffee, contains no caffeine or other active principle, but it does provide color and flavor.

Tea is made by steeping the dried and fermented tea leaves in hot water for 3 to 5 minutes. This extracts water-soluble compounds, primarily caffeine and products made by the oxidation of phenols (catechins), from the leaves.

265

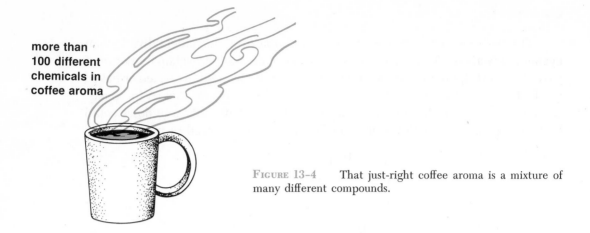

more than 100 different chemicals in coffee aroma

FIGURE 13-4   That just-right coffee aroma is a mixture of many different compounds.

Because of the way tea is made, it ordinarily does not contain any insoluble matter suspended in it, as coffee does. Tea absorbs flavors and odors readily, and this fact can be used to advantage in the preparation of specially flavored teas. It also means that tea should not be stored near onions, garlic, or soaps since it will also absorb these aromas.

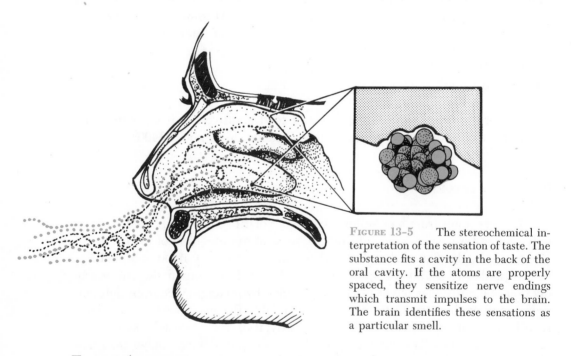

FIGURE 13-5   The stereochemical interpretation of the sensation of taste. The substance fits a cavity in the back of the oral cavity. If the atoms are properly spaced, they sensitize nerve endings which transmit impulses to the brain. The brain identifies these sensations as a particular smell.

FLAVOR ADDITIVES

Most *flavor* additives originally came from plants. The plants were crushed and the compound extracted with various solvents such as ethanol and carbon tetrachloride. Sometimes a single compound was extracted; more often, a mixture of several compounds occurred in the residue. By repeated efforts, relatively pure oils were obtained. Oils of wintergreen, peppermint, orange, lemon, and ginger, among others, are still obtained this way. These oils, alone or in combination, are then added to foods to obtain the desired flavor. Gradually analyses of the oils and flavor components of plants revealed the active compounds responsible for the flavor. Today, the synthetic preparation of flavors actively competes with natural sources.

The Food and Drug Administration has recently banned some of the flavoring agents which formerly were used, including safrole, the primary root-beer flavor.

FLAVOR ENHANCERS

Flavor enhancers have little or no taste of their own but amplify the flavors of other substances. They exert synergistic and potentiation effects. Potentiators were first used in meats and fish. Now they are also used to intensify the flavor and cover unwanted flavors in vegetables, bread, cakes, fruits, nuts, and beverages. Three common flavor enhancers are monosodium glutamate (MSG), 5'-nucleotides (similar to inosinic acid; Chapter 12), and maltol.

MSG is the best known and most widely used flavor enhancer. More than 220 million pounds were sold world-wide in 1966. Glutamic acid was isolated in 1866 by the German chemist Ritthausen and converted to a sodium salt, monosodium glutamate, by another chemist. Neither had any interest in flavor. In 1908, a Japanese chemist at the University of Tokyo, Dr. Kikunae Ikeda, discovered the flavor enhancing properties of MSG. Japanese cooks had used the seaweed *Laminaria japonica* for centuries to improve the flavor of soups and certain other foods. Dr. Ikeda discovered that the ingredient in the seaweed that made the difference was MSG, and that it had an unusual ability to enhance or intensify the flavor of many high protein foods.

MSG imparts no flavor of its own in the concentrations allowed in foods. There are several theories on how MSG, or any enhancer or potentiator, works. Some chemists say that it acts on certain nerve endings to make the taste buds more sensitive and, therefore, increase the flavor of food. Others claim that it increases salivation and that this leads to increased flavor perception. The detailed chemical and biological action of flavor enhancers is only partially understood. They are very effective and we use them without apparent harm, unless the sodium intake must be limited, as in low-salt diets. In some people, MSG causes the so-called "Chinese restaurant syndrome," an unpleasant reaction that includes headaches, sweating, and other symptoms, usually occurring after an MSG-rich Chinese meal. Tomatoes and strawberries affect some individuals in the same way.

MSG causes brain damage when injected in very high dosage under the skin in 10-day old mice. When these laboratory results were reported, considerable discussion ensued concerning the merits of MSG. National investigative councils have suggested that it be removed from baby foods since infants do not seem to appreciate enhanced flavor. However, in the absence of hard evidence that MSG is harmful in the amounts used in regular food, no recommendations were made relative to its use.

The flavor enhancing property of the 5'-nucleotides was discovered in 1913 by Dr. Shintara Kodama of Tokyo University, an associate of Dr. Ikeda. The 5′-nucleotides were discovered to be the ingredients responsible for the flavor enhancing qualities of bonita tuna. They were approved by the FDA in 1962. While MSG is effective in enhancing the flavor of foods in parts per thousand, the 5′-nucleotides significantly enhance the flavor of foods in concentrations of parts per billion and even less. In liquid foods they create a sense of increased viscosity.

BRIGHT COLORS IN FOODS

Food colors are generally large organic molecules having several double bonds and aromatic rings. Some such structures can absorb certain wavelengths of light and pass the rest, giving the substance its characteristic color. β-Carotene is an orange-red substance that occurs in a variety of plants, the carrot in particular, and is commonly used as a food color. It has a conjugated system of delocalized electrons. A conjugated system has alternate double and single bonds along the carbon chain. This allows for smaller energy-level differences for the electrons of the double bonds, and these electrons are readily excited by some of the visible light quanta.

β-*Carotene*

There is little general agreement about which artificial colors are safe. The accepted lists vary widely from country to country. The World Health Organization (UN) has evaluated more than 140 kinds of coloring matter and declared a number of them to be unsafe. The FDA has recently banned some red and yellow dyes made from coal tar.

FOODS THAT FEEL GOOD

LEAVENED BREAD

Sometimes chemical changes in food release carbon dioxide, and the gas causes breads and pastries to rise. Yeast has been used since ancient times to make bread rise, and remains of bread made with yeast have been found in Egyptian tombs and the ruins of Pompeii. The metabolic processes of the yeast furnish gaseous carbon dioxide, which creates bubbles in the bread and makes it rise. When the bread is baked, the CO_2 expands even more to produce a light airy loaf.

Carbon dioxide in cooking can be generated by other processes. For example, baking soda (which is simply sodium bicarbonate, $NaHCO_3$, a base) can react with acidic ingredients in a batter to produce CO_2 (Figure 13–6):

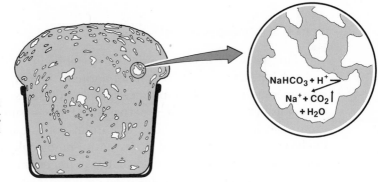

FIGURE 13-6 The carbon dioxide released from baking soda, $NaHCO_3$, expands and causes the bread to rise.

$$NaHCO_3 + H^+ \longrightarrow Na^+ + H_2O + CO_2(gas)$$

Baking powders contain sodium bicarbonate and an added acid salt or a salt which hydrolyzes to produce an acid. Some of the compounds used for this purpose are potassium hydrogen tartrate, $KHC_4H_4O_6$, monocalcium hydrogen phosphate, $Ca(H_2PO_4)_2 \cdot H_2O$, and sodium acid pyrophosphate, $Na_2H_2P_2O_7$. The reactions of these white, powdery salts with sodium bicarbonate are similar, although the compounds all have somewhat different appearances. For example:

$$KHC_4H_4O_6 + NaHCO_3 \longrightarrow KNaC_4H_4O_6 + H_2O + CO_2(gas)$$

PLEASANT CONSISTENCY

Stabilizers and thickeners improve the texture and blends of foods. The action of carrageenin (a polymer from edible seaweed) is shown in Figure 13–7. Most of this group of food additives are polysaccharides (Chapter 9) having numerous hydroxyl groups as a part of their structure. The hydroxyl groups form hydrogen bonds with water, thereby preventing the segregation of water from

FIGURE 13-7 The action of carrageenin to stabilize an emulsion of water and oil in salad dressing. An active part of carrageenin is a polysaccharide, a portion of which is shown above. The carrageenin hydrogen-bonds to the water, which keeps it dispersed. The oil, not being very cohesive, disperses throughout the structure of the polysaccharide. Gelatin (a protein) undergoes similar action in absorbing and distributing water to prevent ice crystals in ice cream.

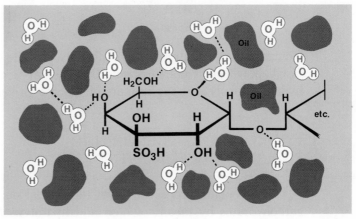

the less polar fats in the food and providing a more even blend of the water and oils throughout the food. Stabilizers and thickeners are particularly effective in icings, frozen desserts, salad dressing, whipped cream, confectionery, and cheeses.

SURFACE ACTIVE AGENTS

Surface active agents are similar to stabilizers and thickeners in their chemical action. They cause two or more normally incompatible (nonpolar and polar) substances to mix. If the substances are liquids, the surface active agent is called an emulsifier. If the surface active agent has a sufficient supply of hydroxyl groups, such as cholic acid has, the groups form hydrogen bonds to water. The cholic acid and its associated group of water molecules are distributed throughout egg in a manner quite similar to that of carrageenin and water in salad dressing.

Some surface active agents have both hydroxyl groups and a relatively long nonpolar hydrocarbon end. Examples are diglycerides of fatty acids, polyoxyethylene(20) sorbitan monopalmitate, and sorbitan monostearate. The hydroxyl groups on one end of the molecule anchor via hydrogen bonds in the water, and the nonpolar end is held by the nonpolar oils or other substance in the food. This provides tiny islands of water held to oil. These islands are distributed evenly throughout the food.

POLYHYDRIC ALCOHOLS

Polyhydric alcohols are allowed in foods as humectants (water retainers), sweetness controllers, dietary agents, and softening agents. Their chemical action is based on their multiplicity of hydroxyl groups that hydrogen-bond to water. This holds water in the food, softens it, and keeps it from drying out. Tobacco is also kept moist by the addition of polyhydric alcohols. An added feature of polyhydric alcohols is their sweetness. Two particularly effective alcohols added to sweeten sugarless chewing gum are mannitol and sorbitol.

$$\underset{\text{D-}Sorbitol}{HOCH_2-\overset{\overset{\displaystyle OH}{|}}{\underset{\underset{\displaystyle H}{|}}{C}}-\overset{\overset{\displaystyle H}{|}}{\underset{\underset{\displaystyle OH}{|}}{C}}-\overset{\overset{\displaystyle OH}{|}}{\underset{\underset{\displaystyle H}{|}}{C}}-\overset{\overset{\displaystyle OH}{|}}{\underset{\underset{\displaystyle H}{|}}{C}}-CH_2OH}
\qquad
\underset{\text{D-}Mannitol}{HOCH_2-\overset{\overset{\displaystyle H}{|}}{\underset{\underset{\displaystyle OH}{|}}{C}}-\overset{\overset{\displaystyle H}{|}}{\underset{\underset{\displaystyle OH}{|}}{C}}-\overset{\overset{\displaystyle OH}{|}}{\underset{\underset{\displaystyle H}{|}}{C}}-\overset{\overset{\displaystyle OH}{|}}{\underset{\underset{\displaystyle H}{|}}{C}}-CH_2OH}$$

Compare the structures of these alcohols with the structure of glucose presented in Chapter 9. It is not surprising that the very similar structures of these polyhydric alcohols and the isomers of glucose produce a similar taste sensation.

COOKING AND PRECOOKING—
PRELIMINARY "DIGESTION"

The cooking process involves the partial depolymerization of proteins or carbohydrates by means of heat and hydrolysis (Figure 13–8). The polymers

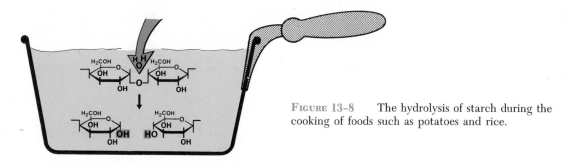

FIGURE 13-8   The hydrolysis of starch during the cooking of foods such as potatoes and rice.

which must be degraded if cooking is to be effective are the carbohydrate cellular wall materials in vegetables and the collagen or connective tissues in meats. Both types of polymers are subject to hydrolysis in hot water or moist heat. In either case, only partial depolymerization is required. A pan of glucose instead of baked bread or a skillet of amino acids in place of broiled steak is not the goal of cooking. However, the partial hydrolysis of starch releases some glucose, which gives the food a sweeter taste. The partial depolymerization into smaller fragments also breaks many of the bonds which must be broken before food can be digested. As a consequence, cooked food is easier to digest—that is, to break down into its monomer monosaccharide or amino acid units. These basic units were discussed in Chapter 9. Of course, overcooking can and does destroy some of the molecular structures that we need, such as the vitamins.

In recent years several precooking additives have become popular; the *meat tenderizers* are a good example. These are simply enzymes which catalyze the breaking of peptide bonds in proteins via hydrolysis at room temperature. As a consequence, the same degree of "cooking" can be obtained in a much shorter heating time. Meat tenderizers are usually plant products such as papain, a proteolytic (protein-splitting) enzyme from the unripe fruit of the papaw tree. Papain has considerable effect on connective tissue, mainly collagen and elastin, and shows some action on muscle fiber proteins. On the other hand, microbial protease enzymes (from bacteria, fungi, or both) have considerable action on muscle fibers. A typical formulation for the surface treatment of beef cuts contains 2 per cent commercial papain or 5 per cent fungal protease, 15 per cent dextrose, 2 per cent monosodium glutamate (MSG), and salt.

ENRICHMENT

Food additives also resupply foods with some of the vitamins and minerals that were lost in food processing. Rickets, pellagra and other deficiency diseases can be eliminated by the addition of nutrients to existing foods, even when the existing foods do not constitute a balanced diet by themselves. Enriched flour is an example; it often contains additions of thiamin, riboflavin, niacin, iron, calcium, and vitamin D.

QUESTIONS

1. Describe some of the chemical changes that occur during the cooking of a carbohydrate such as potato starch.

271

2. What causes fat in foods to become rancid? How can this be avoided?

3. What causes bread to rise?

4. What are the pros and cons of eating "natural" foods as opposed to foods containing chemical additives?

5. Why were cyclamates taken off the market?

6. Do you think it is wise to use animals in safety tests for drugs and food additives? Should mental patients and prisoners be used for this purpose?

7. Many consumer products are almost identical in chemical composition but are sold at widely different prices under different trade names. Do you think the products should be identified by their chemical names or their trade names? Why?

8. See what you can find out about the correlation between taste and smell. Are they the same sensation? Are they independent of each other?

9. A label on a brand of breakfast pastries contains the following additives: dextrose, glycerine, citric acid, potassium sorbate, Vitamin C, sodium iron pyrophosphate, and BHA. What is the purpose of each substance?

10. Choose a label from a food item and try to identify the purpose of each additive.

11. Bixin (annatto extract from the seeds of the tropical tree *Bixa orellana*) is added to food to give it a yellow color. What part of the structure is primarily responsible for the color? If white light is incident on the substance, why is the substance yellow?

Bixin

12. Suggest a way that citric acid complexes metals.

13. Using the structure of a protein given in Chapter 9, show how cooking can affect its structure.

14. What is a synergist? Give an example.

15. What part, if any, does hydrogen bonding play in the activities of surface active agents, humectant action of polyhydric alcohols, and stabilizing effect of gelatin in ice cream?

SUGGESTIONS FOR FURTHER READING

Amoore, J. E., Johnston, J. W., Jr., and Rubin, M., "The Stereochemical Theory of Odor," *Scientific American*, Vol. 210, No. 2, p. 42 (1964).

"Chinese Restaurant Syndrome," *Chemistry*, Vol. 42, No. 8, p. 4 (1969).

"Flavor of a Potato Chip," *Chemistry*, Vol. 43, No. 7, p. 2 (1970).

Furia, T. E. (ed.), "Handbook of Food Additives," The Chemical Rubber Co., Cleveland, 1968. (The GRAS chemicals are listed and discussed on pages 565 to 751.)

Giddings, J. C., and Monroe, M. B., editors; "Our Chemical Environment," Canfield Press, San Francisco, 1972.

Kermode, G. O., "Food Additives," *Scientific American*, Vol. 226, No. 3, p. 15 (1972).

"New Artificial Sweeteners," *Chemistry*, Vol. 43, No. 6, p. 23 (1970).

Pyke, M., "Man and Food," McGraw-Hill Book Co., New York, 1970.

Pirie, N. W., "Orthodox and Unorthodox Methods of Meeting World Food Needs," *Scientific American*, Vol. 216, No. 2, p. 27 (1970).

Sanders, H. J., "The Tasteless Condiment," *Chemistry*, Vol. 40, No. 1, p. 23 (1967).

14
MEDICINE

The average life expectancy in the United States has risen from age 49 in 1900 to 70 in 1970. It is expected to go to beyond the age of 79 by the year 2000. The major contributing factor is the widespread use of a large assortment of new medicinal compounds. Thanks to sulfa drugs, antibiotics, vaccines, and numerous other medicines, quarantine signs warning of scarlet fever have disappeared, polio epidemics no longer exist, and the scourge of smallpox is gone. Tuberculosis, high blood pressure, mental illness, pneumonia, and diabetes are but a few of the many other common afflictions that can now be treated successfully with new drugs.

The contents of the medicine cabinet have changed drastically in the past few years. As the parade of new drugs continues, the indispensable drugs of one decade frequently become obsolete in the next. A survey of physicians shortly before World War I revealed the ten most essential drugs (or drug groups) to be ether, opium and its derivatives, digitalis, diphtheria antitoxin, smallpox vaccine, mercury, alcohol, iodine, quinine, and iron. When another survey was made at the end of World War II, at the top of the list were sulfonamides, aspirin, antibiotics, blood plasma and its substitutes, anesthetics and opium derivatives, digitalis, antitoxins and vaccines, hormones, vitamins, and liver extract. Today there is such a wide array of medicinal chemicals that similar surveys have not been significant statistically, but drugs for (1) reducing fever, (2) relieving pain, and (3) fighting infection still head the list in all areas of medical practice.

TABLE 14-1   DRUG SALES IN THE UNITED STATES FOR SEVERAL IMPORTANT THERAPEUTIC GROUPS

THERAPEUTIC GROUP	1960[a]	1970[a]
Analgesics (pain reducers)	$ 65	$250
Antacids	41	95
Antibiotics	375	660
Cough and cold preparations	65	200
Hormones	141	405
Tranquilizers	140	490
Sulfonamides	44	50

[a]Manufacturers' sales, in millions of dollars.

Medicines are big business (Table 14–1), as anyone knows when he has to buy them. Still, the research required to put a new drug on the market is becoming increasingly more complex because of recent government regulations, and this alone is some justification for its price. In 1970, for example, drug producers in this country tested 3620 new drugs although that same year only 16 new products came onto the market.

In this chapter we are going to look at several of the therapeutic groups given in Table 14–1. The antacids are the first group we shall consider, and they are the simplest in chemical action.

NO ACID REBOUNDS, PLEASE!

The walls of a human stomach contain thousands of cells which secrete 0.1 M hydrochloric acid, the main purposes of which are to suppress growth of bacteria and to aid in the hydrolysis (digestion) of certain foodstuffs. Normally, the stomach's inner lining is not harmed by the presence of this hydrochloric acid, since the *mucosa*, the inner lining of the stomach, is replaced at the rate of about a half million cells per minute. When too much food is eaten the stomach often responds with an out pouring of acid which lowers the pH to a point where discomfort is felt, and the stomach actually begins to digest itself. Fortunately, the mucosa is protected by cells which contain a thick fatty layer.

When presented with this problem of minor stomach upset most people respond by taking one of the commonly available *antacids*. What are these compounds, and how do they work?

Antacids are compounds used to decrease the amount of hydrochloric acid in the stomach. The normal pH of the stomach ranges from 1.2 to 0.3. Various compounds can accomplish this by one or more of the following processes: neutralization, buffering, absorption, or retention in ion exchange resins. These

Antacid

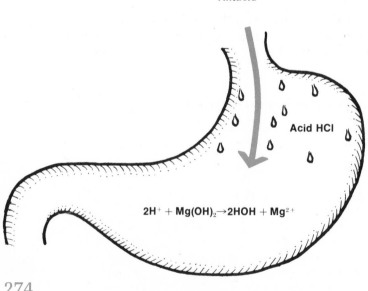

Acid HCl

$$2H^+ + Mg(OH)_2 \rightarrow 2HOH + Mg^{2+}$$

FIGURE 14-1   The chemical action of milk of magnesia. This antacid, which is magnesium hydroxide, neutralizes acid in the stomach. (Courtesy of Pfizer *Spectrum* in J.A.M.A.)

TABLE 14-2   The Chemistry of Some Antacids

Compound	Reaction in Stomach	Comments
Magnesium oxide, MgO	$MgO + 2H^+ \longrightarrow Mg^{2+} + H_2O$	MgO is white and tasteless
Milk of magnesia $Mg(OH)_2$ in water	$Mg(OH)_2 + 2H^+ \longrightarrow Mg^{2+} + 2H_2O$	The water suspension has an unpleasant chalky consistency
Calcium carbonate $CaCO_3$	$CaCO_3 + 2H^+ \longrightarrow Ca^{2+} + H_2O + CO_2$	Calcium carbonate is purified limestone
Sodium bicarbonate $NaHCO_3$	$NaHCO_3 + H^+ \longrightarrow Na^+ + H_2O + CO_2$	Baking soda, like $CaCO_3$, produces gas in the stomach
Aluminum hydroxide $Al(OH)_3$	$Al(OH)_3 + 3H^+ \longrightarrow Al^{3+} + 3H_2O$	$Al(OH)_3$ is a clear gel
Dihydroxyaluminum sodium carbonate $NaAl(OH)_2CO_3$	$NaAl(OH)_2CO_3 + 4H^+ \longrightarrow Na^+ + Al^{3+} + 3H_2O + CO_2$	Sold as Rolaids, will not ordinarily cause pH to go above 5
Sodium citrate $Na_3C_6H_5O_7 \cdot 2H_2O$	$Na_3C_6H_5O_7 + 3H^+ \longrightarrow 3Na^+ + H_3C_6H_5O_7 + 2H_2O$	Mild

compounds are widely used in the treatment of gastric hyperacidity (excess stomach acid) and peptic ulcers. Since they treat *symptoms*, and not the underlying causes, they should be used with some caution by laymen.

Some compounds used for this purpose and their modes of action are given in Table 14–2.

Looking at Table 14–2, we see that antacids are generally insoluble bases or salts with weakly basic anions. All of these bases are strong enough to neutralize a portion of the hydrogen ions in the stomach. The preferred antacids are those which do not reduce the stomach acidity very much. If the reduction of acidity is too great, the stomach responds by secreting an excess of acid. This is called "acid rebound."

PAIN-KILLERS CAN BE KILLERS

One of the largest and most important groups of therapeutics are the *analgesics* or pain killers. Most people have need of these compounds at one time or another. When we have a headache we take aspirin. When we have a tooth filled or extracted, the dentist uses Novocain. Intense suffering requires a strong pain-killer like codeine or morphine. While these compounds are immensely useful, they are nevertheless dangerous if taken or used improperly. Most of them can even become killers if taken in overdose.

Early man may well have used *opium*. Although not all opium derivatives have therapeutic value, most of them are very efficient pain-killers. Their chief disadvantages are their addictive properties.

Opium is obtained from the opium poppy by scarring it with a sharp instrument. From this scar flows a sticky mass which contains about 20 different compounds called *alkaloids* (organic, generally nitrogenous, bases). About 10 per cent of this mass is the alkaloid *morphine*, which is primarily responsible for opium's effects.

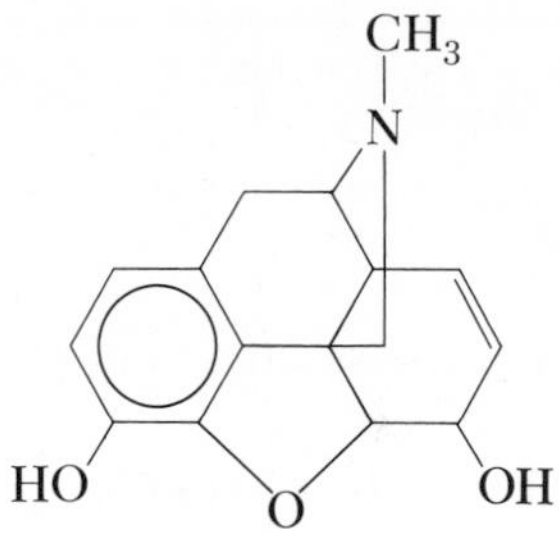

Morphine

Two derivatives of morphine are of interest. One of these is *codeine*, the methyl ether of morphine, which is less addicting than morphine and about as powerful an analgesic. The other compound is *heroin*, the diacetate ester of morphine. Heroin is much more addictive than morphine and for that reason finds no medical uses.

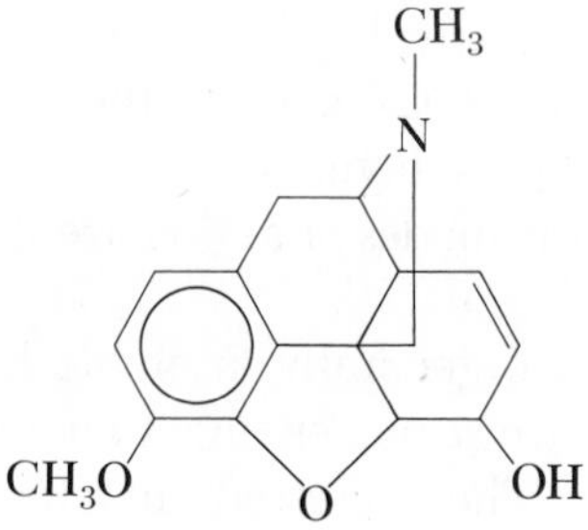

Codeine

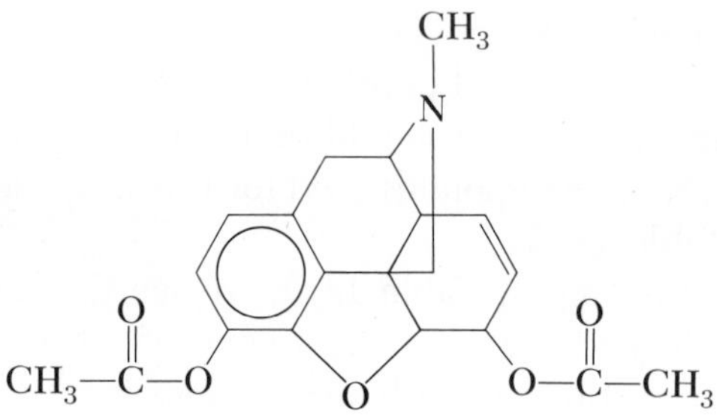

Heroin

The action of morphine, codeine, and other powerful analgesics is not completely understood. Over the years, numerous chemically similar drugs have been synthesized in trying to test various theories of the mechanisms of these drugs.

Many of the *local analgesics*, or local anesthetics, are nitrogen compounds, a similarity they share with the alkaloids. Local analgesics include the naturally occurring *cocaine*, derived from the leaves of the cocoa bush of South America, and the familiar Novocaine (procaine) (see Table 14–3). All of these drugs act by some mechanism of blockage of the nerves which transmit pain. This was discussed in Chapter 10 (see Figure 10–10). Acetylcholine appears to be the "opener" of the gate for sodium (Na^+) and potassium (K^+) ions to flow into a nerve cell (Figure 14–2). The positive ions neutralize the charge on the cell membrane and prepare the nerve cell for another impulse. The anesthetics have a nonpolar part of the molecule that fits into the nonpolar fatty tissue of the cell wall. In the process they constrict the pores, reduce the Na^+ flow, and nullify the effect of acetylcholine. Higher concentrations may also affect the senses of touch, heat, and cold.

There are times when milder, general analgesics are required, and few

TABLE 14-3   SOME LOCAL ANALGESICS

Cocaine		Addicting, probably the first used local analgesic
Procaine (Novocain)		Often used in dental work
Lidocaine (Xylocaine)		More potent than procaine, can be applied to the skin

compounds work as well for as many people as *aspirin*. The history of aspirin goes back to 1763, when Edward Stone noticed that the bark of the willow, when chewed, helped relieve the symptoms of malaria. In 1838 Raffaele Pivia isolated salicylic acid from the active ingredient in the willow. It turned out to be this compound which had the analgesic and antipyretic (fever reducing) properties attributed to the willow bark.

Salicylic acid

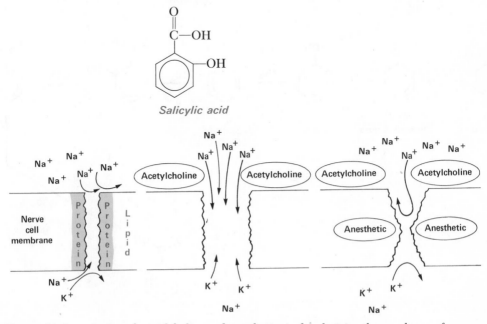

FIGURE 14-2   Action of acetylcholine and anesthetics in depolarizing the membrane of a nerve cell.

277

TABLE 14-4 PRODUCTION OF ASPIRIN IN THE UNITED STATES IN 1966

PRODUCER	LOCATION	CAPACITY (*million lbs./year*)
Monsanto Chemical Co.	Michigan	12.0
Dow Chemical Co.	Michigan	10.0
Sterling Drugs	New Jersey	10.0
Miles Laboratories	Michigan	2.0
Norwich Pharmaceutical Co.	Connecticut	2.0
Rexall Drugs	New Jersey	1.0

Attempts to administer salicylic acid and its salts to patients proved unsuccessful because of their disagreeable tastes. By 1893, Felix Hofmann, a chemist working for the Bayer Company in Germany, discovered a way to attach an acetyl group $(\text{—O}\overset{\text{O}}{\overset{\|}{\text{C}}}\text{CH}_3)$ in place of the hydroxyl hydrogen to make acetylsalicylic acid. The importance of the acetyl group is that it renders the molecule relatively tasteless and reduces the acidity enough so that it can be taken orally.

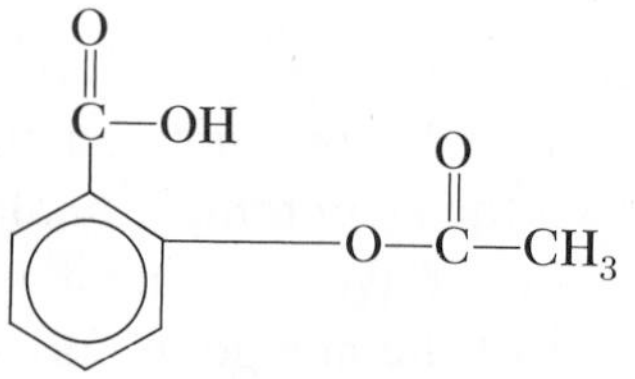

Acetylsalicylic acid (aspirin)

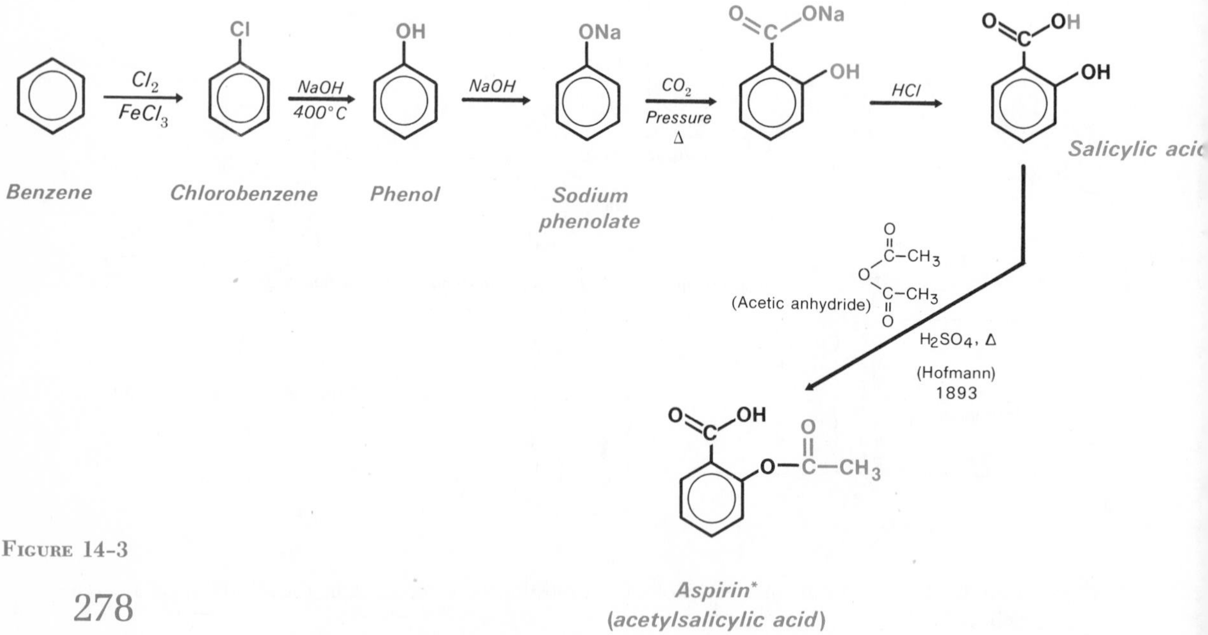
Benzene
Cl2
FeCl3
Chlorobenzene
NaOH
400°C
Phenol
NaOH
Sodium phenolate
CO2
Pressure
Δ
HCl
Salicylic acid
(Acetic anhydride)
H2SO4, Δ
(Hofmann)
1893
Aspirin*
(acetylsalicylic acid)

FIGURE 14-3

Today, aspirin is manufactured on a large scale (Table 14–4) using synthesis methods developed in the last century in Germany. About 30 million pounds are manufactured in the United States each year.

The synthesis steps required to produce aspirin are outlined in Figure 14–3. Each structure represents the beginning or end of a step. Other products, such as NaCl or water, are often formed, but are omitted for simplicity.

Aspirin Down the Hatch

Taking an aspirin tablet appears to be a simple act; yet much research goes into making a tablet which will cause a minimum of bad side-effects for the user. The greatest danger presented by aspirin is that of stomach bleeding, caused when an undissolved aspirin tablet lies on the stomach wall. As the aspirin molecules pass through the fatty layer of the mucosa, they appear to injure the cells, causing small hemorrhages. The blood loss for most individuals taking two five-grain tablets is between 0.5 ml and 2 ml. Some people are more susceptible. Early aspirin tablets were not particularly fast-dissolving, which aggravated this problem greatly. Today, aspirin tablets are formulated to disintegrate and dissolve quickly, although dissolving the tablet in a little water might not be a bad idea.

Aspirin and Headaches

Americans spend more than $250 million a year on headache remedies. More than 200 kinds of tablets and powders to relieve the problem are on the market. According to the National Health Service, 1 out of every 12 Americans has severe headaches regularly.

The basic ingredient of most headache formulations is aspirin. Commercially available headache remedies include caffeine, antacids, extra pain-killers, antihistamines, vitamins, and tranquilizers.

Headaches are triggered by emotional problems (tension), heredity (migraine), and, less frequently, by eye strain, acute sinus conditions, inflammation of the lining of the brain, infection of a cranial nerve, carbon monoxide poisoning, and poorly positioned teeth.

In tension headaches, muscles are tightened and strained. An overworked head muscle can ache just as much as an overworked arm or leg muscle. The tight muscle may also squeeze arteries and reduce blood flow through the muscle, adding to the pain (Figure 14–4).

Experimentation has established that aspirin produces its effect on the central nervous system, and that salicylic acid from the hydrolysis of aspirin is the active chemical. Besides relieving pain, the salicylic acid is an antipyretic (fever reducer). The hypothalamus gland, attached to the pituitary gland near the center of the brain, is the thermostat of the body. Fever is thought to be produced by a chemical secreted by white blood cells when they engulf bacteria. The chemical enters lymph vessels, migrates to the brain, and resets the thermo-

279

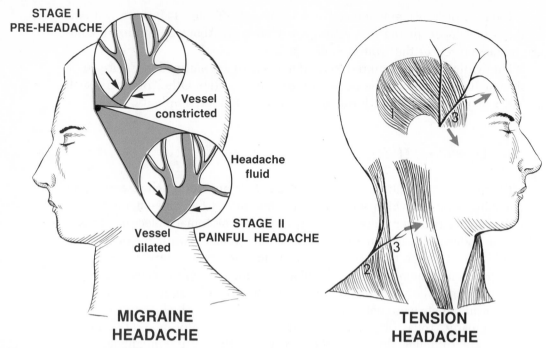

STAGE I
PRE-HEADACHE

Vessel
constricted

Headache
fluid

Vessel
dilated

STAGE II
PAINFUL HEADACHE

**MIGRAINE
HEADACHE**

**TENSION
HEADACHE**

FIGURE 14-4   The physiological action of a headache. In a tension headache the muscles constrict the arteries. The muscles tire and ache, and this causes pain which spreads throughout the head. In a migraine headache the arteries constrict (this is painless) and then expand (this is painful). (In McGraw-Hill Yearbook of Science—Adapted by permission from The New York Times.)

stat to raise the body temperature. Aspirin in some way adjusts the body temperature back to normal.

The pain of a migraine headache appears to be caused by a headache "fluid" and a constriction of the arteries in the head followed by expansion. Each time the heart pumps, the arteries expand further and more pain from the pulsing is produced. The headache fluid contains two small proteins, bradykinin and

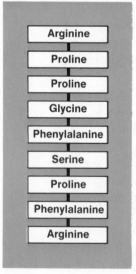

Arginine

Proline

Proline

Glycine

Phenylalanine

Serine

Proline

Phenylalanine

Arginine

FIGURE 14-5   The amino acid sequence in bradykinin, a pain-causing protein. The structures of the amino acids are given in Table 9–1.

neurokinin, which are believed to make nerves sensitive to pain (Figure 14–5). When the substances are extracted from the headache fluid and injected elsewhere in the body, the person will sense pain in the new area, according to Dr. Arnold P. Friedman, physician-in-chief at the Montefiore Headache Unit in New York and a recent chairman of the World Commission for the Study of Headache. The constriction and subsequent expansion of the arteries may be initiated by the release of serotonin, which is known to contract smooth muscle. Normally the serotonin is bound, but its sudden release into the bloodstream could cause the initial contraction of the arteries. An enzyme, monoamine oxidase, metabolizes the serotonin, causing the arteries to relax and expand. When the normal supply of serotonin is depleted, the arteries remain expanded until a new supply is made available. Australian scientists reported in 1967 that the blood content of serotonin does fall sharply at the onset of a migraine attack.

The migraine headache can be relieved by ergotamine tartrate if it is taken early enough. It acts by constricting the muscles of the blood vessels, preventing painful stretching. This property of ergotamine tartrate has been known for over 40 years. A newer drug, methysergide, is effective in about 70 per cent of patients. It is taken between headaches. Neither drug is always effective; both are sometimes associated with serious side-effects when used for long periods.

The hereditary nature of migraine headaches is indicated by a study made by Dr. Adrian M. Ostfield of the University of Illinois. Dr. Ostfield reports that there is a 70 per cent chance that the child will have migraine if both parents have migraine. If one parent has migraine, there is a 45 per cent chance; if neither parent has migraine but there is a history of it in the family, a 25 per cent chance.

THE SULFA DRUGS

Sulfa drugs represent a group of compounds discovered in a conscious search for materials which could control bacterial infections in living animals. In 1904, the German chemist Paul Ehrlich (1854–1915; Nobel prize in 1908) realized that infectious diseases could be conquered if toxic chemicals could be found that attacked parasitic organisms within the body to a greater extent than they did host cells. Ehrlich achieved some success toward his goal; he found that certain dyes which were used to stain bacteria for microscopic examination could also kill the bacteria. This led to the use of dyes against African sleeping sickness and arsenic compounds against syphilis. The mass outbreak of typhus during World War I, the loss of many wounded due to secondary bacterial infection, and the great influenza epidemic of 1917–1918 prepared the medical world for discoveries that might eradicate infectious diseases.

After experimenting with several drugs, Gerhard Domagk, a pathologist in the I. G. Farbenindustrie Laboratories in Germany, found, in 1935, that prontosil, a coloring matter or dye, was active against bacterial infection in mice. Prontosil as such is not effective in killing bacteria, but it is reduced to sulfanilamide, which is effective.

281

$$H_2N-\!\!\bigcirc\!\!-N\!\!=\!\!N-\!\!\bigcirc\!\!-SO_2NH_2 \xrightarrow{H_2O} H_2N-\!\!\bigcirc\!\!-SO_2-NH_2$$
$$\overset{|}{NH_2}$$

Prontosil   *Sulfanilamide*

This discovery led to the synthesis and testing of a large number of related compounds in the search for drugs which were more effective or less toxic to the infected animal. By 1964, more than 5000 sulfa drugs had been prepared and tested. Some of the more effective ones are listed in Table 14–5.

Sulfa drugs inhibit bacteria by preventing the synthesis of folic acid, the vitamin essential to their growth. The drugs' ability to do this apparently lies in their structural similarity to a key ingredient in the folic acid synthesis, para-aminobenzoic acid.

Typical sulfa drug   *p-Aminobenzoic acid*

The close structural similarity of sulfanilamide and p-aminobenzoic acid permits sulfanilamide to be incorporated into the enzymatic reaction sequence instead of p-aminobenzoic acid. By bonding tightly, sulfanilamide shuts off the production of the essential folic acid, and the bacteria die of vitamin deficiency. Not all bacteria are susceptible to sulfa drugs. However, the drugs are effective on streptococci, staphylococci, many gram-negative and gram-positive bacteria, and protozoa such as coccidia. In man and the higher animals, p-aminobenzoic acid is not necessary for folic acid synthesis, and so sulfa drugs have no effect on this mechanism. But sulfa drugs must be buffered (e.g., $NaHCO_3$) to prevent digestive upset, and they attack and pass through several kinds of protein linings. They also are suspected of modifying the base-sequence of nucleic acids (by incorporation) and acting as potential mutagens (Chapter 21).

TABLE 14–5   SOME OF THE MORE EFFECTIVE SULFA DRUGS

NAME	FORMULA
Sulfapyridine	$H_2N-\!\!\bigcirc\!\!-SO_2-N$ (pyridine ring)
Sulfathiazole	$H_2N-\!\!\bigcirc\!\!-SO_2-N$ (thiazole ring)
Sulfisoxazole	$H_2N-\!\!\bigcirc\!\!-SO_2-N$ (isoxazole ring with CH_3 groups)

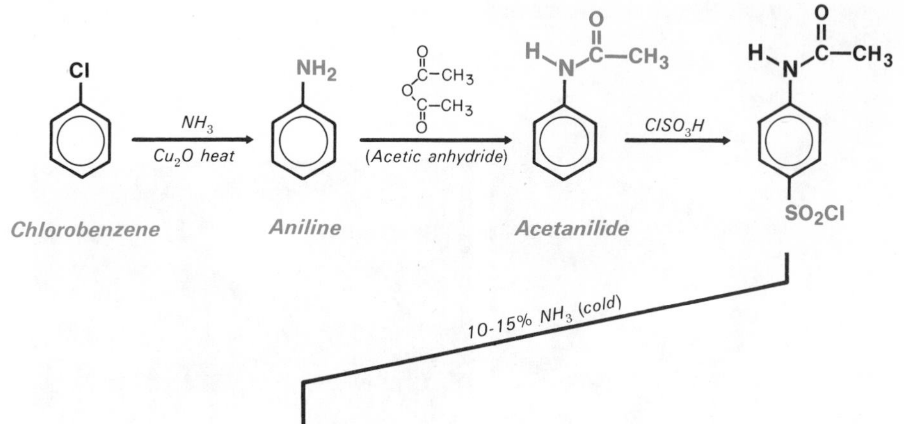

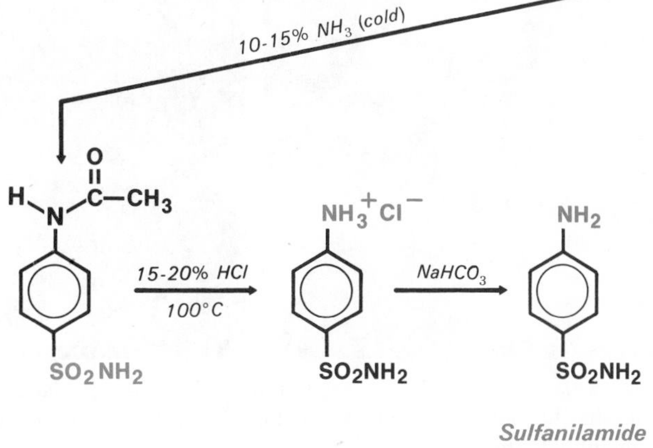

FIGURE 14-6   Outline of the synthesis of sulfanilamide from chlorobenzene. NH$_3$ is ammonia, Cu$_2$O is cuprous oxide, HCl is hydrochloric acid, NaOH is sodium hydroxide, and ClSO$_3$H is chlorosulfonic acid.

Like aspirin, sulfanilamide and many of the sulfa drugs can be prepared from chlorobenzene. Figure 14–6 outlines this synthesis, which is often carried out in the college chemistry laboratory. Great care must be exercised in handling the chlorosulfonic acid, ClSO$_3$H, since it is very corrosive and can cause severe burns. The synthesis employs an often-used technique in organic synthesis. The amine group in aniline is first "blocked" with an acetyl group prior to adding the chlorosulfonic acid. Blocking of the —NH$_2$ group is necessary to prevent its reaction with chlorosulfonic acid. With the NH$_2$ blocked, the acid reacts at the para position of the ring, substituting a —SO$_2$Cl group for a hydrogen; the hydrogen reacts with an OH of the chlorosulfonic acid to form a molecule of water as a coproduct. The technique of blocking one group while another is attacked is necessary in many organic syntheses.

ANTIBIOTICS

In our own time, the quest for drugs to wipe out disease has virtually been fulfilled by the antibiotics. Since these drugs are so efficient, they were the first of what came to be called "miracle" drugs. Their job generally is to aid the white blood cells by stopping bacteria from multiplying. When a person is sick or is killed by a disease, it means that the bacteria have multiplied faster than the white blood cells could devour them, and that the bacterial toxins increased more rapidly than the antibodies could neutralize them. The action of the white blood cells and antibodies plus an antibiotic is generally enough to repulse an attack of the germs.

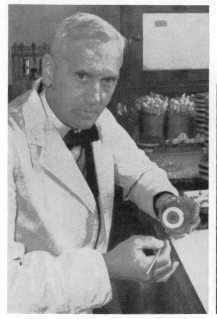

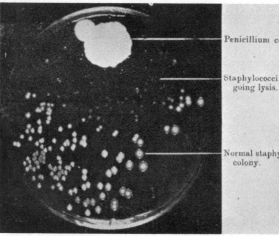

Penicillium colony

Staphylococci under-
going lysis.

Normal staphylococcal
colony.

FIGURE 14-7   A. Sir Alexander Fleming B. A culture-plate showing the dissolution of staphylococcal colonies in the neighborhood of a colony of penicillium. (From article by Fleming, British Journal of Experimental Pathology, June, 1929.)

The first of the antibiotics was penicillin, and the story of its discovery is almost as fantastic as its antibacterial action. In 1928, Alexander Fleming, a bacteriologist at the University of London, was working with cultures of *Staphylococcus aureus,* a germ that causes boils and some other types of infections. In order to examine the cultures with a microscope, he had to remove the cover of the culture plate for a while. One day as he started work he noticed that the culture was contaminated by a blue-green mold. Many people would have discarded the contaminated culture, but Fleming's trained eye had noticed something else.

For some distance around the mold growth, the bacteria colonies were being destroyed. Upon further investigation, Fleming found that the broth in which the mold was grown had an inhibitory or lethal effect against many pathogenic organisms. The mold was later identified as *Penicillium notatum* (the spores sprout and branch out in pencil shapes, hence the name).

Penicillin, the name given to the antibacterial substance produced by the mold, apparently had no toxic effect on animal cells, and its activity was selective. Because Fleming's extracts from the mold were crude, clinical results were discouraging, and his brilliant discovery made little impact on the medical world for almost a decade. Eventually, as a result of the interest and further research by Howard Flory and Ernst Chain, penicillin was developed as a practical drug. In 1941, penicillin was used for the first time on a human being, a London policeman who was hospitalized with a serious case of blood poisoning contracted from a shaving cut. Since he could not recover by normal means, the doctors decided to try the new drug. The effect was immediately favorable. Because of wartime needs, a large supply of penicillin was urgently needed.

Through cooperation between American and British firms, the supply was provided and thousands of lives and limbs were saved.

Fleming, Flory, and Chain shared a richly deserved Nobel Prize in medicine and physiology in 1945.

The structure of penicillin has now been determined.

Penicillin G

Many different penicillins exist, differing in the structure of the R group. Penicillin G is the most widely used in medicine.

The development of a series of antibiotics followed soon after the success of penicillin. In 1937, following collaboration with René Dubos, Selman Waksman isolated a compound from a soil organism, *Streptomyces griseus*, which came to be known as *streptomycin* and was released to physicians in 1947. This compound (Figure 14–8) was quite successful in controlling certain types of bacteria but later had to be withdrawn, owing to its bad side-effects.

In 1945, B. M. Duggar discovered that a gold-colored soil fungus, *Streptomyces aureofaciens*, produced a new type of antibiotic, *aureomycin*, the first of the *tetracyclines*. Research then went at a fever pitch. Pfizer laboratories tested 116,000 different soil samples before they discovered the next antibiotic, which they named *Terramycin*.

Today, the use of antibiotics exceeds that of the sulfa drugs, with penicillins being the most often used.

The fact that antibiotics generally destroy bacterial cells and not human cells is the basis of their success. The action of streptomycin was discussed in the section on antimicrobial action of food additives. Its blocking action of the ribosome surface of bacteria is typical of the specificity of the antibiotics. In each case in which the mode of action is known, the antibiotic interferes with some specific metabolic process in specific bacteria.

Several antibiotics, such as penicillin and bacitracin, are known to prevent cell-wall synthesis. The cell wall of some pathogenic bacteria is composed of mucoprotein. Mucoprotein is a combination of proteins and mucopolysaccha-

285

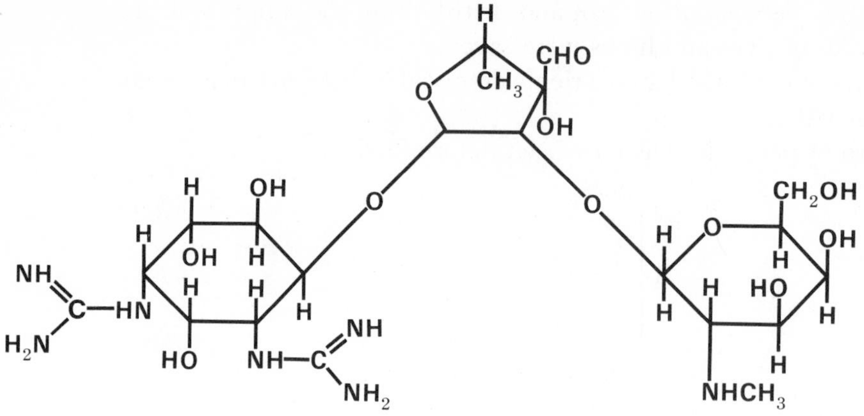

Tetracyclines
Chlortetracycline
("Aureomycin")

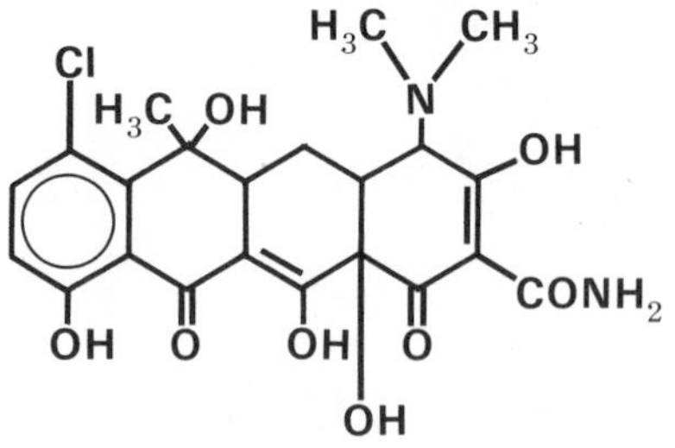

Oxytetracycline
("Terramycin")

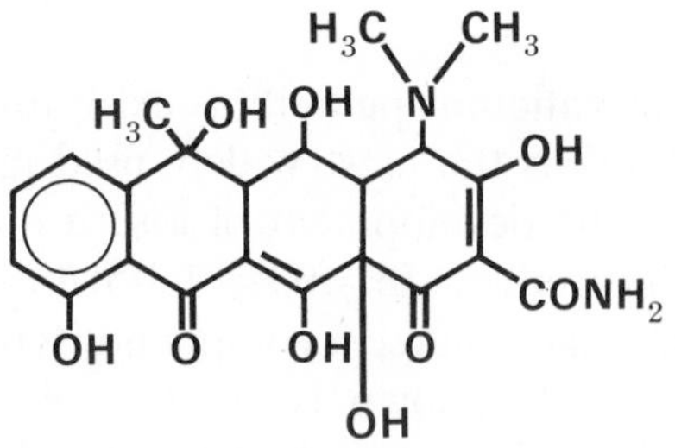

FIGURE 14-8

rides, in which some of the monosaccharide units (usually glucose or galactose) contain a —$NHCOCH_3$ group substituted for a hydroxyl group (—OH). Only bacterial cells have mucopeptide walls.

Penicillin interferes with the synthesis of the mucoprotein cell wall of the bacteria by interfering with the formation of cross links between layers of the wall. The cell wall protects and supports the delicate cell components enclosed within it. The cytoplasmic membrane, immediately inside the cell wall, regulates the flow of nutrients and water in and out of the cell. The layers are reinforced by a series of chemical crosslinks connecting one layer to another. When the crosslinks are not formed properly, the weakened wall, unable to hold its size and shape, expands as water comes into the cell by osmosis. Eventually the cytoplasmic membrane bursts, causing the cell to die.

Of several thousand antibiotics that have been prepared or discovered, only a very few have been found to be relatively nontoxic. Even penicillin is toxic to some people. Streptomycin, which was on the market for more than ten years, has been removed because its toxicity is greater than some equally effective, newer antibiotics. The search continues for antibiotics that are less toxic, yet

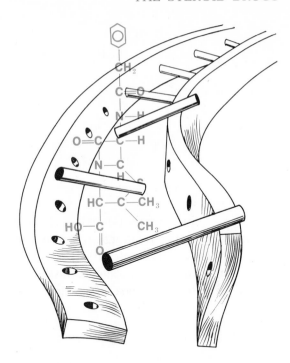

FIGURE 14-9 Penicillin kills bacteria by interfering with the formation of crosslinks in the cell wall.

more effective on more kinds of bacteria. The success of antibiotics in obliterating disease is already renowned. The antibiotics are truly "miracle drugs."

THE STEROID DRUGS

A large and important class of naturally occurring compounds have the tetracyclic structure given below.

These compounds are known as *steroids*, and they occur in all plants and animals. The most abundant animal steroid is *cholesterol*, $C_{27}H_{46}O$. The human body does not synthesize cholesterol but readily absorbs it through the intestines from food. Its association with gallstones and hardening of the arteries was discussed in Chapter 10.

Biochemical alteration or degradation of cholesterol leads to many steroids of great importance in human biochemistry. When *cortisone*, one of the adrenal cortex hormones, is applied topically or injected into a diseased joint, it acts as an anti-inflamatory agent and is of great use in treating arthritis.

287

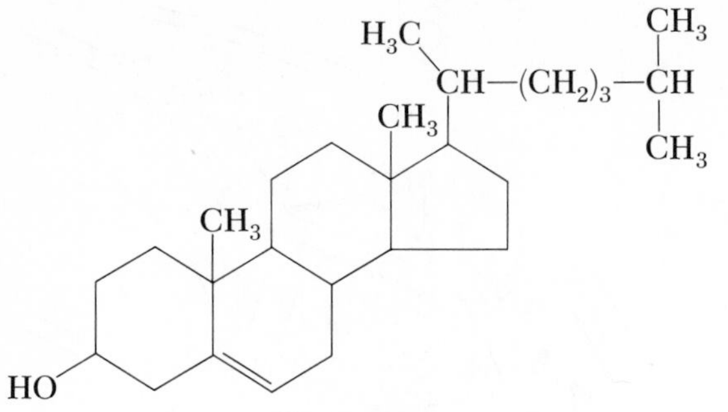

Cholesterol

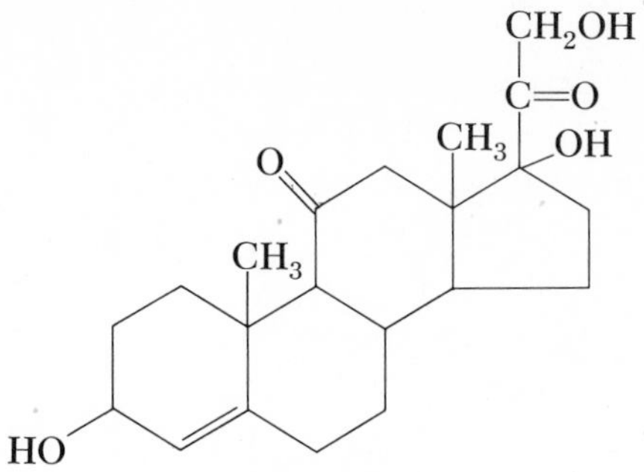

Cortisone

Structurally related to cholesterol and cortisone are the so-called sex hormones. The relationship between gene activation and the female hormones was discussed in Chapter 10. The female sex hormone, *progesterone,* can be converted into the male hormone, *testosterone,* remarkably, by way of a biochemical oxidative decarboxylation (loss of CO).

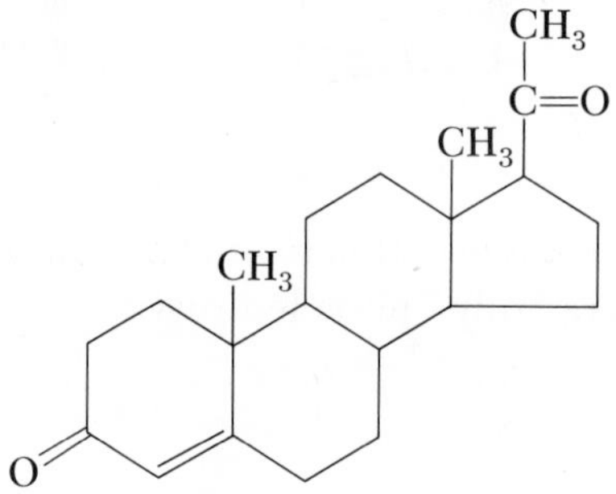

Progesterone

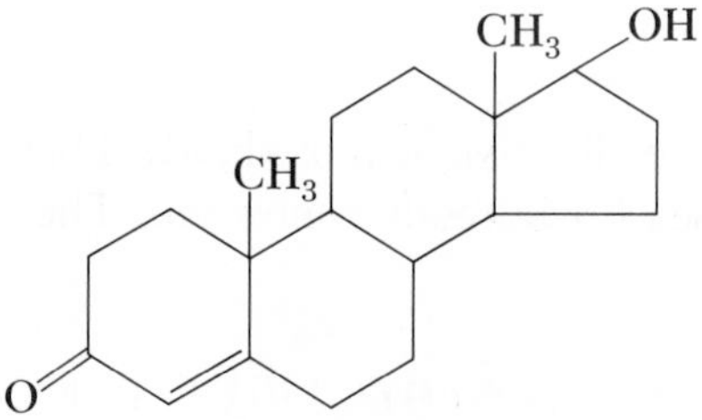

Testosterone

The other female hormones are estradiol and estrone, called *estrogens.* The estrogens differ from the other steroids discussed earlier in that they contain an aromatic A ring.

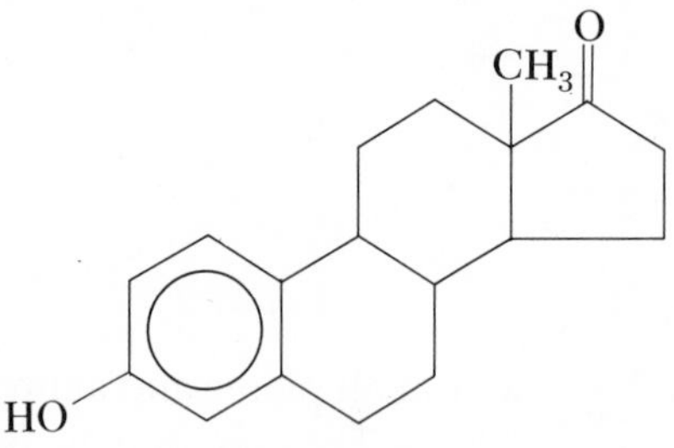

Estrone

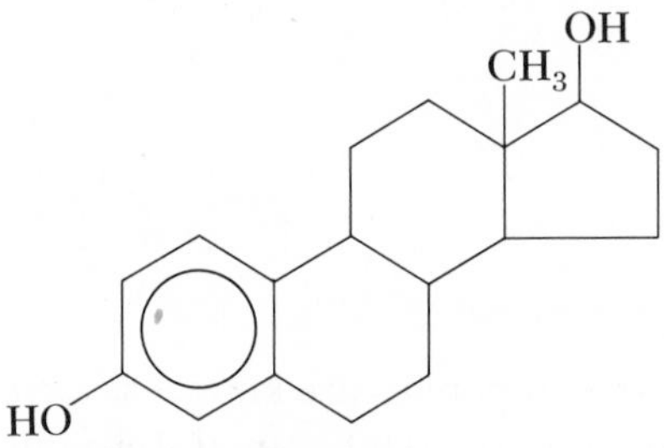

Estradiol

BIRTH CONTROL PILLS

One of the most revolutionary medical developments of the 1960's was the worldwide introduction and use of The Pill. An estimated 8.5 million women in the United States use birth control pills. The basic feature of oral contraceptives for women is their chemical ability to simulate the hormonal processes resulting from pregnancy and, in so doing, prevent ovulation. Ovulation, the production of eggs by the ovary, ceases at the onset of pregnancy because of hormonal changes. (See Chapter 10.) This same result can be produced by the administration of a variety of steroids, some of which are effective when taken orally, although the mechanism of their action and their long-term effects are not known in detail.

The active ingredients of The Pill are the hormones progesterone and estrogen, or their derivatives. Enovid, a product of this sort made by G. D. Searle and Co., is a mixture of Norethindrone and Mestranol. Notice the structural similarities between these compounds and progesterone and estrogen:

Norethindrone
(*major constituent*)

Mestranol
(*minor constituent*)

Do these pills cause cancer? After the first 10 years of usage, there was no conclusive evidence that oral contraceptives are carcinogenic. Nor was there evidence that they cause diabetes, sterility, eye disorders, mental illness, or any of a number of other diseases to which they have been linked by critics. Since the latency period for cancer is thought to range from 10 to 20 years, it seems probable that if oral contraceptives do cause cancer, this will become evident during the 1970's.

There is, however, one serious disorder that has been linked with The Pill: thromboembolic (blood clotting) disease. Such clots are potentially lethal. If they block a major blood vessel in a limb, amputation may be necessary; if they block a vessel to the lungs or brain, death may result. Studies done in Great Britain and the United States indicate that blood clotting is the cause of about 3 deaths per 100,000 users each year. Statistics indicated that this risk was considerably less than the risk of thromboembolism that would accompany the number of pregnancies averted. However, because there is a danger, all women who use The Pill should have medical checkups at least once every 6 months.

Perhaps the major problem facing our world is the population explosion. As the human population doubles every 40 years and associated problems intensify, wider efforts are being made to study procedures for controlling human fertility. Antifertility drugs for males are under active study, and there is every reason to believe that these will be on the market in the near future. Some of the first drugs of this sort to be studied act by temporarily suppressing the formation of sperm cells in the male.

289

DRUGS IN COMBINATION

Drugs, like some food additives, can have enhanced effects when placed in certain chemical environments. Sometimes the effects are harmful; sometimes helpful. Take the case of an aging business executive who took an antidepressant and then ate a meal that included aged cheese and wine. The antidepressant is an inhibitor of monoamine oxidase, an enzyme that helps to control blood pressure. Both the aged cheese he ate and the wine he drank contained pressor amines, which raise blood pressure. Without the controlling effect of the enzyme, these amines skyrocketed his blood pressure and caused a stroke. Neither the amines nor the antidepressant alone would likely cause the stroke, but the combination did.

Likewise, people who take digitalis both for heart trouble and to cut down on the sodium level should take aspirin only under medical supervision. Aspirin can cause a 50 per cent reduction in salt excretion for three or four hours after it is taken.

Alcohol increases the action of many antihistamines, tranquilizers, and drugs such as reserpine (for lowering blood pressure) and scopolamine (contained in many over-the-counter nerve and sleeping preparations), making such combinations extremely dangerous. Staying away from dangerous alcohol-drug combinations is not as easy as it may seem. Many people fail to realize that a large number of over-the-counter preparations—such as liquid cough syrup and tonics—contain appreciable amounts of alcohol.

The alcohol molecule can easily pass through the mucosa lining of the stomach. Experiments have shown that aspirin can more easily pass through the mucosa in the presence of alcohol. Therefore it would appear cocktails and aspirin would be a bad mixture since the passage of aspirin through the stomach walls too quickly can cause excessive bleeding.

Not all drug combinations are bad. Doctors have been highly successful in prolonging the lives of leukemia and other cancer victims with combinations of drugs that individually could not do the job. Resistant kidney disease has also responded to drug combinations in cases in which single drugs were ineffective.

Perhaps the best advice is to take drugs only when you are sick, making sure that your doctor knows what you are taking and approves of it.

QUESTIONS

1. What weight of MgO is required to neutralize all the acid in a stomach which contains 0.5 mole of hydrochloric acid?

2. Calculate the weight of $Al(OH)_3$ required to neutralize the acid in Question 1.

3. Assume a 60% overall yield and calculate the amount of benzene necessary to produce 12 million pounds of aspirin.

4. What are some dangers inherent in all analgesics? Give one specific example for morphine, procaine, and aspirin.

5. Someone asks you how aspirin works to relieve a headache. What would be your answer?

6. Briefly, how do sulfa drugs act to kill bacteria? Why do they not harm humans?

7. What is the mechanism by which penicillin kills bacteria?

8. Describe the methods for testing drugs and medicines for safety.

9. How is the action of antibiotics different from the action of sulfa drugs in limiting bacterial infection?

10. Many drugs are almost identical in chemical composition but are sold at widely different prices under different trade names. What are some of the reasons for this? Do you believe drugs should be sold under *generic* names only?

11. There are several examples in which the medicine is effective because it mimics a chemical required by a germ. Give two such examples.

12. Draw the steroid structure and list four compounds that have this as a basic part of their structure.

13. In the synthesis of aspirin,

    a. what by-product is formed along with chlorobenzene?
    b. identify the groups on salicylic acid.
    c. which hydrogen atom is the more acidic (more easily ionizable) on salicylic acid?

14. Write a chemical equation for the hydrolysis of aspirin to salicylic acid.

SUGGESTIONS FOR FURTHER READING ▰▰▰▰▰▰▰▰▰▰▰▰▰

Collier, H. O., "Aspirin", *Scientific American*, Vol. 209, No. 5, p. 97 (1963).
Davenport, H. M., "Why the Stomach Does Not Digest Itself," *Scientific American*, Vol. 226, No. 1, p. 86 (1972).
"How Penicillin Kills Bacteria," *Chemistry*, Vol. 41, No. 7, p. 44 (1968).
Marks, G. and Beatty, W., "The Chemical Garden," Charles Scribner's Sons, New York, 1971.
Sates, M., "Analgesic Drugs," *Scientific American*, Vol. 215, No. 5, p. 131 (1966).
Shideman, F. E., "R_x . . . Take as Directed," The Chemical Rubber Co., Cleveland, Ohio, 1967.
Solmssen, U. V., "The Chemist and New Drugs," *Chemistry*, Vol. 40, No. 4, p. 22 (1967).

15

THE CHEMISTRY OF CLEANING (OR, GETTING THE DIRT OUT)

Dirt has been defined as matter in the wrong place. Tomato catsup is esteemed as a palatable and nutritious food, but on your shirt it is dirt. There are a large number of cleansing, or surface-active, agents capable of removing the dirt without harm to the shirt. Indeed, radio and television advertising might lead us to believe that the soaps and detergents we have today are unique and vastly superior to the products of a year or a century ago. This is not always so. Soap, for example, has always been made by a time-tested recipe that dates back at least to the second century of the Christian era. Galen, the great Greek physician, mentions that soap was made from fat, ash lye, and lime. Moreover, Galen stated that soap not only served as a medicament but also removed dirt from the body and clothes.

What *is* new is the greater purity of soap, the improvement in its cleaning action by numerous additives, and the advent of the relatively new synthetic detergents. The soap-making industry was revolutionized by two events. The first was the discovery of the process of making soda ash (Na_2CO_3; its water-softening and alkaline properties will be discussed later) from ordinary salt (NaCl) by Nicolas LeBlanc in 1791, and the second was the epoch-making work of the celebrated French chemist Chevreul, whose researches into the chemical constitution of the natural fats extended from 1813 to 1823. As a result of these two advances, the soap-makers of the 19th century were provided with ample quantities of sodium carbonate at a reasonable price and were armed with the knowledge of the true nature of fats and fatty acids. Accordingly, they made rapid progress; their products improved steadily and grew in diversity, until little by little the soap industry attained its present gigantic proportions.

GRANDMA'S LYE SOAP

Fundamental soap chemistry is fairly simple. A strong base is first needed. This can be furnished by extracting wood ashes with a small amount of water, as in Grandma's time. The solution known as "lye" contained KOH, Na_2CO_3, and K_2CO_3. Today sodium hydroxide (NaOH) is also known as *lye*. In aqueous solution, the basic carbonates produce hydroxide (OH⁻) ions.

$$CO_3^{2-} + H_2O \rightleftharpoons HCO_3^- + OH^-$$

The other fundamental ingredient in soap-making is a suitable *fat,* an ester (Chapter 7) of a fatty acid and glycerin. The actual soap-making process is one in which the component parts of the fat molecule are separated by reaction with a strong base. The hydroxide (OH⁻) ions play a key role in the process, which is called *saponification.*

Fatty acid ions   Glycerin

The R-groups may vary as will be discussed later.

Grandma made soap by saponification of animal fats (tallow or lard) mixed with the extract of ashes of hard wood in a heated kettle. Since she cared little for the chemistry involved and wanted no grease in her soap, she often added too much lye. The resulting soap was fairly harsh, to say the least, since the excess lye was often retained by the soap.

The actual soap is the potassium or sodium salt of the fatty acid, and it precipitates from the solution by "salting out." (See Figure 15–1.)

The typical lard used in Grandma's lye soap contained mostly stearic acid units, although other fatty acids were present as well.

$$CH_3-(CH_2)_{16}-COOH$$
Stearic acid

Modern soaps make use of other fatty acids. Some render the soap soft to the touch, while others render the soap more water-soluble with high lather ability. Several sources of fats, and the soaps they produce, are shown in Table 15–1. The length and number of double bonds in the fatty acid chain influence the solubility of a soap. The major disadvantage of soaps is their reactivity toward double positive metal ions (M^{2+}) found in most water. More will be said about this later.

293

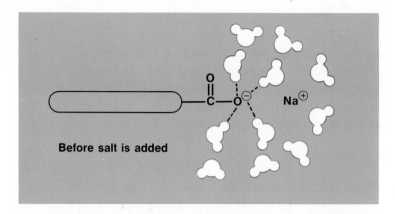

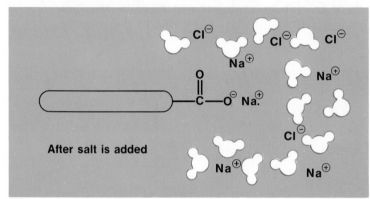

FIGURE 15-1 Salting out (shown diagrammatically). Without salt, there is sufficient water of hydration to keep the soap dissolved. When the salt is added, the equilibrium between the soap anions and the positive ions is shifted toward more neutral soap molecules, which decreases the hydration and renders the soap less soluble.

HOW SOAP CLEANS

The cleansing action of a soap can be explained in terms of its molecular structure. Material that is water-soluble can be readily removed from the skin or other surface simply by washing with an excess of water. To remove a sticky sugar syrup from one's hands, the sugar is dissolved in water and rinsed away. Many times the material to be removed is oily and water will merely run over the surface of the oil. Since the skin has natural oils, even substances, such as ordinary dirt, which are not oily themselves, can cover the skin in a greasy layer

TABLE 15-1 SOME FATTY ACIDS AND THEIR SOURCES

SOURCE	PRINCIPAL FATTY ACID	PROPERTIES OR USES OF THE SOAP PRODUCED
Animal fat, beef and mutton	CH_3—$(CH_2)_{16}$—COOH *stearic acid*	Hard soap; usually mixed with other soaps
Coconut oil	CH_3—$(CH_2)_{10}$—COOH *lauric acid*	Very water-soluble, good lather
Palm oil, Olive oil	CH_3—$(CH_2)_7$—CH=CH—$(CH_2)_7$—COOH *oleic acid*	Used in toilet soaps
Cottonseed oil	CH_3—$(CH_2)_4$—CH=CH—CH_2—CH=CH—CH_2—$(CH_2)_6$—COOH *linoleic acid*	Cheap source of a fatty acid

FIGURE 15-2   The cleansing action of soap. (*a*) A piece of glass coated with grease inserted in water gives evidence for the strong adhesion between water and glass at 1, 2 and 3. The water curves up against the pull of gravity to wet the glass. The relatively weak adhesion between oil and water is indicated at 4 by the curvature of the water away from the grease against the force tending to level the water. (*b*) A soap molecule, having oil soluble and water soluble ends, will orient at an oil-water interface such that the hydrocarbon chain is in the oil (with molecules that are electrically similar-nonpolar) and the COO^-Na^+ group is in the water (highly charged polar groups interacting electrically). (*c*) In an idealized molecular view, a grease particle, 1, is surrounded by soap molecules which in turn are strongly attracted to the water. At 2 another droplet is about to break away. At 3 the grease and clean glass interact before the water moves between.

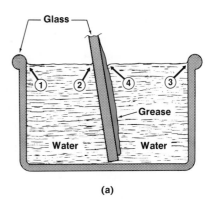

(a)

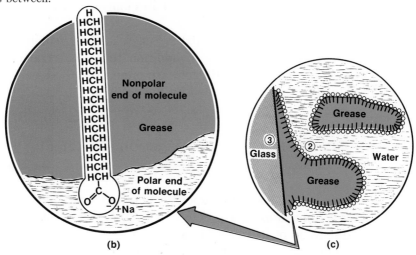

(b)   (c)

held by the skin oils. The cohesive forces (forces between like molecules tending to hold them together) within the water layer are too large to allow the oil and water to intermingle (Figure 15–2). When present in an oil-water system, a soap such as sodium stearate,

$$CH_3CH_2CH_2CH_2CH_2CH_2CH_2CH_2CH_2CH_2CH_2CH_2CH_2CH_2CH_2CH_2CH_2C \begin{matrix} O \\ \diagdown \\ O^-Na^+ \end{matrix}$$

will move to the interface between the two liquids. The carbon chain, which is nonpolar, will readily mix with the nonpolar grease molecules, whereas the highly polar —COO^-Na^+ group enters the water layer. The soap molecule will then tend to lie across the oil-water interface. The grease is broken up into small droplets, each surrounded by hydrated soap molecules. The surrounded oil droplets cannot come together again since the exterior of each is covered with —COO^-Na^+ groups which strongly interact with the surrounding water. If enough soap and water are available, the oil will be swept away as an emulsion, leaving a clean, wettable surface. This is what is meant by "detergent" action. A fat-derived soap is one kind of detergent.

TYPES OF SOAP

Soaps come in a variety of forms. Toilet soaps, for example, often contain much of the glycerin released in saponification and little of the free alkali so they will be soft and mild to the skin. Perfumes, dyes, and medicinal agents may be added prior to casting the soap into a solid form. Floating soaps have air beaten into them as they solidify. A hard soap is obtained if there is a high percentage of a sodium salt of a relatively long-chain fatty acid, such as stearic acid, present. A soft or liquid soap is obtained by saponification with potassium hydroxide; liquidity increases as the chain length of the fatty acid decreases. Fatty acids with chains as short as C_{12} or shorter are not used because the resultant soaps irritate the skin. These soaps are more volatile and create an odor problem. Fatty acids with chains longer than stearic acid tend to give very insoluble soaps.

Shampoos are often mixtures of several ingredients designed to satisfy a number of requirements. In addition to soaps, condensation products from diethanolamine and lauric acid are often used. These are essentially a type of synthetic detergent obtained by the reaction

$$HN(CH_2CH_2OH)_2 + CH_3(CH_2)_{10}COOH \xrightarrow{\Delta} CH_3(CH_2)_{10}\overset{\overset{\displaystyle O}{\|}}{C}-\overset{\overset{\displaystyle H}{|}}{N}(CH_2CH_2OH)_2$$

Diethanolamine   *Lauric acid*   *An amide* (detergent)

Shampoos also contain compounds to prevent the calcium or magnesium in hard water from forming a precipitate; EDTA is often used for this purpose. (For a discussion of its chelating action, see Chapter 13.) Lanolin and mineral oil are often added, and these give the product a cloudy appearance, in addition to keeping the scalp from drying out and scaling.

"SYNDETS"—THE MODERN SOAPS

Synthetic detergents ("syndets") are derived from organic molecules which have been designed to have the same cleansing action as soaps, but not the same reaction with the cations found in hard water, such as Ca^{+2}, Mg^{+2} and Fe^{+3}. As a consequence, synthetic detergents are equally effective in hard and soft water, while a soap used in hard water will give a precipitate of the Ca^{+2}, Mg^{+2} or Fe^{+3} salt of the long-chain fatty acid. Since such precipitates have no cleansing action and tend to stick to laundry, their presence is very undesirable.

$$2CH_3(CH_2)_{16}COO^- + Ca^{2+} \longrightarrow [CH_3(CH_2)_{16}COO^-]_2Ca^{2+}$$

Stearate ion   *Bath tub ring*

Water softeners can remove the hard water ions, replacing them with ions such as Na^+, which do not interfere with the ordinary soap's action, but this is often an expensive process.

The first true synthetic detergent was made by a Belgian chemist named

Reychler, who, as early as 1913, reported the laboratory preparation of several compounds with definite soaplike properties. Although he reported his work in a scientific journal, it received little attention. No great need was seen to improve on a well-established product. The impetus for the development of detergents came when the Allied Forces blockaded Germany during World War I. Germany, cut off from access to imported fats and oils, sought to relieve the situation by developing soap substitutes not requiring the use of fats. The first patent for a synthetic detergent was filed on October 23, 1917, by Dr. Fritz Gunther of the company then known as the Badische Anilin und Soda Fabrik, the same company that developed the commercial production of ammonia via the Haber process. The raw materials were derived from coal tar, with which Germany was well supplied, and from other nonfat sources. The product was finally marketed in 1925 under the trade name "Nekal." It is still being sold today and is the prototype of the largest class of synthetic detergents, the aryl-alkyl sulfonates. The major component of Nekal is sodium alkyl naphthalene sulfonate.

$$\text{naphthalene}-(CH_2)_x-\overset{\displaystyle O}{\underset{\displaystyle O}{S}}-O^-, Na^+$$

The values of x are generally 12 to 14.

There are an enormous number of different synthetic detergents on the market today. Their molecular structures consist of a long oil-soluble (hydrophobic) group and a water-soluble (hydrophilic) group. The hydrophilic groups include the sulfate ($-OSO_3-$), sulfonate ($-SO_3-$), hydroxyl ($-OH$), ammonium ($-NH_3^+$), and phosphate [$-OPO(OH)_2$] groups.

The great majority of synthetic detergents were originally of the sodium alkyl sulfate type. The preparation of sodium lauryl sulfate, the essential ingredient of some early common household synthetic detergents, illustrates the chemical processes involved. The principal starting material is a suitable vegetable oil, such as cottonseed or coconut. The first step is to treat the dry oil, dissolved in toluene, with sodium dispersed in toluene:

$$\text{Coconut oil} + \underset{\substack{\uparrow \\ \textit{Dispersion}}}{\text{Na}} \longrightarrow \text{Glycerol} + \text{Alcohols} + H_2$$

(Water free in toluene) [Mainly lauryl alcohol, $CH_3(CH_2)_{11}OH$]

The second step involves putting the more polar hydrogen sulfate group on the end of the hydrocarbon chain. This is accomplished by treating the lauryl alcohol with sulfuric acid.

$$CH_3(CH_2)_{11}OH + H_2SO_4 \longrightarrow CH_3(CH_2)_{11}OSO_3H + H_2O$$

Lauryl alcohol   Sulfuric acid   Lauryl hydrogen sulfate

The final step involves neutralizing the acidic lauryl hydrogen sulfate with sodium hydroxide:

$$CH_3(CH_2)_{11}OSO_3H + NaOH \longrightarrow CH_3(CH_2)_{11}OSO_3Na + H_2O$$

Sodium lauryl sulfate

Another group of synthetic detergents (also called "surfactants," from *surface active agents*) is the alkylbenzene sulfonates. These are prepared by putting large alkyl groups on a benzene ring and then sulfonating the benzene ring with sulfuric acid or a related reagent. Before use, they are transformed into their sodium salts. The reactions involved are:

*Sodium alkyl-
benzenesulfonate*

These molecules, like all others in this class, consist of a long hydrophobic chain and a highly polar group which interacts strongly with water.

What's In Your Detergent?

Your synthetic laundry detergent actually contains little real detergent. Most solid syndets are formulated like the one shown in Table 15–2. Liquid syndets are somewhat different since they contain different active detergent molecules.

The trisodium tripolyphosphate is known as the *builder* and functions in two important ways. By reacting with water molecules, the tripolyphosphate

TABLE 15–2   COMPOSITION OF A TYPICAL SOLID DETERGENT
The Chemistry of the Two Main Ingredients Is Discussed in the Text

	COMPOSITION BY WT. %
Sodium alkylbenzene sulfonate, *detergent*	18
Dedusting agent	3
Foam Booster	3
Sodium Tripolyphosphate, *builder*	50
Anticorrosion agent	6
Optical brightener	0.3
Water and Inorganic Filler	19.7
	100.0

ions produce OH⁻ ions and keep the wash water slightly basic, thus helping to emulsify grease particles.

$$\left[\begin{array}{ccc} O & O & O \\ | & | & | \\ P & P & P \\ \end{array} \right]^{5-} + 2H_2O \longrightarrow HP_2O_7{}^{3-} + H_2PO_4{}^- + OH^-$$

Tripolyphosphate ion   *Makes solution basic*

In addition, the tripolyphosphate ions can tie up Ca^{2+} and Mg^{2+} ions which cause hardness in the water.

Other builders are used, such as carbonate ions (CO_3^{2-}) and silicate ions (SiO_3^{2-}). Each of these has serious drawbacks, however. Carbonates are too basic in solution and silicates cause skin sensitization and inhalation problems.

Recently, the magnitude of the detergent use in this country ($1.5 billion per year sales) and the phosphorus content of these detergents have combined to produce an interesting conflict of interests. The phosphorus run-off into our water supplies amounted to some 500 million pounds per year, creating massive overgrowth of certain algae in streams and lakes (*eutrophication*). To counteract this problem detergent manufacturers began to switch over to other builders. One of these was nitrilotriacetic acid (NTA).

$$N\!\!-\!\!\begin{array}{l} CH_2COOH \\ CH_2COOH \\ CH_2COOH \end{array}$$

NTA

After an almost complete switch-over by the detergent makers in the late 1960's and early 1970's, evidence came to light linking NTA to cancer and deformed offspring in laboratory animals. At present, detergent manufacturers are searching for other "builders"; the need is urgent since entire metropolitan areas have banned the sale of phosphate-containing syndets.

ENZYMES EAT DIRT

In 1958, a Danish pharmaceutical firm discovered a protein-destroying enzyme which, when used as a presoak, thoroughly removed certain stains from work clothes. By 1967, 75 per cent of all U.S. laundry detergents contained enzymes of some type. For example, lipases promote the hydrolysis of relatively insoluble fats to mixtures of glycerol and fatty acids. These have been used in laundering cotton for some time. The glycerol is water-soluble, and the fatty acids are converted to water-soluble soaps by fillers in the detergents. Some stains, such as blood, are proteins. Proteases are included in detergents and presoaks for decomposing protein stains into smaller fractions that either are water-soluble or can be mechanically washed away because their attachment to the fabric has been broken. A mechanism whereby enzymes can facilitate the hydrolysis of fats and proteins is described in Chapter 11.

Enzyme detergents and presoaks are being examined very carefully for

299

possible health hazards. Several cases of dermatitis (irritation and breaking out of the skin) on the hands have been tentatively associated with the use of enzyme detergents. At this time, there is no conclusive evidence that enzyme detergents present a hazard to the consumer which would justify label warnings or other action under the Federal Hazardous Substances Act.

DOWN THE DRAIN, TO WHERE?

After a detergent molecule has been used for cleansing, it is washed out with rinse water and enters a sewage system or river. If the normal microorganisms present in such systems can transform the detergent molecule to CO_2 and H_2O in a reasonable time, the detergent is said to be *biodegradable*. Soaps, having biological origin, are typical biodegradable molecules. Many of the newer synthetic detergents, however, are not readily biodegradable. We find that straight chain hydrocarbon derivatives are easily biodegraded, but when we build up substituents on the chain, the biodegradability is decreased. For example, the molecule,

$$CH_3-CH-CH_2-CH-CH_2-CH-CH_2-CH-\underset{CH_3}{\overset{}{\bigcirc}}-\overset{O}{\underset{O}{\overset{\|}{S}}}-O^-, Na^+$$

with CH_3 groups on the carbons.

while it might be an excellent detergent, would be a very poor candidate for biodegradability. The enzymes present in bacteria are simply not suited to break down such molecules. Aromatic rings within the detergent molecule also reduce biodegradability, since they cannot be attacked by microorganisms.

When such detergents are widely used, extreme foaming is often observed in rivers and lakes (Figure 15–3).

Since the middle 1960's the soap and detergent industry has voluntarily curbed the use of non-biodegradable detergents in this country.

WHITES WHITER

Bleaching agents are compounds which are used to remove color from textiles. Most commercial bleaches are oxidizing agents such as sodium hypochlorite. Optical brighteners are quite different, since they act by converting a portion of the invisible ultraviolet light, which impinges on them, into visible blue or blue-green light, which is emitted. Together or separately, these two classes of compounds find their way into commercial laundry and cleaning preparations, since, by appearance, they seem to be making clothes cleaner.

In earlier times textiles were bleached by exposure to sunlight and air. In 1786, the French chemist Berthollet introduced bleaching with chlorine, and subsequently this process was carried out with sodium hypochlorite, an oxidizing agent prepared by passing chlorine into aqueous sodium hydroxide:

$$2Na^+ + 2OH^- + Cl_2 \longrightarrow \underline{2Na^+ + OCl^-} + Cl^- + H_2O$$

Sodium hypochlorite

FIGURE 15-3   Foam in natural waters. A high concentration of organic molecules dissolved in natural water causes foam because of the lowered surface tension of the water. If the water is badly polluted, the foam may exist miles from the pollution source. (From Singer, S. F.: Federal interest in estuarine zones builds. Environmental Science & Technology, 3:2, 1969.)

Shortly after this, hydrogen peroxide was introduced as a textile bleach. Subsequently, a number of other compounds which contain oxidizing agents based on chlorine were developed and introduced.

One way to decolorize materials is to remove or immobilize those electrons in the material which are activated by visible light. The hypochlorite ion is capable of removing electrons from many colored materials. In this process, the hypochlorite is reduced to chloride and hydroxide ions.

$$ClO^- + H_2O + 2e^- \longrightarrow Cl^- + 2OH^-$$

As stated above, optical brighteners are compounds which transform incident ultraviolet light into emitted visible light; this is a type of fluorescence. When optical brighteners are incorporated into textiles or paper, they make the material appear brighter and whiter (Figure 15–4).

WHY ARE DYES NOT REMOVED BY SOAP?

The answer to our question is that effective dyes are bonded strongly to the cloth. Such bonding can be accomplished in several ways, as is shown in Figure 15–5.

301

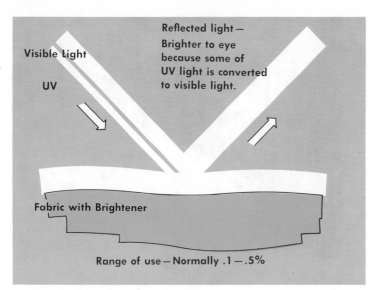

FIGURE 15–4 An optical brightener converts ultraviolet energy to visible light; hence, more light can be detected by the eye.

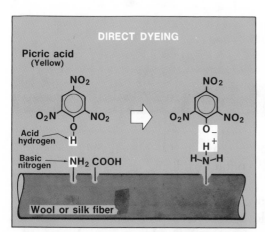

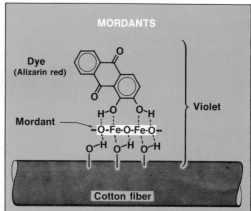

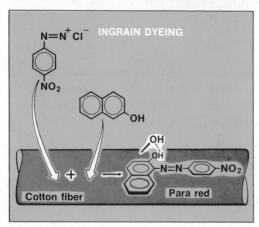

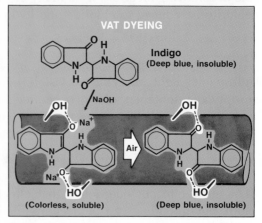

FIGURE 15–5   Some examples of some methods of dyeing cloth. Details are given in the text.

Para red is used to dye cotton by a technique known as ingrain dyeing. In this technique the last step of the synthesis is carried out directly on the fiber rather than by synthesizing the dye first and then trying to apply it to the fiber. This is an effective dyeing technique because smaller molecules can penetrate crevices in the fiber which might be very difficult for a large molecule to enter. After the synthesis, the large molecules are trapped where they are formed, like a ship in a bottle. Besides the physical entrapment of the dye molecule, cotton holds onto para red by means of hydrogen bonds.

Direct dyeing is a chemical reaction in which a reactive part of the dye molecule, usually an acidic or basic group, reacts directly with some group on the fiber itself. Wool and silk are particularly easy to dye in this way since they are composed of proteins (Chapter 9) with available amino ($-NH_2$) and carboxyl ($-COOH$) groups. The amino groups, because they are basic, will react with and bind to the fiber any dye that has an acidic group. Likewise the carboxyl groups, because they are acidic, will react with and bind any dye that has a basic group. Picric acid will dye wool and silk because it has an acidic phenolic hydrogen which reacts with the amino groups of wool and silk.

When a fiber is relatively unreactive to dyes, it can often be dyed by vat dyeing. In this process the cloth is immersed in a vat of the soluble dye in reduced form, and the dye penetrates into the inner crevices of the fiber. While imbedded in the fibers, the dye is rendered insoluble by oxidation to prevent it from being easily washed from the cloth. This method is particularly useful for dyeing cotton with indigo and the phenanthrene dyes.

Some dyes do not adhere very well to a fiber unless it is first treated with a substance known as a mordant (Latin: *mordere* = to bite). Mordants interact with both the fiber and the dye, forming a link between them. Metal oxides are often used as mordants, and anyone who has tried to remove an iron rust stain from cotton can testify to the strong adherence between the two. Reactive centers in a dye, suitable for causing a reaction with mordants, are the acidic carboxyl and phenolic hydroxyl groups. Alizarin with phenolic hydroxyl groups will dye cotton that has previously been treated with a metallic oxide. The mordant frequently changes the color of the dye, since it reacts chemically with it. For example, alizarin is blood-red, but it gives violet colors with iron mordants and brownish red colors with chromium mordants.

OUT, OUT, SPOT

To a large extent, stain removal procedures are based on solubility patterns or chemical reactions. Many stains, such as those due to chocolate or other fatty foods, can be removed by treatment with the typical drycleaning solvents such as tetrachloroethylene, $Cl_2C=CCl_2$.

Stain removers for the more resistant stains are almost always based upon a chemical reaction between the stain and the essential ingredients of the stain remover. A typical example is an iodine stain remover, which is simply a concentrated solution of sodium thiosulfate. The reaction here is

$$I_2 + 2Na_2S_2O_3 \longrightarrow \underline{2\ NaI + Na_2S_4O_6}$$

Soluble in water
(colorless)

TABLE 15-3   SOME COMMON STAINS AND STAIN REMOVERS°

STAIN	STAIN REMOVER
Coffee	Sodium hypochlorite
Lipstick	Isopropyl alcohol, isoamyl acetate, Cellosolve $(CH_2OHCH_2OCH_2CH_3)$, chloroform
Rust and ink	Oxalic acid, methyl alcohol, water
Airplane cement	50/50 amyl acetate and toluene or acetone
Asphalt	Benzol (benzene) or carbon disulfide
Blood	Cold water, hydrogen peroxide
Berry, fruit	Hydrogen peroxide
Grass.	50/50 amyl acetate and benzol or sodium hypochlorite or alcohol
Nail polish	Acetone
Mustard	Sodium hypochlorite or alcohol
Antiperspirants	Ammonium hydroxide
Perspiration	Ammonium hydroxide, hydrogen peroxide
Scorch	Hydrogen peroxide
Soft drinks	Sodium hypochlorite
Tobacco	Sodium hypochlorite

° Before any of these stain removers are used on clothing, the possibility of damage should be checked on a portion of the cloth that ordinarily is hidden.

Iron stains are removed by treatment with oxalic acid, which forms a soluble coordination compound with the iron:

$$Fe_2O_3 + 6H_2C_2O_4 \longrightarrow 3H_2O + 2Fe(C_2O_4)_3^{3-} + 6H^+$$

Oxalic acid   *Soluble in water*

Mildew stains can be removed by hydrogen peroxide or laundry bleach (sodium hypochlorite), which oxidizes the fungus responsible for the mildew. Blood stains on cotton can be removed by hypochlorite solution. Bleach should not be used on wool because it reacts chemically with the nitrogen atoms present in the peptide chains. The chemicals used to remove a few common stains are listed in Table 15–3.

QUESTIONS

1. A detergent label indicates that the material contains 20 per cent $Na_2CO_3 \cdot 10H_2O$. What is the percentage of water in the detergent from this source?

2. Why are detergents better cleansing agents than soaps in regions where the water supply contains calcium or magnesium salts?

3. (a) Write an equation for the chemical reaction between the calcium ion (Ca^{2+}) and soap.
   (b) Write an equation for the chemical reaction between the calcium ion (Ca^{2+}) and the

water softener, Na_2CO_3. (c) A washtub contains 40 liters of water. The water has run through a limestone region and picked up 0.1 mole calcium ion per liter. What is the maximum amount (in grams) of soap powder [$NaOCO(CH_2)_{16}CH_3$, molecular weight: 306] required to precipitate the calcium ions?

4. Why is a soap from coconut oil more soluble in water than a soap made from palm oil?

5. Suggest ways of removing the following from clothing: (a) motor oil, (b) iodine stain, (c) lard, (d) copper sulfate.

6. Explain why vinegar is able to remove some stains which are soluble in weak acids.

7. Why was the discovery of cheap Na_2CO_3 important to the development of the soap industry?

8. What is castile soap?

9. Explain why Grandma's lye soap produced rough, red hands.

10. Explain how an optical brightener in a detergent works.

11. Search recent issues of *Science, Chemical and Engineering News,* and other scientific journals for the present status of enzyme detergents as regards their safety and extent of use.

12. Do the same for the present status of NTA and other builders in laundry detergents.

SUGGESTIONS FOR FURTHER READING

Bennett, H., "The Chemical Formulary," Chemical Publishing Co., New York, 1933–1965. (Twelve volumes of formulas for making soaps, cosmetics, perfumes, and so forth.)

"Enzymes in Detergents," *Chemistry,* Vol. 43, No. 2, p. 25 (1970).

"Phosphates in Detergents and Eutrophication of America's Water," U.S. Government Printing Office, Washington, D.C., 1970.

"The Smell of Detergents," *Chemistry,* Vol. 40, No. 5, p. 10 (1967).

16

THE CHEMICALS WE USE ON OURSELVES— COSMETICS

People find many reasons for applying various chemical preparations (*cosmetics*) to their skin and hair. We wish to be clean, beautiful, healthy, and pleasing to others. Often we find that a cosmetic can do something easier than we can do it ourselves, such as take off unwanted hair or hold hair in place on a windy day.

Considerable progress has been made in producing chemical products which color hair and skin, disinfect our body surfaces, and keep down unwanted body odor. In this chapter we are going to look at a few of them in some detail. The purpose is not so that you can make face creams or hair sprays, but so that you can more fully understand the basic chemistry involved. In fact, amateur chemical preparations should be avoided, since toxic reactions often are encountered when impure or otherwise harmful concoctions are used without proper prior testing. Even the professionals have their problems. Scarcely a cosmetic producer is in business today who hasn't received a letter of complaint saying, in effect, "Your shampoo caused my hair to fall out!"

SKIN AND HAIR

The work of a cosmetic begins the moment it is applied to the skin or hair. These are the two most important protein structures on our bodies.

Our skin, like the other organs of the body, is not uniform tissue. Rather it is composed of layers, each parallel to the surface. The outermost layer, called the *stratum corneum,* or *corneal* layer, is where most cosmetic preparations for the skin act. The corneal layer is composed of essentially dead cells with low moisture content and a surface pH of about 4, slightly acid. Depending upon its location on the body, the corneal layer is populated with as many as a million microorganisms per square centimeter. While these play a role in our lives (see later), perhaps the chemistry of the corneal cell protein *keratin* is of greatest importance. Keratin is composed of about 22 different amino acids. Its structure renders it insoluble in, but slightly permeable to, water. This is important since we do not wish to dissolve in rain, but we do need to perspire! In order to control the moisture content of the corneal layer so it does not dry out and slough off too fast, moisturizers are added to the skin.

Hair is composed principally of keratin. An important difference between hair keratin and other proteins is its high content of the amino acid *cystine*. About 16 to 18 per cent of hair protein is cystine, while only 2.3 to 3.8 per cent of the keratin in corneal cells is cystine. This amino acid plays an important role in the structure of hair, as we will see shortly.

$$HOOC-\underset{\underset{NH_2}{|}}{CH}-CH_2-S-S-CH_2-\underset{\underset{NH_2}{|}}{CH}-COOH$$

Cystine

The toughness of both skin and hair is due to the bridges between different protein chains, such as hydrogen bonds (Chapter 3) and —S—S— linkages (called *disulfide* linkages) between one part of a cystine molecule on one chain and the other part on another chain.

$$\begin{array}{cc} O{=}\overset{|}{C} & \overset{|}{N}{-}H \\ H{-}\overset{|}{C}HCH_2{-}S{-}S{-}CH_2\overset{|}{C}{-}H \\ H{-}\overset{|}{N} & \overset{|}{C}{=}O \end{array}$$

Another type of bridge between two protein chains which is important in keratin as in all proteins is the *salt* bridge. Consider the interaction between a lysine —NH$_2$ group and a carboxylic —COOH group of glutamic acid on a neighboring protein chain. At pH 4.1, the formation of a —NH$_3^+$ group and a —COO$^-$ group is most favorable for keratin. These two groups then interact as unlike charges, attracting, and forming an ionic bond.

$$H\overset{|}{C}CH_2CH_2CH_2CH_2NH_2 + HOOCCH_2CH_2\overset{|}{C}H \longrightarrow$$

Lysine   *Glutamic acid*

at pH 4.1

$$H\overset{|}{C}CH_2CH_2CH_2CH_2NH_3^+ \quad {}^-OOCCH_2CH_2\overset{|}{C}H$$

As the pH rises above 4, keratin will swell and become soft as these salt bridges are broken. This is an important aspect of hair chemistry.

SHAMPOOS AND GENERAL HAIR CARE

We discussed soaps and detergents in a general way in Chapter 15; however, the cleaning of hair presents several interesting items which can best be discussed

307

here. Shampoos are designed to clean hair of dirt, oils, and loose corneal cells. After their use they must leave the hair manageable. In addition, we as consumers demand a nice smell, and lots of foam, although this has nothing to do with the effectiveness of the product.

Most shampoos contain anionic detergents, since these are the least damaging to the eyes and still have good foaming and detergent properties. Sodium lauryl sulfate is an often used example of this class of detergent.

$$CH_3(CH_2)_{11}OSO_3^-Na^+$$
Sodium lauryl sulfate

The chief drawback to these detergents is their high sensitivity to calcium ions, which tend to precipitate and form a scum which dulls the hair. Often a shampoo will contain a "builder" such as phosphate ion (PO_4^{3-}), which can complex the calcium ions.

Another major problem with anionic detergents is their accumulation at free positive centers in the keratin structure after the bulk of the shampoo is rinsed away. Since negative centers such as —NH_2 and —C=O are common in keratin they greatly predominate after some of the positive centers are tied up by the anionic detergent. These negative centers then begin to repel each other and make the hair hard to manage. A *rinse* contains a cationic detergent (about 1%) which neutralizes the anions causing the problem. Caution should be exercised with these solutions, since the cationic detergents are damaging to the eyes.

$$CH_3(CH_2)_{11}OSO_3^-$$
$$CH_3(CH_2)_{11}OSO_3^-$$

$$\overset{+}{H}OOC-CH_2-CH_2-\overset{|}{\underset{|}{C}}-H$$

Free negative center

CHANGING THE SHAPE OF HAIR

From the time of ancient Egypt, Greece, and Rome, there have always been women who considered their hair too straight or too curly. The properties of hair in this respect are determined by the extent and nature of the cross linking. This is dictated as the messenger-RNA constructs the protein structure from the amino acids available (see Chapter 12).

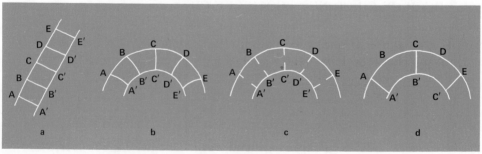

FIGURE 16–1   A schematic diagram of a permanent wave.

308

When hair is wetted it can be stretched to one and a half times its dry length, because water (pH 7) destroys some of the salt bridges and causes swelling of the keratin. Imagine the disulfide crosslinks remaining between two protein chains in hair as (a) in Figure 16–1. Winding the hair on rollers causes tension to develop at the crosslinks (b). In "cold" waving, these crosslinks are broken by a suitable chemical (c), relaxing the tension. Next, another chemical re-forms the cross links (d) and the hair holds the shape of the roller.

The basic chemical process in cold permanent waving is the disruption of

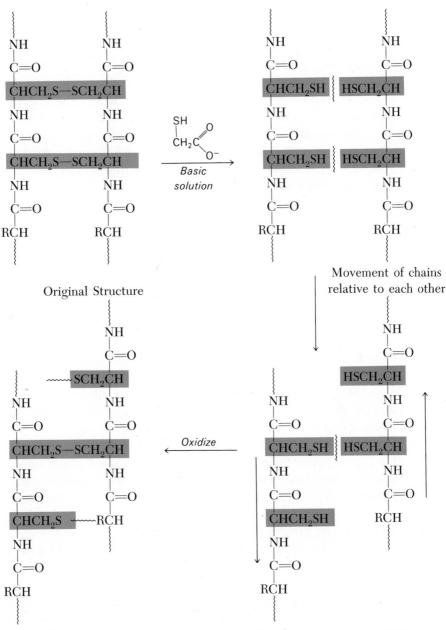

FIGURE 16–2   Structural changes that occur in hair during a permanent wave.

309

the disulfide linkages by a chemical reducing agent, and the subsequent re-forming of at least some of these bonds by an oxidizing agent. While the structure of the hair is disrupted, it is shaped on curlers and then reoxidized. The chemical reactions in simplified form are shown in Figure 16–2.

$$CH_2-C\underset{OH}{\overset{O}{\diagup}}$$
$$|$$
$$SH$$

Thioglycolic acid

The most commonly used reducing agent is thioglycolic acid. The common oxidizing agents used include hydrogen peroxide, perborates ($NaBO_2 \cdot H_2O_2 \cdot 3H_2O$), and sodium or potassium bromate ($KBrO_3$). A typical neutralizer solution contains one or more of the oxidizing agents dissolved in water. The presence of water and strong base in the oxidizing solution also helps to break and re-form hydrogen bonds between adjacent protein molecules.

Various additives are present in both the oxidizing and the reducing solutions in order to control pH, odor, and color, and for general ease of application. A typical waving lotion contains 5.7 per cent thioglycolic acid, 2.0 per cent ammonia, and 93.3 per cent water.

Hair can be straightened by the same solutions. It is simply "neutralized" (or oxidized) while straight (no rolling up).

HAIR COLORING AND BLEACHES

Hair contains two pigments: brown-black melanin and an iron-containing red pigment. The relative amounts of each actually determine the color of the hair. In deep-black hair melanin predominates, while in light-blond, the iron pigment predominates. The depth of the color depends upon the size of the pigment granules. A number of other factors affect the color of hair. These include the person's age, the amount of sunlight striking the hair, and whether or not permanent waves are used.

In recent years hair bleaches and colors have become very popular, largely because of newer formulations which can produce a much more uniform coloration of human hair. The formulations vary from temporary coloring (removable by shampoo), which is usually achieved by means of a water-soluble dye, which acts on the surface of the hair, to semi-permanent dyes, which penetrate the hair fibers to a great extent. These often consist of cobalt or chromium complexes of dyes dissolved in an organic solvent. Permanent dyes are generally "oxidation" dyes. They penetrate the hair, and then are oxidized to give a colored product which is permanently attached to the hair by chemical bonds or which is much less soluble than the reactant molecule. Permanent dyes are similar to the fabric dyes used in ingrain dyeing, discussed in Chapter 15. These hair dyes generally are derivatives of phenylenediamine.

310

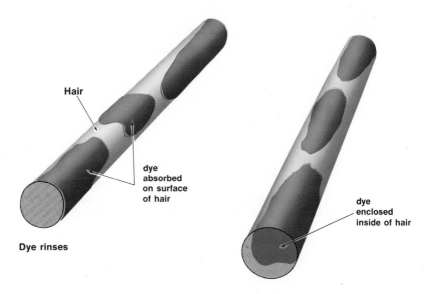

Hair

dye
absorbed
on surface
of hair

Dye rinses

dye
enclosed
inside of hair

Semi-permanent dye

FIGURE 16–3   Methods of dyeing hair.

NH$_2$

NH$_2$

This compound itself dyes hair black. A blond dye can be formulated with
p-aminodiphenylamine-sulfonic acid,

NH$_2$—⟨benzene⟩—NH—⟨benzene⟩—SO$_3$H

or p-phenylenediamine-sulfonic acid.

NH$_2$

SO$_3$H

NH$_2$

One blond formulation contains p-phenylenediamine (0.3 per cent),
p-methylaminophenol (0.5 per cent), p-aminodiphenylamine (0.15 per cent),
o-aminophenol (0.15 per cent), pyrocatechol (0.25 per cent), resorcinol (0.25 per
cent), and inert solvent (98.40 per cent). These compounds are applied in an
aqueous soap or detergent solution containing ammonia to make the solution
basic. The dye material is then oxidized, using hydrogen peroxide to develop
the desired color. The amines are oxidized to nitro compounds.

311

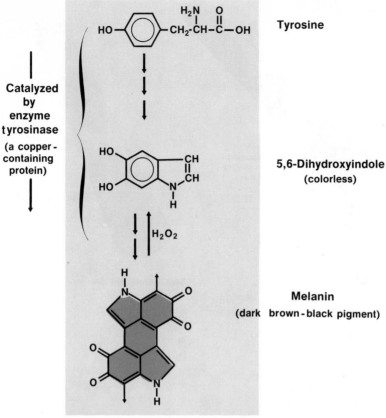

FIGURE 16-4   Bleaching of the hair by hydrogen peroxide. There are several chemical intermediates between the amino acid—tyrosine—and the hair pigment—melanin, which is a protein. Hydrogen peroxide oxidizes melanin back to colorless compounds, which are stable in the absence of tyrosinase (found only in the hair roots). Melanin is a high molecular weight polymeric material of unknown structure. The structure shown here is only a segment of the total structure.

$$-NH_2 + 3H_2O_2 \xrightarrow{\text{oxidation}} -NO_2 + 4H_2O$$

Amine   *Nitro*

Hair can be bleached by hydrogen peroxide, which destroys the hair pigments by oxidation. The solutions are made basic with ammonia to enhance the oxidizing power of the peroxide. Parts of the chemical process are given in Figure 16-4. This drastic treatment of hair does more than just change the color. It can destroy sufficient structure to render the hair brittle and coarse.

HAIR SPRAYS

Hair sprays are essentially solutions of plastic in a very volatile solvent whose purpose, when sprayed on hair, is to furnish a film with sufficient strength to hold the hair in place after the solvent has evaporated. After early experiments with shellac, the introduction of the aerosol can allowed the use of a wider variety of resins and solvents and provided greater control over the application of the product.

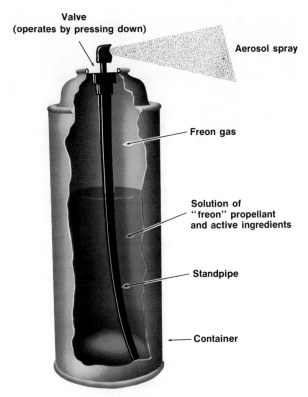

Valve
(operates by pressing down)

Aerosol spray

Freon gas

Solution of
"freon" propellant
and active ingredients

Standpipe

Container

FIGURE 16-5   Cross section of a typical aerosol spray can.

A very common resin in hair sprays is the addition polymer, polyvinyl-pyrrolidone (PVP).

Polyvinylpyrrolidone, PVP

The resin is blended in hair spray formulations with a plasticizer, a water repellant, and a solvent-propellant mixture. The plasticizer makes the plastic more pliable, as described and illustrated in Chapter 8. The solvent-propellant system is a solvent such as anhydrous ethyl alcohol mixed with a liquefied propellant, such as Freon 12.

$$Cl-\overset{\displaystyle F}{\underset{\displaystyle Cl}{C}}-F$$

Dichlorodifluoromethane (Freon 12)

313

FIGURE 16-6 Film of hair spray. Hair spray was allowed to dry on white surface and was then pulled up to reveal film.

The resin concentration of hair sprays is of the order of 1.5 per cent with a ratio of 30 per cent ethanol to 70 per cent propellant for the liquid phase.

Other additives are often put into hair sprays to give the hair a sheen (silicone oils) and to render the plastic more pliable (plasticizers).

A typical hair spray formulation contains the following ingredients and amounts.

PVP	(Resin)	4.70 parts by weight
Dimethyl phthalate	(Plasticizer)	0.20
Silicone oil	(Sheen)	0.10
Ethanol	(Solvent)	25.00
Freon 11	(Propellant)	45.00
Freon 12	(Propellant)	25.00
Perfume	(For effect)	
		100.00 parts by weight

Since PVP tends to pick up moisture, other, less hydroscopic polymers are beginning to replace PVP in hair sprays. For example, significantly better moisture control is obtained with a copolymer made from a 60/40 ratio of vinyl pyrrolidone and vinyl acetate.

KILLING GERMS

Germs, or microorganisms, as they should be called, are always with us in large number. Fortunately, there are so many non-pathogenic microorganisms among them that there is little room left for the pathogenic kinds.

Certain things we do to our skin, such as shaving, upset the balance between harmful and harmless bacteria and promote infection by pathogenic microorganisms. *Disinfectants* are chemicals used to kill these organisms before they can overcome the skin's defenses. Most commonly used are the alcohols. These are the only disinfectants used in many after-shave preparations. Maximum effectiveness of ethyl alcohol is at a concentration of 70%, while isopropyl (rubbing) alcohol is most effective at a concentration of 50%. These alcohols kill germs apparently by hydrogen bonding with water, which dehydrates the cellular structure of the germ.

The oldest known disinfectant is phenol: its aqueous solution is known as carbolic acid. The acidic nature of the —OH group attached to the benzene ring was discussed in Chapter 7. Phenol was discovered in 1867 by Sir Joseph Lister, who introduced it into surgery. It appears that phenol kills bacteria by

OH

Phenol

denaturing their cellular proteins. Today, about one-third of all toilet soaps sold contain some derivative of phenol. One famous compound in this class is commonly known as *hexachlorophene*. This compound is extremely effective against staphylococci and streptococci bacteria which cause most so-called "body odors" and small infections.

Hexachlorophene

In 1971 it became known that hexachlorophene was related in some way to damage in the brain cells of test baby monkeys and in December of that year the FDA warned against the use of products containing hexachlorophene. Later, in 1972, the FDA placed a ban on hexachlorophene in all but prescription uses.

There are still other phenol derivatives which find wide use in soaps, deodorants, facial creams, and other cosmetics.

DEODORANTS

The 2,000,000 sweat glands on the body surface are primarily used to regulate body temperature via the cooling effect produced by the evaporation of the water they secrete. The evaporation of the water leaves the solid constituents, mostly sodium chloride, plus smaller amounts of other materials, including some organic compounds. The odor results largely from amines and hydrolysis products of fatty oils (fatty acids, acrolein, etc.) emitted from the body and from bacterial growth on the body. Sweating is both normal and necessary for the proper functioning of the human body.

There are three kinds of deodorants: those which directly "dry up" perspiration or act as astringents, those which have an odor to mask the odor of sweat, and those which remove odorous compounds by combining with them. Among those which have astringent action are hydrated aluminum sulfate, hydrated

315

aluminum chloride, ($AlCl_3 \cdot 6H_2O$), aluminum chlorohydrate [actually aluminum hydroxy chloride, $Al_2(OH)_5Cl \cdot 2H_2O$ or $Al(OH)_2Cl$ or $Al_6(OH)_{15}Cl_3$], and alcohols. Those compounds which act as deodorizing agents include zinc peroxide, essential oils and perfumes, and a variety of mild antiseptics. Zinc peroxide removes odorous compounds by oxidizing the amines and fatty acid compounds. The essential oils and perfumes absorb or otherwise mask the odors, and the antiseptics are generally oxidizing or reducing agents.

The most widely advertized deodorants contain aluminum salts as the active ingredients. Other materials are added to assist the application or to provide a fragrance. Aluminum salts are astringents in that they can reduce or close up the openings of the sweat glands by affecting the hydrogen bonds that hold protein molecules together or (less often) by precipitating skin proteins. The ones that precipitate skin proteins give skin irritating effects. A typical spray deodorant will have the aluminum salt and minor ingredients dissolved in an alcoholic solution. A typical cream antiperspirant deodorant contains aluminum hydroxy chloride (astringent; 20 per cent), sorbitan monostearate (hydrogen bonds water and absorbs it; 5 per cent), polyoxyethylene sorbitan monostearate (hydrogen-bonds water and absorbs it; 5 per cent), stearic acid (precipitates amines; 15 per cent), propylene glycol (precipitates fatty acids; 5 per cent), and water (for desired consistency; 50 per cent).

SKIN PREPARATIONS—MOSTLY BEAUTY

Healthy skin is important if we are to ward off bacterial invasion and disease. To remain healthy, the moisture content of skin must stay near 10 per cent; if it is higher, microorganisms grow too easily; if lower, the corneal layer breaks down. Washing skin takes away fats which tend to retain the right amount of moisture. If dry skin is treated with a fat after washing it will be protected until enough natural fats have been regenerated.

Lanolin is an excellent skin softener (emollient) and is a component of many cosmetics. It is a complex mixture of esters from hydrated wool fat. The alcohols in the esters have up to 33 carbon atoms, and the fatty acids have up to 37 carbon atoms. Cholesterol, a common alcohol in lanolin, is found both free and in esters (Chapters 10 and 14). Cholesterol appears to endow fat mixtures with the property of absorbing water. This is one factor that makes lanolin an excellent emollient (a substance applied to soften or soothe irritated skin). With its high proportion of free alcohols, particularly cholesterol, and hydroxy acid esters, lanolin has the structural groups (—OH) to hydrogen-bond water (to keep the skin moist) and to anchor within the skin (the fatty acid and ester hydrocarbon structures). See Figure 16–7.

Creams

Creams are generally emulsions of either an oil-in-water type or a water-in-oil type. An emulsion is simply a colloidal suspension of one liquid in another. The oil-in-water emulsion has tiny droplets of an oily or waxy nature dispersed throughout a water solution (homogenized milk is an example). The water-in-oil

316

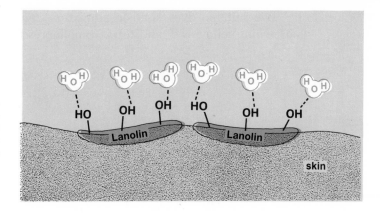

FIGURE 16-7 The hydroxyl groups of lanolin form hydrogen bonds with water and keep the skin moist. The fat parts of the molecule are "soluble" in the protein and fat layers of the skin.

emulsion has tiny droplets of a water solution dispersed throughout an oil (natural petroleum and melted butter are examples). An oil-in-water emulsion can be washed off the hands with tap water, while a water-in-oil one gives the hands a greasy, water-repellant surface.

Cold cream originally was a suspension of rose water in a mixture of almond oil and beeswax. Subsequently, other ingredients were experimented with in order to get a more stable emulsion. An example of a modified composition is the following cold cream: almond oil, 35 per cent; beeswax, 12 per cent; lanolin, 15 per cent; spermacetti (from whale oil), 8 per cent; and strong rose water, 30 per cent. Other oils can be substituted for some or all of the almond oil. The lanolin stabilizes the emulsion.

Vanishing cream is a suspension of stearic acid in water, to which a stabilizer is added to make the suspension more stable. The stabilizer may be a soap, such as potassium stearate. These creams do not actually vanish, they merely spread as a smooth, thin covering over the skin.

Creams of various sorts may be used as the base for other cosmetic preparations, in which case further ingredients are added, to give additional properties to the creams. As an example, hydrated aluminum chloride can be added to prepare a cream deodorant.

Lipstick

The skin on our lips is covered by a very thin corneal layer which is free of fat and easily dried out. In order to protect lips from the drying effects of wind and air, women use lipstick. (They may have a few other purposes in mind also.)

Lipstick consists of a solution or suspension of coloring agents in a mixture of high-molecular-weight hydrocarbons or their derivatives, or both. The material must soften to produce an even application when pressed on the lips, yet the film must not be too easily removed, nor may the coloring matter run. Lipstick is perfumed to give it an odor and pleasant taste. The color usually comes from a dye or "lake" from the eosin group of dyes. A *lake* is a precipitate of a metal ion (Fe^{3+}, Ni^{2+}, Co^{3+}) with an organic dye. The metal ion enhances the color or changes the color of the dye.

Two suitable dyes, used in admixture and with their lakes, are dibromo-fluorescein (yellow-red) and tetrabromofluorescein (purple):

Tetrabromofluorescein
(*sodium salt*)

The ingredients in a typical formulation include:

Dye	Furnishes color	4–8%
Castor oil, paraffins or fats	Dissolves dye	50%
Lanolin	Emollient	25%
Carnauba wax }	{ Makes stick stiff by	18%
Beeswax }	{ raising the melting point	18%
Perfume	Imparts pleasant taste	1.5%

Carnauba wax and beeswax are high-molecular-weight esters (Chapter 9).

In the manufacture of lipstick, the dye is first dispersed in the castor oil, and then the waxes, lanolin, and perfume are added, while the whole is heated and stirred to obtain a homogeneous mixture. The molten mass is then cast into suitable forms and subsequently inserted into holders, passed momentarily through a gas flame to obtain a smooth surface, and then packaged.

Suntan Lotions

One of the agents most harmful to white skin is the short wavelength, ultraviolet light from the sun. Various preparations have been used in the past to screen out all ultraviolet radiation and thus protect the skin when it must be exposed to the sun for long periods of time. Today, it is considered desirable to exclude the shorter, more harmful wavelengths, while transmitting enough longwave ultraviolet to permit gradual tanning.

The variety of suntan products ranges from lotions, which selectively filter out the higher energy ultraviolet rays of the sun, to preparations which essentially dye the skin a tan color. The lotions which filter out the ultraviolet rays are more accurately described as sunscreens, and their ingredients are often mixed with other materials, to give a lotion which both screens and tans. A common ingredient in preparations used to *prevent* sunburn is *p*-aminobenzoic acid.

$$H_2N-\langle\bigcirc\rangle-COOH$$

p-Aminobenzoic acid

Like most aromatic compounds, it absorbs strongly in the ultraviolet region of the spectrum (Figure 16–8). The *p*-aminobenzoic acid is emulsified with a mixture of alcohols, an oil, and water by a high-molecular-weight fatty acid ester. A leading suntan lotion contains monoglyceryl *p*-aminobenzoate (3 per cent), mineral oil (25 per cent), sorbitan monostearate and polyoxyethylene sorbitan monostearate (10 per cent; used as emulsifiers), and water (62 per cent). Perfume is added to give the material a pleasant odor.

Newer suntan lotions contain 2-ethoxyethyl-*p*-methoxycinnamate as the absorber of ultraviolet light. It is said to absorb the peak burning wavelength, which is at 308 nanometers. These lotions, like all the others, do not prevent sunburn completely. They merely prolong the possible length of exposure to the sun before severe sunburn occurs.

$$CH_3O-\langle\bigcirc\rangle-CH=CHCOOCH_2CH_2OC_2H_5$$

2-ethoxyethyl-p-methoxycinnamate

Tanning is a process by which the skin is stimulated to increase its production of the pigment *melanin* (see *Hair Coloring and Bleaching* in this chapter). At the same time, the skin thickens and becomes more resistant to deep burning.

Preparations for the relief of the pain of sunburn are solutions of local anesthetics (such as benzocaine), plus other ingredients for color, odor, and

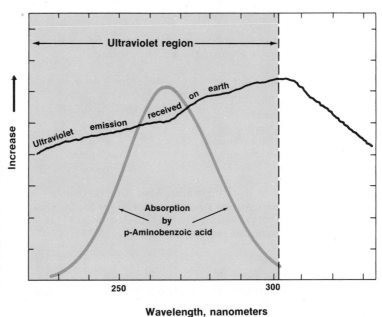

FIGURE 16–8 Absorption spectrum of p-aminobenzoic acid. Maximum absorption occurs at 265 nanometers, although it absorbs at other wavelengths as shown. The maximum of the deep-burning ultraviolet radiation received on earth is at about 308 nanometers. One nanometer is 10^{-9} meter.

Wavelength, nanometers
(1 nanometer = 10^{-9} meter)

319

softening of burned tissue. As you might expect, lanolin is especially suited for this purpose and it is used often. By softening the tissue, the emollient provides a ready access of the anesthetic and of blood plasma.

QUESTIONS

1. You read in the newspaper about a new compound which will break disulfide bonds in proteins. What potential use might it have?

2. (a) What is the purpose of an emulsifier? (b) In which of the following cosmetics is an emulsifier important: suntan lotion, hair spray, cold cream?

3. Which of the following properties would be appropriate for a hair spray propellant?

   (a) High or low boiling point?
   (b) Soluble or insoluble in the active ingredients?
   (c) Capable of chemical reaction with the active ingredients?
   (d) Odor?
   (e) Toxicity?

4. What is the monomer unit in polyvinylpyrrolidone?

5. What is the purpose of the following:

   (a) Freon in hair sprays?
   (b) Polyvinylpyrrolidone in hair sprays?
   (c) Aluminum chloride in deodorants?
   (d) p-Aminobenzoic acid in suntan lotion?

6. What specific substance is broken down during the bleaching of hair?

7. Use the structures of the constituents of lanolin to justify its ability to emulsify face creams.

8. Hydrogen-bonding is a very handy theoretical tool. Name three applications of hydrogen bonding in consumer products.

9. If you were going to formulate a suntan lotion, what particular spectral property would you look for in choosing the active compound?

10. A typical hair spray for women *or* men has a formula:

resin (like PVP)	3.0%
plasticizer	0.2%
ethyl alcohol	26.0%
propellant (Freon 11/12)	70.8%
perfume (to suit application)	negligible
	100.0%

Costs on these ingredients vary with the number of cans of hair spray a manufacturer produces. Assume the resin costs $.90 per lb., the plasticizer, $1.10 per lb., the ethyl alcohol, $0.15 per lb., and the Freon 11/12 mixture $0.45 per lb. Compute the costs of the ingredients of a 16 oz. (wt.) can of hair spray.

SUGGESTIONS FOR FURTHER READING

Bennett, H., "The Chemical Formulary," Chemical Publishing Co., New York, 1933–1965. (Twelve volumes of formulas for making soaps, cosmetics, perfumes, and so forth.)

Bennett, H., "Chemical Specialities," Chemical Publishing Company, New York, 1969. (A detailed practical guide on how to set up a chemical specialities business to manufacture and sell cosmetics, herbicides, and so forth.)

Gleason, M. N., Gosselin, R. E., Hodge, H. C., and Smith, R. P., "Clinical Toxicology of Commercial Products," Third Edition, The Williams & Wilkins Co., Baltimore, 1969. (Contains a wealth of information on the toxic aspects of various commercial products as well as their composition.)

Harry, R. G., "Modern Cosmetology," Chemical Publishing Co., New York, 1947. (A source of formulas and manufacturing techniques for almost all standard types of cosmetics.)

Jellinek, J. S., "Formulation and Function of Cosmetics," John Wiley and Sons, New York, 1970.

17

THE CHEMISTRY OF PAINTS—PROTECTING WHAT YOU HAVE

Most of us have taken up the brush, roller, or spray can at one time or another to apply a coating of what we call paint. We do this either to protect or beautify, or both. The things we paint are made of wood, metal, masonry, plastic, paper or ceramic. About $3 billion is spent on paints of all types in a typical year in the United States. In this chapter we will look at some of the chemistry of paints. From time to time we shall borrow on our knowledge of polymers from Chapter 8, since, as we will see, a paint forms a protective film which is much like a typical polymer film.

FIGURE 17-1 Photographers photographing a painter painting—in this picture the John Quincy Adams birthplace in Quincy, Massachusetts. The Sears Great American Home Series of advertisements illustrates the importance of surface coatings as preservatives. (Courtesy of Sears, Roebuck and Co., Chicago, Illinois.)

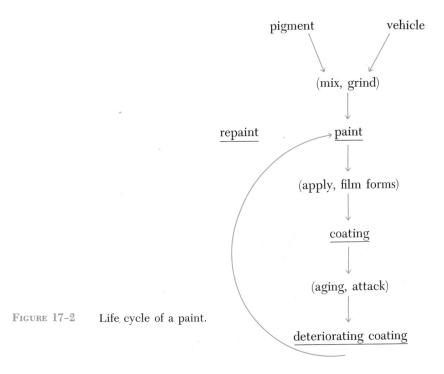

pigment   vehicle

(mix, grind)

repaint   paint

(apply, film forms)

coating

(aging, attack)

deteriorating coating

FIGURE 17-2   Life cycle of a paint.

BASIC COMPONENTS OF PAINTS

Paints are generally composed of two basic components: *pigment* and *vehicle*. The pigment supplies the desired color and, for exterior paints, provides a shield against the harsh ultraviolet radiation of the sun. The vehicle is the liquid part of the paint (in earlier years, linseed oil). Nowadays the vehicle often consists of a solvent and some sort of dissolved polymer, which act as a binder.

Unfortunately, most paint films break down in use, owing to attack by oxygen, sunlight, or pollutants, as well as through physical abuse. Much of the research in paints, and the interest the user has in paints, has to do with extending their life (Figure 17–2).

FILM FORMATION—A COVERING

In order to hold out moisture and air, and to hold the pigment in place, the binder of a paint must be capable of forming a solid film, which adheres to the painted object. Films can be formed by simply allowing the solvent to evaporate, or by more complicated processes in which polymerization reactions take place during the "drying" process.

For years, many farmers have used whitewash around the farm to brighten things up and protect wooden structures. Tom Sawyer's whitewash was a simple kind, composed of hydrated (slaked) lime ($Ca(OH)_2$) and water. (About 50 pounds of hydrated lime to 6 gallons of water will do the job.) After application, the lime slowly was converted to calcium carbonate (chalk) by carbon dioxide (CO_2) in the air. The resulting white color looked nice but protected the surface

$$Ca(OH)_2 \quad + CO_2(\text{in air}) \xrightarrow{\textit{Slow}} CaCO_3 + H_2O$$

Hydrated lime   *Chalk*

underneath very little. A film-forming, weather-resistant whitewash can be formed by adding hydrated lime to casein (milk protein, skimmed milk will do) and formaldehyde, HCHO. As we saw in Chapter 9, some proteins contain the amino acid lysine, which contains an amine group at the end of the hydrocarbon chain (Figure 17–3).

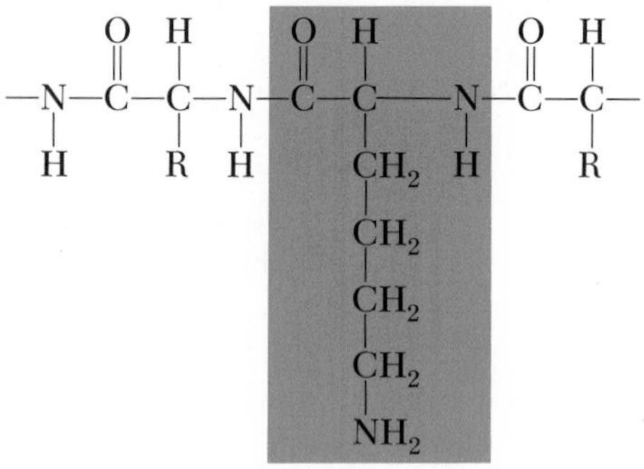

FIGURE 17-3 A protein chain with a lysine amino acid unit designated.

In the presence of formaldehyde, the casein undergoes *crosslinking* (see Chapter 8), to produce a rigid film. The reaction is actually a condensation polymerization.

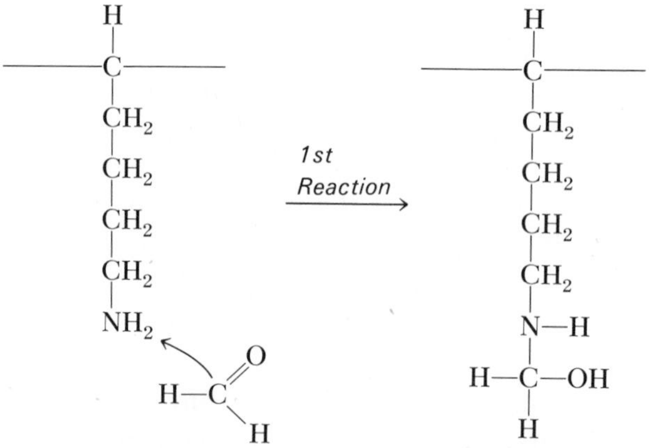

Then, crosslinking occurs between two chains.

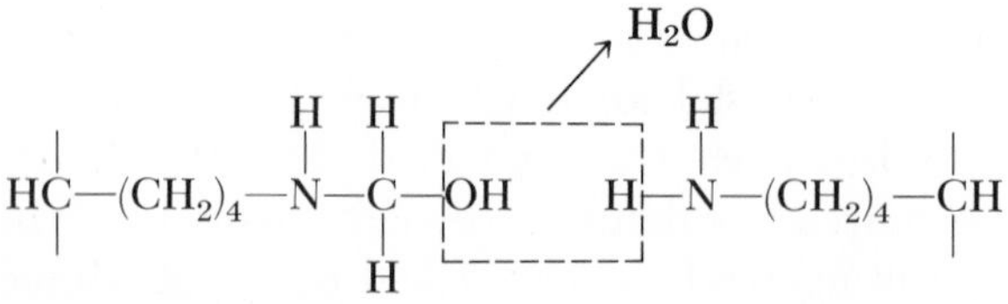

The resulting casein-formaldehyde polymer is a dense, three-dimensional web which holds the calcium carbonate very tightly and prevents it from washing off.

Not everyone wants to whitewash his buildings, so let's look at some other types of paints.

OIL-BASED PAINTS

For outside painting, few paints excel white lead $(Pb(OH)_2 \cdot 2PbCO_3)$ dispersed in linseed oil and turpentine. This paint was used by the early Greeks and Romans. Of course, the lead is toxic, and lead-based paints are now banned in interior applications. In addition, lead reacts with hydrogen sulfide fumes present in industrial atmospheres to form black lead sulfide, PbS.

$$Pb(OH)_2 + H_2S \longrightarrow PbS + 2H_2O$$

(*White*) *Hydrogen sulfide* (*black*)

The drying of an oil-based paint is a fairly complicated chemical process we shall look at later. However, the solvent mentioned above, turpentine, is a hydrocarbon mixture of two isomers of $C_{10}H_{16}$, α- and β-pinene—appropriately named since turpentine is obtained from pine trees. As a turpentine-based paint dries, the solvent simply evaporates, leaving a characteristic odor.

α-Pinene   *β-Pinene*

Modern paints overcome the odor problem by using less-obnoxious hydrocarbons such as mineral spirits, a petroleum fraction boiling between 93 and 150°C.

The pigment of a modern oil-based paint contains titanium dioxide (TiO_2), "titanium calcium" (30 to 50 per cent TiO_2 on the surface of calcium sulfate particles), zinc oxide (ZnO), and various extender pigments. Titanium dioxide has greater opacity and whiteness and is less toxic than white lead. The different crystalline forms of TiO_2 catalyze the decomposition of the binder (such as linseed oil) at different rates. The deterioration of the binder forms a white powder, which is easily rubbed off. This process is called chalking. The rate of chalking of the dried paint can be controlled with different crystalline forms

325

of TiO_2. One form, rutile, chalks slowly while another, anatase, chalks more rapidly. Controlled chalking is desirable because it gives a constantly renewed surface over a period of years.

Zinc oxide is a hardener in oil-based paints; zinc and other divalent ions (e.g., Ca^{+2}) form insoluble soaps with the fatty acids from the linseed oil, making the dried surface more durable. Extender pigments do not contribute much to opacity but are added to control the hardness of the paint film and to reduce the cost of the paint. If the relative amount of the total pigment is too low, the film is likely to be too soft. Talc (magnesium silicate, $Mg_3Si_4O_{10}(OH)_2$), limestone (calcium carbonate, $CaCO_3$) and silica (silicon dioxide, SiO_2) are examples of extender pigments. A typical outdoor white oil-based paint could contain the following ingredients:

	Percent by weight
Mineral spirits	34.2%
Titanium dioxide (TiO_2)	29.0%
Linseed oil	19.6%
Calcium carbonate ($CaCO_3$)	12.0%
Zinc oxide (ZnO)	2.2%
Drier	1.7%
Silica (SiO_2)	1.3%

Interior and exterior oil paints are very similar. Titanium dioxide is added in greater amounts to give whiter whites. Natural resins or greater concentrations of pigments are added to give a quicker drying film. Many additives are used in both exterior and interior oil-based paints to hasten the drying of the paint, provide better dispersion of the pigment, resist mildew and fungi formation, and prevent flooding, settling, skinning, yellowing, and destruction by ultraviolet light. For mildew resistance, mercury phenyl acetate, $Hg(OCOCH_2C_6H_5)_2$, is used to the extent of about 0.5 per cent by weight. The mercury ion can be the cause of slight darkening of the paint, since it reacts, like lead, with the pollutant hydrogen sulfide to form black mercuric sulfide.

$$Hg^{2+} + H_2S \longrightarrow \underset{black}{HgS} + 2H^+$$

"Flooding" in paints results when the color components separate on drying and concentrate in streaks or patches in the surface of the film. Antiflooding agents are generally wetting agents, such as aluminum stearate or stearic acid. The polar end of the wetting agent bonds to the pigment particle, and the nonpolar hydrocarbon chain entangles itself compatibly in the oil film.

The Drying Mechanism

The drying of a modern oil-based paint involves much more than the evaporation of the solvent. A principal ingredient of these paints is a drying oil. Many drying oils are used such as soybean, castor, coconut, and linseed (obtained from the seed of the flax plant). All of these oils are esters of glycerin

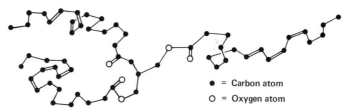

FIGURE 17-4   Structure of a triglyceride of linolenic acid.

and various fatty acids, which were discussed in Chapter 15 in the making of soaps. The most important component of these oils in relation to their drying properties are the unsaturated fatty acids. The most important of these are oleic, linoleic, and linolenic acids.

$$CH_3CH_2CH_2CH_2CH_2CH_2CH_2CH_2CH=CHCH_2CH_2CH_2CH_2CH_2CH_2CH_2\overset{\overset{\displaystyle O}{\|}}{C}OH$$

Oleic acid

$$CH_3CH_2CH_2CH_2CH_2CH=CHCH_2CH=CHCH_2CH_2CH_2CH_2CH_2CH_2CH_2\overset{\overset{\displaystyle O}{\|}}{C}OH$$

Linoleic acid

$$CH_3CH_2CH_2CH_2CH=CHCH=CHCH=CHCH_2CH_2CH_2CH_2CH_2CH_2CH_2\overset{\overset{\displaystyle O}{\|}}{C}OH$$

Linolenic acid

The most widely used drying oil is linseed oil, which contains varying proportions of these unsaturated acids. Figure 17–4 gives a perspective view of the structure of a triglyceride drying oil molecule.

In one film-forming mechanism of drying, oxygen from the air attacks the unsaturated fatty acid chains at a carbon atom adjacent to the double bond, producing first a hydroperoxide radical ($\cdot$OOH) which then attaches itself to the fatty acid chain.

$$—CH_2—CH_2—CH=CH—CH_2—CH_2—$$
$$\nwarrow O_2 \qquad \downarrow$$

$$—CH_2—\underset{\cdot}{CH}—CH=CH—CH_2—CH_2—$$
$$\cdot OOH \qquad \downarrow$$

$$—CH_2—CH—CH=CH—CH_2—CH_2—$$
$$\underset{OOH}{|}$$

When a hydroperoxide group has attached itself to a fatty acid, it quickly forms a bridge to a neighboring fatty acid chain, again at a carbon atom adjacent to a double-bonded carbon atom. This crosslinking mechanism joins together two adjacent molecules by an ether linkage.

Another possible mechanism produces two $\cdot$OH radicals, which continue the process by other free-radical activation.

327

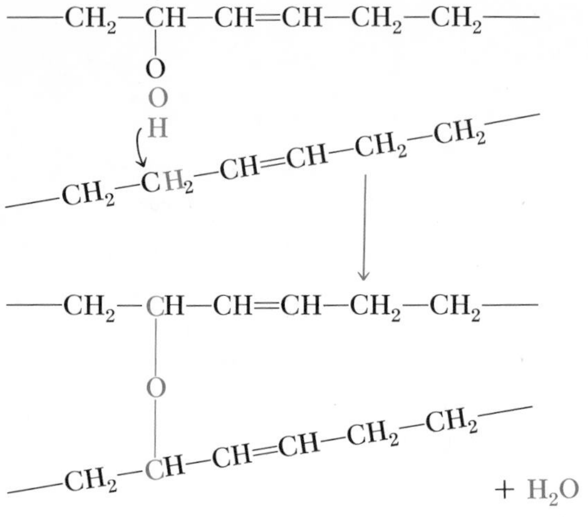

$$-\!-CH_2-CH-CH=CH-CH_2-CH_2-\!-$$

$$\underset{\underset{\displaystyle CH_2-CH_2-CH=CH-CH_2-CH_2-}{|}}{\overset{\displaystyle O}{\underset{O}{|}}}$$

$$-\!-CH_2-CH-CH=CH-CH_2-CH_2-\!-$$

$$\underset{CH_2-CH-CH=CH-CH_2-CH_2-}{\overset{O}{|}}$$

$$+ \ H_2O$$

FIGURE 17-5   Crosslinking by ether formation.

Both water and hydrogen peroxide are formed when oil paints dry. The hydrogen peroxide can be formed by the reaction of two hydroxy free radicals.

$$2 \cdot OH \longrightarrow H_2O_2$$

The result of these cross-linking reactions taking place throughout the drying paint is the production of a web of linked molecules, a polymeric solid. Since the triglyceride molecules prior to drying were relatively small, some penetrated the openings in the painted surface. After drying, good adhesion results. The solid nature of the film also holds the pigment particles in place.

To accelerate the decomposition of the hydroperoxide intermediates discussed above, metallic salts of fatty acids are added. One of the best metal ions is Pb^{2+}. This explains why lead salts were used for such a long time in paints until their toxic nature was discovered. Today salts of Zn, Co, Fe, Mn, and Ca are added to paints. These metallic ions catalyze the decomposition of peroxides. They also precipitate fatty acids as insoluble soaps.

Paint Film Breakdown—The Unwanted Kind

Although oxidation appears to cease when the paint film becomes solid, closer examination reveals that it continues at a slower rate over the lifetime

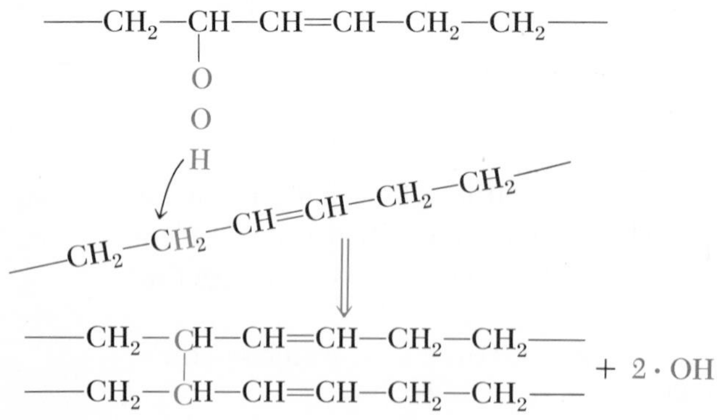

$$-\!-CH_2-CH-CH=CH-CH_2-CH_2-\!-$$

$$\underset{\underset{\displaystyle CH_2-CH_2-CH=CH-CH_2-CH_2-}{|}}{\overset{\displaystyle O}{\underset{O}{|}}}$$

$$-\!-CH_2-CH-CH=CH-CH_2-CH_2-\!-$$
$$-\!-CH_2-CH-CH=CH-CH_2-CH_2-\!-$$
$$+ \ 2 \cdot OH$$

FIGURE 17-6   Crosslinking of two fatty acid chains by formation of a carbon-carbon bond.

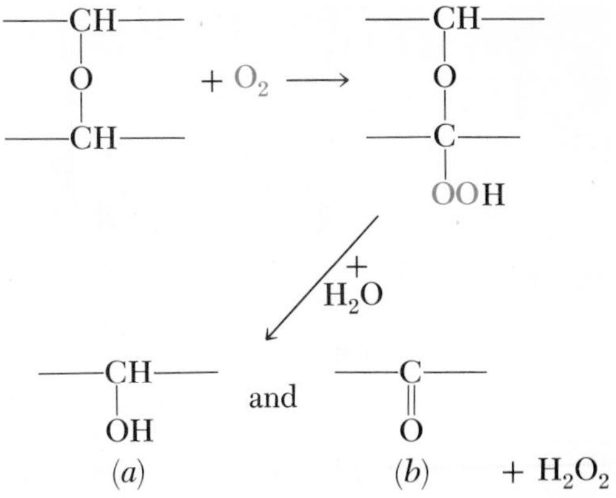

$$—CH— \qquad\qquad —CH—$$
$$| \qquad\qquad\qquad\qquad |$$
$$O \quad + O_2 \longrightarrow \quad O$$
$$| \qquad\qquad\qquad\qquad |$$
$$—CH— \qquad\qquad —C—$$
$$\qquad\qquad\qquad\qquad\qquad ||$$
$$\qquad\qquad\qquad\qquad\qquad OOH$$

$$\overset{+}{H_2O}$$

FIGURE 17-7   Breakdown mechanism for an ether linkage between two fatty acid chains. Oxidation first produces a hydroperoxo group. Water then reacts with this group producing (a) and (b).

$$—CH— \qquad\qquad\text{and}\qquad —C—$$
$$| \qquad\qquad\qquad\qquad\qquad ||$$
$$OH \qquad\qquad\qquad\qquad\quad O$$
$$(a) \qquad\qquad\qquad\qquad (b) \qquad + H_2O_2$$

of the film. For example, the oxidation of ether linkages can lead to a breakdown of the crosslinked polymer structure.

Ultraviolet radiation can cause paint to crack, fade, and depolymerize (Figure 17–8). Protection against ultraviolet rays can be afforded by (1) pigmentation with suitable colors, which absorb the rays without being broken down (e.g., zinc oxide, titanium dioxide, carbon blacks, and iron oxides), (2) pigmentation with substances which reflect the rays (e.g., aluminum flakes), or (3) incorporation of small amounts of ultraviolet absorbers in the first coat (e.g., derivatives of benzotriazoles or benzophenone). Good absorbers have little color and absorb without being broken down themselves. It should be added, however, that black paints really last.

FIGURE 17-8   The results of paint film breakdown are familiar. In order to prevent damage to surface underneath, repainting is suggested before peeling takes place.

329

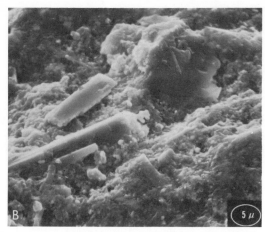

FIGURE 17-9   The surface of a linseed oil solvent base paint after four years of exterior exposure. A, before, and B, after washing with water and detergent to remove chalk; magnified 2000 times. (Courtesy of L. H. Princen, Oilseed Crops Laboratory, U.S. Department of Agriculture.)

The yellowing of paint is observed often, particularly with interior paints. It has been confirmed experimentally that paints tend to yellow more in the dark and in the presence of moisture. Darkness favors the condensation of atmospheric moisture on surfaces; light bleaches and tends to evaporate the moisture. The role (or roles) of moisture in the atmospheric yellowing of paints is not clearly understood. Water may hydrolyze linseed oil into acids that have stronger colors. It may catalyze the oxidation of the oils to other compounds that are highly colored.

Paint Film Breakdown—The Wanted Kind

Often it is necessary to remove paint for one reason or another. Films produced from drying oils are most easily removed by strong alkali such as solutions of lye (NaOH), or of trisodium phosphate (Na_3PO_4). Recalling what we learned in Chapter 15 about saponification of fats with lye to make water-soluble soaps, it should be easy to understand why alkali literally eats up these paints. Drying oils are glycerides, and still contain ester linkages in the polymer film. Lye simply breaks down the polymer to a point at which it loses its structural integrity. When using lye or other strong alkali, lots of water should be used to wash the alkali away.

LATEX PAINTS

A useful method of addition polymerization is called *emulsion polymerization*. This process is used to polymerize styrene and butadiene to make SBR rubber (see Chapter 8). By stabilizing the mixture of styrene and butadiene in water with a soap, the reaction heat is carried off easily by the water. In making tire rubber the emulsion must be broken down. If not destroyed, the emulsion can be used as a paint.

Latex paints have a styrene content as high as 85 per cent with the butadiene

FIGURE 17-10   Clean-up of latex paints is easy if you don't wait too long.

about 15 per cent. An advantage of these paints is that the polymer is already formed in a readily applicable form, the water emulsion.

Immediately after application with brush or roller, the water begins to evaporate. When only a part of the water is gone the emulsion breaks, and the remaining water quickly evaporates, leaving the dry film of paint. Drying times may be as short as 30 minutes. Since no hydrocarbon solvents are involved, odor

TABLE 17-1   ADDITIVES USED IN EMULSION PAINTS

(1) Dispersing agents for pigments	Example: tetrasodium pyrophosphate ($Na_4P_2O_7$). Same principle of like-charged particles repelling as in oil-based paints.
(2) Protective colloids and thickeners	A thicker paint is slower to settle and drips and runs less. A protective colloid tends to stabilize the organic-water interface in the emulsion. Examples: sodium polyacrylates, carboxymethylcellulose, clays, gums. (Same mechanism as soap-dispersing oil in water; Chapter 15.)
(3) Defoamers	Foaming presents a serious problem if not corrected. The surface tension of the water must be decreased. Chemicals used: tri-n-butylphosphate, n-octyl alcohol and other higher alcohols.
(4) Coalescing agents	As the water evaporates and the paint dries, a coalescing agent is needed to stick the pigment particles together. As the resin film forms, the coalescing agent evaporates. Coalescing agents must volatilize very slowly. Examples: hexylene glycol and ethylene glycol.
(5) Freeze-Thaw Additives	Freezing will destroy the emulsion. Antifreezes such as ethylene glycol are used.
(6) pH Controllers	The effectiveness of the ionic or molecular form of the emulsifier depends on the acid or alkaline conditions (pH). The wrong pH will break down the emulsion. Most paints tend to be too acidic. Ammonia, NH_3, is added to neutralize the acid.

331

is at a minimum, and there is no fire hazard. Interior painting with these paints allows quick occupancy after painting.

Since these paints are emulsions based on water, the painting tools can be easily cleaned with soap and water. This fact alone has had an important impact on their acceptance.

In the past few years, paint manufacturers have begun to blend linseed oil emulsions with latex emulsions in order to take advantage of the penetrating ability of the triglyceride molecules in the linseed oil. Presently, some "latex" paints contain as much as 75 per cent linseed oil emulsion, but still have the desirable characteristics of latex. Table 17–1 gives a summary of various additives to emulsion paints and the rationale for their use.

BAKED-ON PAINTS

If you have ever had a car repainted, perhaps you have had the opportunity to see the baking oven in which the paint is dried. Automobile finishes, and those on major appliances such as refrigerators, washing machines, and stoves, require very tough adherent paints in order to withstand the abuse they get.

The chemistry of coatings such as these involves producing extensively cross-linked polymers. A popular type of baked-on paint is the so-called *alkyd*. The term alkyd comes from a mixture of *al*cohol and ac*id*. In Chapter 8, we saw that long-chain polyesters are produced by the reaction of difunctional (two-functional group) alcohols and acids. In alkyds, *poly*functional (more than one reactive group) alcohols and acids are used to form cross-linked films. Consider, for example, the possible reactions of the diacid, phthalic acid, and glycerin, a *tri*alcohol. Note the functional groups left unused.

$$
\begin{array}{c}
\text{COOH} \\
\text{COOH}
\end{array}
+
\begin{array}{c}
\text{HO—CH}_2 \\
\text{HO—CH} \cdot \\
\text{HO—CH}_2
\end{array}
\longrightarrow
\begin{array}{c}
\text{COOH} \\
\text{C—O—CH}_2 \\
\parallel \\
\text{O} \quad \text{HO—CH} \\
\text{HO—CH}_2
\end{array}
$$

Phthalic acid   *Glycerin*

FIGURE 17–11   The high temperatures in a drying oven cause numerous crosslinking reactions to take place, which increase the surface strength of an alkyd paint.

A second as well as a third molecule of phthalic acid can react with a molecule of glycerin.

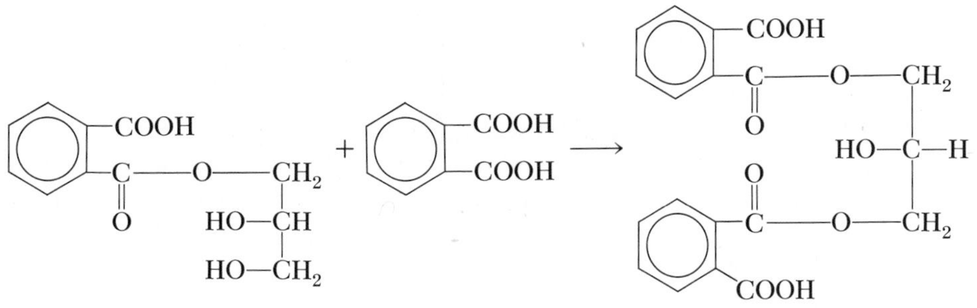

With more reactants, extensive cross-linking becomes possible. The resulting polymer is called a *resin* (Figure 17–12).

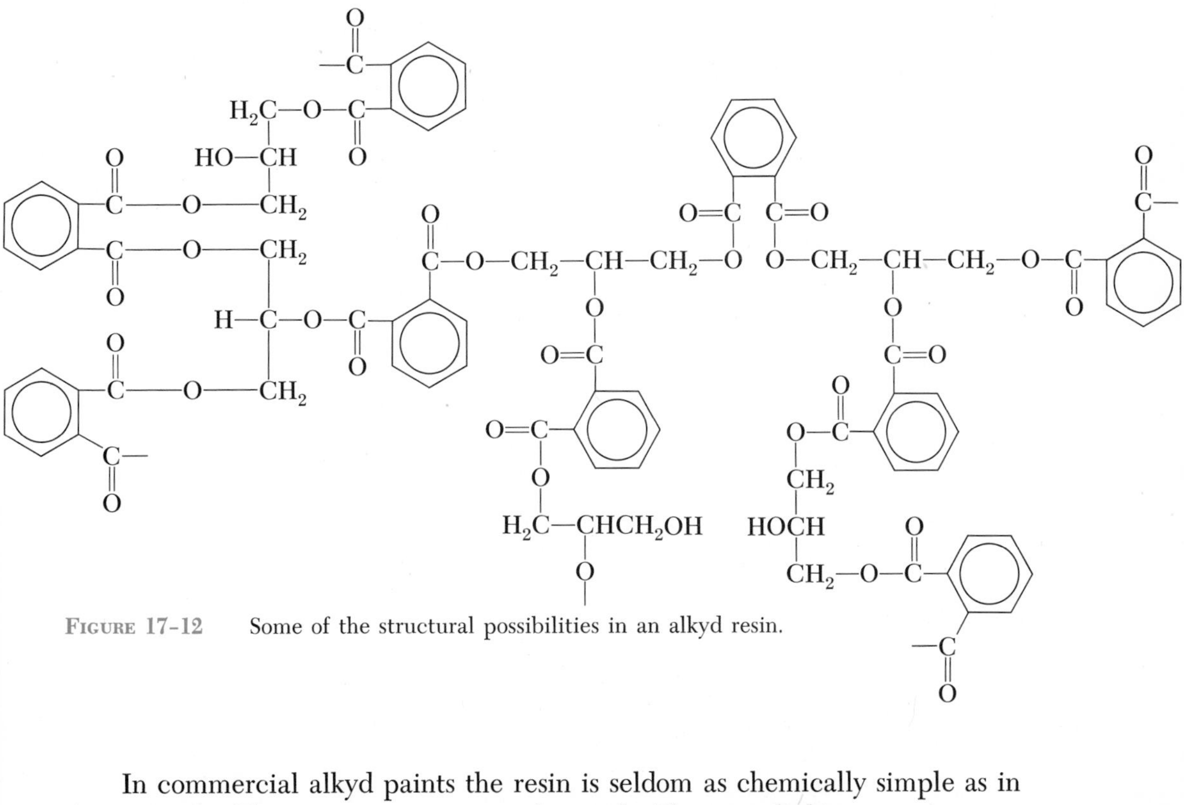

FIGURE 17–12   Some of the structural possibilities in an alkyd resin.

In commercial alkyd paints the resin is seldom as chemically simple as in this example. These resins are extremely tough. The cross-linking reactions are usually begun at the paint factory and carried to completion after the paint has been applied by heating the paint surface, a process known as baking.

Quite often a different type of resin is used along with an alkyd resin to give it more strength. This resin is one formed when urea reacts with formaldehyde. Recall a similar reaction in the casein-formaldehyde white wash discussed earlier.

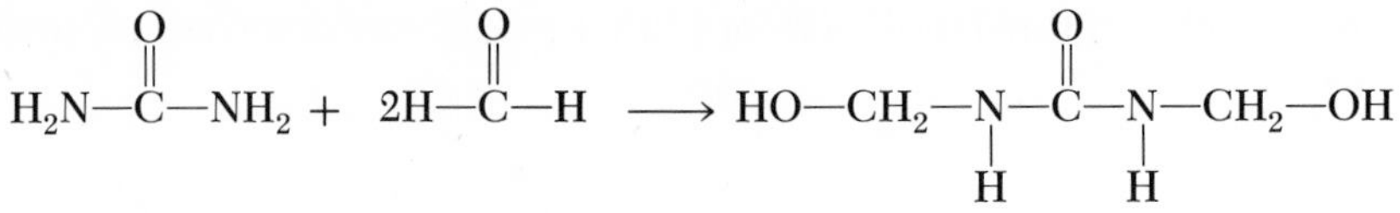

$$H_2N-\overset{\overset{\displaystyle O}{\|}}{C}-NH_2 \; + \; 2H-\overset{\overset{\displaystyle O}{\|}}{C}-H \; \longrightarrow \; HO-CH_2-\underset{\underset{\displaystyle H}{|}}{N}-\overset{\overset{\displaystyle O}{\|}}{C}-\underset{\underset{\displaystyle H}{|}}{N}-CH_2-OH$$

Urea　　　　*Formaldehyde*　　　　　　　*Dimethylol urea*

The dimethylol urea can then polymerize (Figure 17–13).

Many automobile finishes are combinations of alkyd and urea-formaldehyde resins which have been extensively cross-linked by a final baking at temperatures around 130°C for about 1 hour.

In applications like washing machine finishes, alkyds are not as much used as urea-formaldehyde resins since hot soap solution, being alkaline, attacks the ester linkages in the alkyd resin (Figure 17–12).

Paint chemistry is actually an extension of the chemistry of polymers, as we have seen. While it might be possible to make our own paints based on the information we have just studied, numerous other factors, some subtle, go into the making of a good paint. The chemistry is generally refined by a paint chemist so that the paint will protect well, look good, last long and be easy to apply. If it fails in one of these it will be replaced by another.

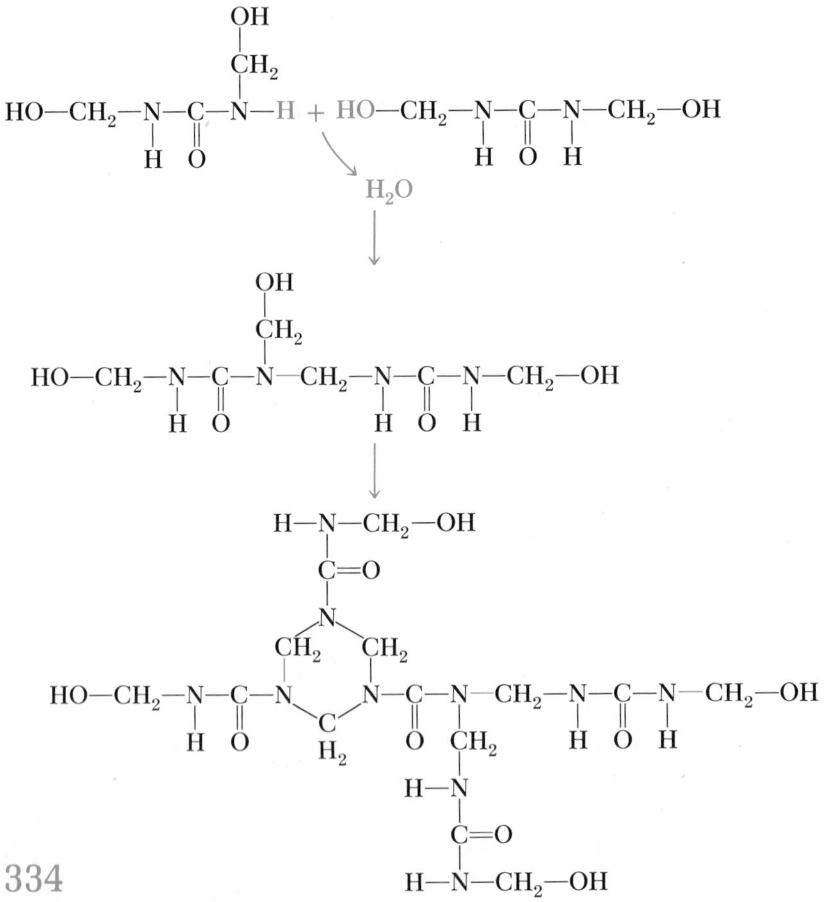

FIGURE 17–13

334

QUESTIONS

1. Name three modes of film formation used in various paints and give an example of each.

2. Do modern latex paints always contain a rubber latex? Explain.

3. What is the chemical cause of chalking?

4. What agents in the environment destroy a paint?

5. What is the simple chemistry of the whiting of whitewash?

6. Which white pigment is banned in interior paints? Explain.

7. Explain why cross-linking increases the toughness of a polymer film.

8. If a white, outside paint becomes black with age, what is probably taking place?

9. What is the source of turpentine? How is it used in paints?

10. What are the two most important ingredients in the drying of an oil-based paint?

11. Could an oil-based paint "dry" in a vacuum? Explain.

12. What is happening when a paint yellows?

13. Lye literally eats up an oil-based paint. What is the chemistry?

14. Would lye be effective on an alkyd paint? Explain.

15. Explain the drying of a latex paint.

16. Give two chief advantages of latex paints over oil-based paints.

17. Why do latex paints have trouble adhering to some surfaces?

18. Acrylics are a popular "latex-type" of paint today. The monomer of polyacrylates is where R is CH_3 or C_2H_5. Draw a representative portion of polyacrylate.

$$CH_2{=}CH$$
$$O{=}C{-}OR$$

SUGGESTIONS FOR FURTHER READING

Hahn, A. V., "The Petrochemical Industry," McGraw-Hill, New York, 1970.
Marteni, C. R., "Alkyd Resins," Reinhold Publishing Co., New York, 1961.
Payne, H. F., "Organic Coating Technology," Vol. I, John Wiley and Sons, New York, 1964.

18

WATER POLLUTION AND PURIFICATION

Water is certainly one of the most important and unusual of all the chemical compounds. Its intimate participation in life processes marks it as indispensable to all forms of life known to man. The rapidly expanding scope of human life, with its associated industrial development, is increasing the demand for water supplies so rapidly that future needs may easily be underestimated. The time has passed when man can take an abundant water supply for granted.

It is obvious that over the centuries to modern times the world water supply has remained essentially constant. Pollution often diminishes the useable water supply to less than comfortable levels in regions in which it would otherwise be quite adequate.

Water pollution arises from two main sources. First, water, being an unusually good solvent, readily dissolves objectionable materials when available. Second, and by far the most serious source, are the activities of man; mercury poisoning in Lake Erie and raw sewage effluents are but two examples.

Not all the chemical knowledge needed to understand water pollution problems and their control is at hand. However, much is known, and research on the chemical problems of polluted water is intense. More important, the technological know-how, much of which is patterned after natural purification processes, is available to solve the water pollution problem. Certainly, purifying polluted water will raise the cost. A wise observer of modern man has stated, "Man will pay for pure water, one way or the other. Either he will pay affordable costs of purification or he will pay the catastrophic costs of an increasingly sick environment."

WATER REUSE

336   It is estimated that 4350 billion gallons of rain (snow and ice) fall on the contiguous United States each day. Of this amount, 3100 billion gallons return

to the atmosphere by evaporation and transpiration. The discharge to the sea and underground reserves amounts to 800 billion gallons daily, leaving 400 billion gallons of surface water each day for personal or commercial use. The 48 states withdrew from natural sources 40 billion gallons per day in 1900, 325 billion gallons per day in 1960, and it is estimated that the demand will be at least 900 billion gallons per day by the year 2000. It is evident then that the *reuse* of surface water is the only way that human needs for water can be met in the future.

CLASSES OF WATER POLLUTION

There was a time when polluted water could be thought of in terms of natural silt and contaminations associated with the natural wastes of animals and humans. As man's use of water has increased and diversified, pollution has become distinctly more chemical in nature. The U.S. Public Health Service has classified water pollutants into eight broad categories, listed in Table 18–1.

A consideration of each of the pollutants in Table 18–1 involves their chemical aspects. Some of these chemical problems will be presented after we have looked at nature's purification processes and the pollutants it is able to control.

NATURAL WATER PURIFICATION

Water is a natural resource which, within limitations, is continuously renewed. The familiar hydrologic or water cycle (Figure 18–1) offers a number of opportunities for nature to purify its water. The world-wide *distillation* process produces rain water with only traces of nonvolatile impurities, along with gases dissolved from the air. *Crystallization* of ice from ocean salt water results in

TABLE 18-1   CLASSES OF WATER POLLUTANTS, WITH SOME EXAMPLES

1. Oxygen-demanding wastes	plant and animal material, reducing agents from paper mills, SO_2.
2. Infectious agents	bacteria and viruses
3. Plant nutrients	fertilizers, such as nitrates and phosphates
4. Organic chemicals	pesticides such as DDT, detergent molecules
5. Other minerals and chemicals	acids from coal mine drainage, inorganic chemicals such as iron salts from steel plants
6. Sediment from land erosion	clay silt on stream bed may reduce or even destroy life forms living at the solid-liquid interface
7. Radioactive substances	waste products from mining and processing of radioactive material, radioactive isotopes after use
8. Heat from industry	cooling water used in steam generation of electricity

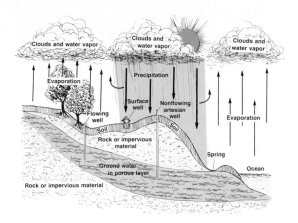

FIGURE 18-1   The water cycle in nature.

relatively pure water in the form of ice floes. *Aeration* of ground water as it trickles over rock surfaces, as in a rapidly running brook, allows volatile impurities, previously dissolved from mineral deposits, to be released to the air. *Sedimentation* of solid particles occurs in slow-moving streams and lakes. *Filtration* of water through sand and other permeable solids rids it of suspended matter, such as silt. Next, and of very great importance, are the *oxidation processes*. Essentially all naturally occurring organic materials—plant and animal tissue, as well as their waste materials—are changed through a complicated series of oxidation steps in surface waters to simple molecules common to the environment. Finally, another process that has been used by nature is dilution. Most, if not all, of the pollutants found in nature are rendered harmless if reduced below certain levels of concentration by dilution with pure water.

Before the advent of the exploding human population and the industrial revolution, natural purification processes were quite adequate to provide ample water of very high purity in all but the arid regions; even in these regions pure water was available from time to time. Nature's purification processes can be thought of as massive but somewhat delicate. In many instances the activities of man push the natural purification processes beyond their limit, and polluted water accumulates.

A simple example of nature's inability to handle man-made pollution is in dragging gravel from stream beds. This excavation leaves large amounts of suspended matter in the water. For miles downstream from a source of this pollutant, aquatic life is nearly all destroyed. Eventually, the solid matter settles, and normal life can be found again in the stream.

A more complex example, one for which there is not nearly so much reason to hope for the eventual solution by natural purification, is the degradability of organic materials. A *biodegradable* substance, as mentioned in Chapter 15 relative to soaps and detergents, is composed of molecules which are broken down to simpler ones in the natural environment. For example, cellulose suspended in water will eventually be converted to carbon dioxide and water.

$$(C_6H_{12}O_6)_x \xrightarrow[\textit{Biochemical catalysts}]{\textit{Oxidizing agents}} CO_2 + H_2O$$

Cellulose   *Carbon*   *Water*
                                    dioxide

The overall oxidation process is a complicated one, involving microorganisms and their enzymatic secretions. Many reaction steps are involved; a detailed study of all the chemistry involved here is yet to be made. Some organic molecules, notably some of those synthetically produced, are not easily biodegradable; these molecules simply stay in the natural waters or are absorbed by life forms and remain unchanged for long periods of time.

Sewage is a very dilute suspension or solution of organic substances in water, if we exclude industrial pollutants. The organic material is approximately 0.05 per cent of the total weight of typical city sewage. The amount of degradable organic material decomposable by natural waters is determined by a delicate balance between the numbers of microorganisms and the supply of oxygen. This is related to the quantity known as biochemical oxygen demand (BOD).

BIOCHEMICAL OXYGEN DEMAND

After many uses, water has an increased concentration of organic material. Even in the natural state, life forms associated with natural waters are constantly putting organic debris into the water. Thus all natural bodies of water supporting life have a *biochemical oxygen demand* (BOD), the amount of oxygen required to oxidize the organic material to simple inorganic molecules. The oxygen is required by microorganisms, such as many forms of bacteria, to utilize organic matter as food in their metabolic reactions. The waste waters of modern society often have very high concentrations of organic material, which have a very large biochemical oxygen demand. In extreme cases, more oxygen is required than is available from the environment. Ultimately, given near normal conditions and enough time, the microorganisms will convert huge quantities of organic matter into the following chemical end-products.

$$\text{Organic carbon} \longrightarrow CO_2$$
$$\text{Organic hydrogen} \longrightarrow H_2O$$
$$\text{Organic oxygen} \longrightarrow H_2O$$
$$\text{Organic nitrogen} \longrightarrow NO_3^-$$

In the extreme case, when the BOD is higher than the oxygen available, radical changes occur in the chemistry and biology of the system. Septic conditions result; fish and other fresh water aquatic life can no longer survive. The aerobic bacteria (those that require oxygen for the decomposition processes) die. As a result of the death of these life forms, even more lifeless organic matter results and the BOD soars. Nature, however, has a back-up system for such conditions. A whole new set of microorganisms (anaerobic bacteria) takes over; these organisms take oxygen from oxygen-containing compounds to convert organic matter to CO_2 and water. Organic nitrogen is converted to elemental nitrogen by these bacteria. Given enough time, enough oxygen may become available, and aerobic oxidation by aquatic life will return.

A stream containing 10 parts per million (ppm) by weight (just 0.001 per cent)° of an organic pollutant from sewage, the formula of which can be represented by $C_6H_{10}O_5$, will contain 0.01 g of this material per liter.

$$?g = 1 \text{ liter of water}$$

$$?g = 1 \text{ liter} \times \frac{1000 \text{ ml}}{1 \text{ liter}} \times \frac{1 \text{ g}}{\text{ml}} = 1000 \text{ g}$$

0.001 per cent of this is the pollutant:

0.001 per cent of 1000 g = (0.00001)(1000 g) = 0.010 g

To transform this pollutant to CO_2 and H_2O, the bacteria present use oxygen as described by the equation:

$$C_6H_{10}O_5 + \quad 6O_2 \quad \longrightarrow 6CO_2 + 5H_2O$$

| *Relative weight* | *Relative weight* |
| 162 | 192 |

° Parts per million is a concentration unit often used by the chemist. It means, just as the name implies, so many parts for every million (10^6) parts. Ten ppm by weight means 10 parts of impurity for every 10^6 parts of the entire sample: so $\frac{10}{10^6} \times 100 = 10^{-3}$ per cent or 0.001 per cent by weight.

FIGURE 18-2   Fish kills can be caused by the lack of a substance necessary for life, such as oxygen, or the presence of toxic materials that interfere with the life processes. A heavy concentration of organic matter in a stream may depress the oxygen concentration below that required to support fish life. (From Tidwell, P.: Anti-pollution: A fish management priority. The Tennessee Conservationist, 37:4, 1971.)

The 0.01 g of pollutant requires 0.012 g of dissolved oxygen.

$$?g \text{ oxygen} = 0.010 \text{ g pollutant} \times \frac{192 \text{ g oxygen}}{162 \text{ g pollutant}} = 0.012 \text{ g oxygen}$$

At 68°F the solubility of oxygen in water under normal atmospheric conditions is 0.0094 g of oxygen per liter.

Since the BOD (0.012 g per liter) is greater than the equilibrium concentration of dissolved oxygen (0.009 g per liter), as the bacteria utilize the dissolved oxygen in this stream, the oxygen concentration of the water will soon drop too low to sustain any form of fish life (Figure 18–2). However, if the water is flowing vigorously in a shallow stream (this facilitates the absorption of more oxygen from the air via aeration), the life forms common to fresh water will likely survive.

In the study of polluted water, it is important to know how much oxygen a given sample of polluted water will require. Simple laboratory measurements give a very good estimate. For example, a known volume of the sewage is diluted with a known volume of standardized sodium chloride solution of known oxygen content. This mixture is then held at 20°C for five days in a closed bottle. At the end of this time the amount of dissolved oxygen remaining in the sample is determined, and the amount of oxygen which has been consumed is taken to be the biochemical oxygen demand.

THERMAL POLLUTION

Thermal pollution results when water is used for cooling purposes and in the process has its own temperature raised. Water has a very high *heat capacity* (the heat required to raise a unit of weight of water one degree) and a high *heat of vaporization* (the heat required to change a unit of weight of water to gaseous water); this combination makes water an ideal cooling fluid for thermal power stations, nuclear energy generators, and industrial plants. The extent to which thermal pollution can occur is illustrated by the Thames River, the center of an industrial complex in England. The yearly average temperature in this river, corrected for changes in climatic conditions, rose from 53°F in 1930 to 60°F in 1950. The heat released into this river in 1930 was at the average rate of 550 megawatts. In 1950 the rate became 3700 megawatts. The river, even with various means of cooling (evaporation, conduction, and discharge to the

TABLE 18-2   HEAT CONTRIBUTORS TO INCREASED HEAT CONTENT
OF THAMES IN 1950

	PERCENT
Fossil fuel power stations	75
Industrial effluents	6
Sewage effluents	9
Fresh water discharge	6
Biochemical activity	4

TABLE 18–3   Solubility of Oxygen in Water at Different Temperatures

Temperature (°C)	Solubility (Parts per Million)
25	8.4
15	10.2
5	12.5

sea), simply could not dissipate the heat as fast as it has been fed in by man. Table 18–2 lists the contributors to this heat inflow.

The most obvious result of thermal pollution is to make the water less efficient for further cooling applications. Far more important, however, are the biological and biochemical implications.

The oxygen content of water in contact with air is dependent on the temperature of the water, and increases as the water temperature is decreased (Table 18–3). Also, the rate at which water dissolves oxygen is directly proportional to the *difference* between the actual concentration of oxygen present and the equilibrium value. This is extremely fortunate, since it means the rate of solution of oxygen increases sharply as the oxygen is used up. Thus, the rate of absorption of oxygen from the air is greatly facilitated in a shallow, cold mountain stream compared to the case of a deep lake behind a dam in a warmer river.

The increased temperature of the Thames River also increased the oxygen deficit by four per cent over what it otherwise would have been. However, the biochemical result of this factor alone could not be determined, since the river was and would have been anaerobic as a result of other pollutants in 1950. Even though we readily admit there is much about the consequences of thermal pollution that we do not know, two conclusions seem obvious: (1) thermal pollution aggravates the problem of oxygen supply, and (2) a significant rise in the temperature of a stream can drastically change or even destroy entire biological populations.

Solutions to thermal pollution involve cooling the water in evaporation towers or storage lakes before returning it to the natural body of water. Cycling the water for reuse after cooling has obvious advantages.

TABLE 18–4   Projection of the Thermal Pollution Problem in the United States°

Year	Per Capita Electrical Energy Consumption in U.S. (48 States, Actual or Projected)	Cooling Water Needs (Billion Gallons per Day)
1950	2,000 Kwh	
1968	6,500 Kwh	
1980	11,500 Kwh	200†
2000		450†

° Notes: (a) Annual surface runoff is 1250 billion gallons per day.
(b) Generation of electrical energy is doubling every 10 years.
(c) 500 new power stations will be needed by the year 2000, at the present growth rate.

† Projected.

FIGURE 18-3   Thermal pollution. Plant effluent causes the Calumet River to "steam." Such heat discharges can change the temperature of a natural body of water enough to vastly alter the ecological balances. (From Chemical and Engineering News ["Technology"], Feb. 24, 1969.)

FERTILIZERS

Contamination of water by fertilizers leads to very undesirable effects. In a large measure, this contamination results from the phosphate (PO_4^{3-}) and nitrate (NO_3^-) present in fertilizers. It is generally believed that phosphate and nitrate encourage the growth of large amounts of algae. It is certain that nitrate in sufficient concentration is toxic to most higher organisms.

Limnology is the study of physical, biochemical, and biological aspects of fresh-water systems. One of the conclusions of limnology is that massive but relatively delicate balances exist in such systems. The growth of a limited kind and quantity of algae is good for a lake; such growth is a source of needed oxygen via photosynthesis:

$$6CO_2 + 6H_2O \xrightarrow[Catalysts]{Sunlight} C_6H_{12}O_6 + 6O_2$$

Carbon   Water   Sugar
dioxide

However, in waters that contain excessive amounts of phosphate and nitrate leached from fertilized farm land, algae growths are sometimes so massive they tend to choke out desirable life forms. Such "blooms" of algae may lead to eutrophication. In this state, the normal oxygen-producing algae actually produce an oxygen-deficient environment, especially at lower depths. This is because the massive quantity of dead plant material causes the BOD to rise sharply.

Recent research studies by Wyandotte Chemical Company suggest that

343

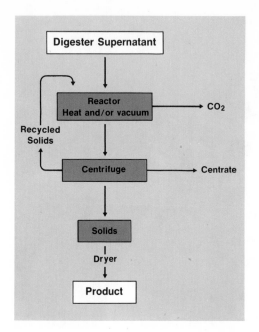

FIGURE 18-4   W. R. Grace process for removing phosphates from sewage.

phosphate and nitrate are not the primary causes of algae blooms. While the issue is not settled to the satisfaction of all concerned, this research suggests that the excessive organic pollutant is the primary cause of algae blooms. Bodies of water in which the phosphate concentrations were kept abnormally low—less than 10 parts per billion—produced algae blooms. It was reasoned that the carbon dioxide at higher-than-normal levels resulting from the bacterial oxidation of organic material allowed the excessive algae growth. This "defense" of higher-than-normal phosphate concentrations in our natural waters has to be judged, at best, not entirely satisfactory since the limnological problem is so complex and our present knowledge is so limited.

In the meantime, W. R. Grace Company, at the end of a ten-year project, has announced a process that would reduce the great bulk of phosphate from municipal sewage. The process (Figure 18–4) can be accomplished in most cases without any addition of chemicals. The Grace process involves heating the sewage to 70°C. The solution becomes alkaline as a result of the decomposition of ammonium bicarbonate and other chemicals.

$$NH_4HCO_3 \longrightarrow NH_3 + H_2O + CO_2$$

$$\text{Ammonium bicarbonate} \qquad \text{Ammonia} \quad \text{Water} \quad \text{Carbon dioxide}$$

The ammonia causes the solution to be basic (alkaline) and the carbon dioxide is removed with a vacuum. Phosphate is then precipitated as magnesium ammonium phosphate.

$$NH_3 + H_2O \longrightarrow NH_4^+ + OH^-$$

Ammonia   Water   Ammonium   Hydroxide
                       ion   ion

$$Mg^{2+} + NH_4^+ + PO_4^{3-} \longrightarrow MgNH_4PO_4$$

Magnesium   Ammonium   Phosphate   Magnesium ammonium
ion   ion   ion   phosphate

The solid product is centrifuged and dried; it is a marketable product. The process also reduces nitrogen, BOD, and COD (chemical oxygen demand) levels.

The presence of nitrate in water can be quite dangerous, and normal procedures used in purifying water for municipal supplies do not remove it. When the concentration of nitrate is over a few ppm, it acts as a poison and destroys the ability of human hemoglobin to carry oxygen, as a result of the reduction of nitrate (NO_3^-) to nitrite NO_2^-. The nitrite interferes with the transport of oxygen by hemoglobin and leads to cyanosis (a condition in which the surface of the body becomes blue because of a lack of oxygen in the blood). This is especially dangerous to infants, who are abnormally susceptible to poisoning (and death) via nitrate in their drinking water.

Fertilizers are made from simple chemicals, such as ammonium nitrate, NH_4NO_3, that are quite soluble in water. During and after a hard rain the surface run-off carries much of the fertilizer away. On the other hand, natural fertilizers are not water-soluble, but release nitrate, phosphate, and metal ions at a rate commensurate with the rate of uptake by the plants. New fertilizer products are now on the market that imitate nature in releasing their soil nutrients at a slower rate.

PESTICIDES

The use of synthetic insecticides increased enormously on a world-wide basis after World War II. As a result, insecticides such as DDT have found their way into lakes and rivers. There is a great variety of pesticides, and their use frequently leads to severe damage to other forms of animal life, such as fish and birds. The widespread use of pesticides was originally justified by the need for more food for a growing human population. However, it has to be admitted that the toxic reactions and peculiar biological side-effects of many of the pesticides were not thoroughly studied or understood prior to their widespread use.

A good case in point is DDT. This insecticide, which has not been shown to be toxic to man in doses up to those received by factory workers involved in its manufacture (400 times the average dosage over years of exposure), does

DDT

have peculiar biological consequences. The structure of DDT is such that it is *not* broken down very rapidly by animals or fish; it is deposited and stored unchanged in the fatty tissues. The biological half-life of DDT is about eight years; that is, it takes about eight years for an animal to metabolize one-half of an amount it assimilates. The enzymes to catalyze the breakup of this molecule are just not present. If ingestion continues at a steady rate, it is evident that DDT will build up within the animal over a period of time. For many animals this is not a problem, but for some predators, such as eagles and ospreys, which feed on other animals and fish, the consequences are disastrous. The DDT in the fish eaten by such birds is concentrated in the bird's body, which attempts to metabolize the large amounts by an alteration in its normal metabolic pattern. This alteration involves the use of molecules which normally regulate the calcium metabolism of the bird and are vital to its ability to lay strong eggs. When these molecules are diverted to their new use, they are chemically modified and are no longer available for the egg-making process. As a consequence, the eggs the bird does lay are easily damaged, and their survival rate decreases drastically. This process has led to the nearly complete extinction of eagles and ospreys in some parts of the United States where formerly they were numerous.

DDT and other insecticides such as *dieldrin* and *heptachlor* are referred to as *persistent pesticides*. Substitutions of other substances with biodegradable structures such as chlordan are now made possible. It is interesting to note that the structural differences between heptachlor (persistent) and chlordan (short-lived) are relatively slight (look at the chlorine atom on the lower five-membered ring).

Dieldrin

Heptachlor

Chlordan

In a study made under the sponsorship of the U.S. Department of Health, Education and Welfare from 1964 through 1967, 10 municipal water supplies were examined for pesticides. These waters were taken from the Mississippi and

Missouri rivers. Over 40 per cent of the samples contained dieldrin, over 30 per cent contained endrin, and 20 per cent contained chlordan. The amounts present were barely detectable and in the parts per billion (ppb) concentration range.

While it might occur to us to forbid completely the use of such insecticides, we would soon find that there are many areas of the United States in which agricultural production would drop precipitously were this to be done. An enlarged human population requires that crops be protected against the ever-present threat of destruction by insects. Furthermore, the use of insecticides is essential if we are to control malaria, plague, and other diseases. It must be recognized that a single insect species, the tsetse fly, alone has retarded the development of over 4,000,000 square miles of Africa by means of the diseases which it transmits to men and cattle.

The choice of solutions to our problems with insects is not an easy one. The use of insecticides introduces them into our environment and our water supplies. A refusal to use insecticides means that man must tolerate malaria, plague, sleeping sickness, and consumption of a large part of his food supply by insects. It is obvious that continuing research is needed on new methods and materials for the control of insect populations by both chemical and biological means. From the chemical point of view it is entirely reasonable to believe that chemical agents can be found which will control pests and then in turn be quickly biodegraded in the environment.

INDUSTRIAL WASTES

Industrial wastes can be an especially vexing sort of pollution problem because they often are not removed or are removed very slowly by naturally occurring purification processes and are generally not removed at all by a typical municipal water treatment plant. Table 18–5 lists some of the large variety of industrial pollutants that have been put into our natural waters on a large-scale basis.

The technology necessary to remove industrial wastes from the water before it is returned to natural waters is available now. The limiting factor is the cost.

TABLE 18-5   A PARTIAL LIST OF INDUSTRIAL WASTES THAT HAVE BEEN DUMPED INTO NATURAL WATERS

Acids
Bases
Salts
Various metal solutions
Oils
Bacteria
Emulsifying agents
Grease
Scrap plant and animal wastes
Dyes
Waste solvents
Poisons such as cyanides and mercury
Numerous chemicals from washing operations

347

Society will simply have to decide whether it is willing to spend a greater fraction of its productivity in order to have an unpolluted environment. The control of this pollution is made very difficult by the lack of national and even international standards for all parties to observe. For example, steel mills have been in the past, and to a lesser extent still are, notorious polluters of water. One company would find it very difficult to compete if it had to clean up all of its wastes while other companies went free. Concerted action, whether voluntary or enforced, is necessary. Very large sums of money have recently been spent by some industrial concerns to control or limit pollution. Dow Chemical Company, for example, built a $1.3 million plant to control wastes in a plant producing organic chemicals at Midland, Michigan. However, with the population explosion and the resulting industrial expansion, our society is still losing ground in industrial water pollution. A much greater effort is needed.

An example will now be given of an industrial pollution problem that can be solved in a classical as well as in a recently developed way. Metal processing and finishing operations produce large quantities of ions in solution. Chromium plating, for example, produces a waste solution containing chromium in the form of chromate (CrO_4^{2-}) and cyanide (CN^-), and other, less objectional ions. Such wastes formerly were emptied into the natural waters. Chemical treatment has been understood for many years and has been used where efforts were made to clean up these wastes. The cyanide can be oxidized by an alkaline chlorine solution,

$$\text{Cyanide} + \text{Chlorine} \xrightarrow{\textit{Base}} \text{Nitrogen} + \text{Bicarbonate} + \text{Chloride}$$

The chromium-containing ion (CrO_4^{2-} [chromate]) is reduced by sulfur dioxide (SO_2).

$$\text{Chromate} + \text{Sulfur dioxide} \longrightarrow \text{Chromium(III) salts} + \text{Sulfate ions}$$

The products of these reactions either are insoluble or are not objectionable, being common to the natural environment. However, it must be admitted that

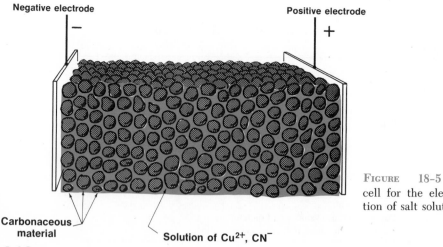

Negative electrode

Positive electrode

Carbonaceous material

Solution of Cu²⁺, CN⁻

FIGURE 18–5 Carbonaceous cell for the electrolytic purification of salt solutions.

TABLE 18-6   PURIFICATION OF COPPER PLATING WASTE SOLUTION BY
CARBONACEOUS ELECTROLYTIC CELL

	CONCENTRATIONS (PPM)	
	Before	After
Copper	40	less than 1
Cyanide	20	less than 0.5

the waste is not chemically pure, since it is a solution of NaCl and other substances, including a high concentration of CO_2.

It has been known for some time that an electrolytic reduction would produce chemically pure water. This method, however, has not been used because water is such a poor conductor of electricity. Excessive amounts of time and power would be required to complete the electrolysis. Recently, Resource Control, Inc., of West Haven, Conn., developed an electrolytic cell packed with an insoluble carbonaceous material (Figure 18–5). In this cell, the carbonaceous material provides good electrical conductivity between the electrodes; carbon is a good conductor of electricity. The electrical resistance is not a function of the salt concentration, which can be reduced essentially to zero in a short period of time with minimal power requirements. While the theory is as yet not well understood, there appear to be many microcells on the surface of the carbon. The positive ions (cations) are reduced at each cathode and the negative ions (anions) are oxidized at each anode. Table 18–6 gives before-and-after results for a copper plating waste solution.

Another industrial pollution problem, which does not have a simple economic and chemical solution, is the acid drainage from coal mines. Iron pyrite (FeS_2) is oxidized in the presence of water to give an acidic solution.

$$2FeS_2 + 7O_2 + 2H_2O \longrightarrow 2Fe^{2+} + 4SO_4^{2-} + 4H^+$$
Sulfuric acid

Groundwater flushes this acid into surrounding streams, and aquatic life in those streams is destroyed. It was once thought that if the coal mines were sealed, the oxygen would be used up and the reaction would stop. However, bacterial action catalyzes an oxidation process (not as yet well understood) in mines in which the oxygen concentration is nil, and the acid continues to flow. The only apparent solution at this time is to treat the drainage to remove the pollutants.

Heavy metals such as lead, mercury, cadmium, chromium, manganese, copper, zinc, etc., are generally toxic to life forms in greater than trace amounts (Chapter 10). Millions of pounds of these metals are "consumed" each year in the United States. For example, the United States government has estimated that 163 million pounds of mercury have been consumed in this country since 1900, with 6 million pounds being the consumption figure in 1969. A breakdown of uses is given in Table 18–7.

We do not know where all of the waste mercury goes, but we do know that widespread mercury pollution has been found in most industrialized regions throughout the United States. One chlorine plant is known to have been dis-

TABLE 18-7   MERCURY CONSUMPTION IN THE UNITED STATES TOPS 6 MILLION POUNDS°

	1969 CONSUMPTION (THOUSANDS OF POUNDS)
Electrolytic chlorine	1,572
Electrical apparatus	1,382
Paint	739
Instruments	391
Catalysts	221
Dental preparations	209
Agriculture	204
General laboratory use	126
Pharmaceuticals	52
Pulp and paper making	42
Amalgamation	15
Other	1082
TOTAL	6035

° U.S. Department of Interior, Bureau of Mines.

charging as much as 200 pounds of mercury a day. Downstream from another plant, silt and clay deposits contained 560 ppm mercury. Four miles downstream the figure was still 50 ppm, and fish in a 35-mile stretch of the river had mercury levels as high as 1.13 ppm. The upper limit in water set by both the United States and Canada for mercury is 0.5 ppm. Water in most polluted regions has much less mercury actually in solution, the figure of 0.03 ppm or less being given.

Mercury in solution is often in the form of dimethyl mercury, $(CH_3)_2Hg$. Evidently the metal either goes into the water in elemental form, or—since it is easily reducible—it is reduced to the element.

$$Hg^{2+} + \text{Reducing agent} \longrightarrow Hg + \text{Oxidized form of the reducing agent}$$

Slime deposits containing elemental mercury evidently serve as a source of methyl mercury. It is generally believed that mercury enters the food chain through small organisms that feed on the bottom. These in turn are food for bottom-feeding fish. Game fish in turn eat these fish and accumulate the largest concentration of mercury, the accumulation of poison building up as the food chain progresses.

Mercury pollution can be stopped. For example, the tiny particles of mercury in the outflow of chlorine plants can be held in settling tanks. The mercury will be deposited on the bottom, and the brine can be concentrated by evaporation and used again.

FRESH WATER FROM THE SEA

As we look for new supplies of water, attention is inevitably drawn to the sea as well as to the vast supplies of brackish water frequently found by drilling in arid lands. Sea water contains 3.5 per cent salts; brackish waters have smaller salt concentrations. It is obvious that processes which can convert brackish and salty water to potable (drinkable) water are potentially very valuable. Of course, the value of such processes is actually determined by the cost of the pure water produced. Thus far, the cost of purification of salt water is high in comparison to the cost of water supplied by a typical European or American municipal water system. We can expect the cost of fresh water purification to increase if pollution continues to increase, and the cost of purifying salt water to decrease as technology is improved.

The purification of salt water can be divided into two general approaches. Water may be separated from the salt by a change in state, either by evaporation to the gas or by freezing to the solid. A second approach is to cause the ions in the salt water to move, under the influence of electrical charge, chemical attraction, or selective membranes, out of the stream of water flow. If this process continues long enough, pure water results. Bringing about a change of state in water is expensive because of the large amount of energy that must move either into or out of the water. Recall that 540 cal is required for the vaporization of 1 g of water and 80 cal must be removed for the freezing of each gram of water. Moving ions out of the water is expensive because it is relatively slow for large flows of water and involves new techniques that are still under development.

DISTILLATION

The use of distillation to obtain fresh water from sea water has been developed to a considerable state of refinement. In view of the large amounts of heat required, various methods have been devised to heat the water with "waste heat" prior to entering the still. One method is to cool the steam with raw water as it enters the still. Since the steam has to lose 540 cal per gram in order to liquify, this energy can be used to heat up the raw water to near the boiling point. Solar energy has also been used to preheat the water. Distillation plants are now the most common method in use to refine sea water and can produce water at a cost of less than $1 per 1000 gal.

FREEZING

When cold sea water is sprayed into a vacuum chamber, the evaporation of some of the water cools the remainder and ice crystals form in the brine. Heat is required to vaporize the cold water, and there is no source for the heat except the cold water and the walls of the chamber. If the chamber is well insulated, the water is cooled to 0°C and some of it freezes. When the crystals of ice form, they tend to exclude the salt ions. Even though the separation of salt and water is not complete in one step, the ice has less salt than the same weight of liquid solution. The ice crystals can be collected on a filter, washed with a small amount of fresh water, and then melted to obtain "pure" water. The process is repeated until the degree of purity desired is achieved. Plants have been built and successfully operated which produce as much as 250,000 gal of pure water per day by this purification technique.

Ion Exchange

In this process, brackish water or sea water is first passed through a cation exchange resin to replace the cations with H^+ and then through an anion exchange resin to replace the anions with OH^-; these two ions then neutralize each other. Unfortunately, the amount of sea water which can be purified by a given amount of ion exchange resin is quite small, and the resulting water is relatively costly.

Modern ion-exchange resins are high-molecular-weight polymers which contain firmly bonded groups which can exchange one ion for another present in solution. *Cation* exchangers, or positive-ion exchangers, swap their hydrogen ions for the positive ions present in solution. They are usually organic derivatives of sulfuric acid, and their action can be depicted as:

$$R\text{—}SO_3^- H^+ + Na^+ + Cl^- \longrightarrow R\text{—}SO_3^- Na^+ + H^+ + Cl^-$$

When the polymer is saturated with cations, it can be regenerated by treatment with strong acid, which reverses the above reaction. Because they generate weakly acidic solutions, *cation* exchange resins themselves do not do a complete job. When they are followed by resins that exchange negative ions (anions), they provide a good route to very pure water. An *anion* exchange resin can replace the anions in solution by the hydroxide ion. These resins are again high-molecular-weight polymers, but now the polymer contains a nitrogen atom bonded to four other groups. They function as shown:

$$\text{Polymer} \quad \text{—N—}^+, OH^- + \underbrace{H^+ + Cl^-} \longrightarrow \text{Polymer} \quad \text{—N—}^+, Cl^- + H_2O$$

From cation exchangers

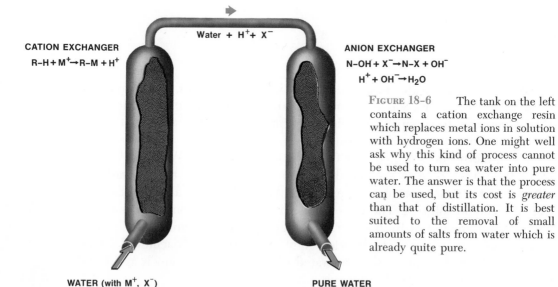

CATION EXCHANGER
$R\text{-}H + M^+ \rightarrow R\text{-}M + H^+$

Water + H^+ + X^-

ANION EXCHANGER
$N\text{-}OH + X^- \rightarrow N\text{-}X + OH^-$
$H^+ + OH^- \rightarrow H_2O$

FIGURE 18-6   The tank on the left contains a cation exchange resin which replaces metal ions in solution with hydrogen ions. One might well ask why this kind of process cannot be used to turn sea water into pure water. The answer is that the process can be used, but its cost is *greater* than that of distillation. It is best suited to the removal of small amounts of salts from water which is already quite pure.

WATER (with M^+, X^-)

PURE WATER

When the anion exchange resin has exchanged all its hydroxide, it can be regenerated by treatment with a strong solution of sodium hydroxide, which reverses the above reaction.

ELECTRODIALYSIS

In the electrolysis of salt water, cations such as Na^+ migrate toward the negative electrode and anions such as Cl^- move toward the positive electrode. If an electrolysis cell is divided into three compartments by semipermeable membranes, one permeable to cations and one permeable to anions (Figure 18–7), the process is termed "electrodialysis." Dialysis is the passage of selected species in solution through membranes while other species are excluded. Electrodialysis is the special case in which the passage of ions is influenced by positive and negative electrodes.

In Figure 18–7, note that cations can move to the left out of the center compartment but cannot move through the anion membrane from the right compartment to the center one. In a similar way, anions move out of the center compartment to the right. Thus, the ionic concentration of the water in the central compartment is reduced. If the process is continued long enough, the water in the central compartment is freed from its salt content.

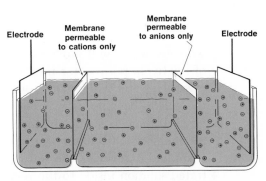

All three compartments filled with brackish water

FIGURE 18–7 The basic electrodialysis process. Each compartment of the cell contains brackish water. Application of an electrical potential across the cell causes the cations to move into the left compartment and the anions to move into the right compartment. The salt content of the water in the center compartment is thus reduced.

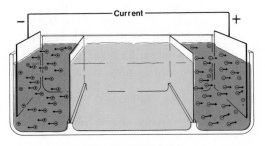

Salts removed from water
in center compartment

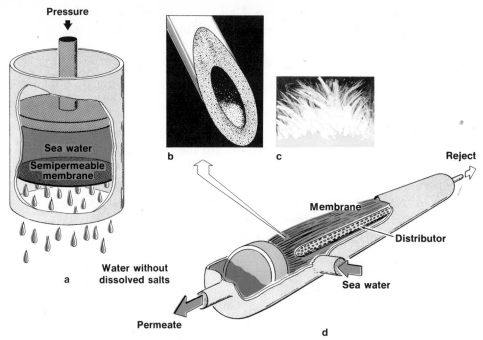

FIGURE 18-8   Reverse osmosis. (*a*) Mechanical pressure forces water against osmotic pressure to region of pure water. (*b*) Enlargement of individual membrane. (*c*) Mass of many membranes. (*d*) Industrial unit; feed water (salt) that passes through membranes collects at left end (permeate). The more concentrated salt solution flows out to the right as the reject.

REVERSE OSMOSIS

The elementary science student has learned that osmosis is the process whereby water will move through a semipermeable membrane from a region of relatively pure water into a region containing a concentrated solution. For example, water will move through a cell membrane into the protoplasm, causing the cell to become turgid. The resulting pressure inside the cell is often very large. If enough mechanical pressure is brought to bear on the solution inside the membrane (in other words, if a pressure greater than the osmotic pressure is applied in reverse), the water can actually be made to flow from the concentrated solution inside to the region outside (Figure 18–8).

Reverse osmosis is a very promising process. Costs are down to 25¢ per 1000 gal for a plant producing 30,000,000 gal of pure water per day. This compares to 7.5¢ for the typical municipal treatment and 11¢ for the activated sludge treatments. Membranes used thus far are mostly cellulose acetate and work very well for salt water. Organic materials in the water tend to foul these films when used for municipal sewage. However, much effort is being made in developing new films for a wide variety of applications of reverse osmosis.

<hr />

QUESTIONS

1. What are some processes that *decrease* the amount of dissolved oxygen in a stream? What are some processes that *increase* the amount of dissolved oxygen in a stream? Which ones are most readily subject to man's control?

2. Explain why each of the following introduces a pollution problem when its wastes are emptied into a stream.

   (a) A slaughterhouse or meat-packing plant.
   (b) A paper mill.
   (c) An electric generating plant burning oil or coal.
   (d) An agricultural area which is intensively cultivated.

3. An old rule of thumb is, "Water purifies itself by running two miles from the source of incoming waste." What processes are active in purifying the water? Is this adage foolproof? Explain.

4. To render a pollutant like DDT harmless, is it necessary to convert the carbon to carbon dioxide, the hydrogen to water, and the chlorine to chloride? Explain.

5. What are some consequences of requiring by law that all water put into our national waters be clinically and chemically pure H_2O? Do you think this is the proper solution to water pollution? If not, what is the proper compromise?

6. What are some ecological consequences of thermal pollution?

7. What pertinent facts would you try to gather if it is your responsibility to vote on a bill to regulate water pollution?

8. Which generally comes first: theoretical solutions to water pollution problems based on an understanding of the reaction of the molecules or technological solutions based on field or laboratory experiences? From the material in this chapter, select an example of each of these two approaches.

9. After thinking about water pollution, at what point do you think pollutants should be removed from used water? Who should be responsible for this removal? Would you distinguish between industrial wastes and household wastes?

10. The major elements in organic compounds are carbon, hydrogen, oxygen, and nitrogen. What are the oxidation products for these elements in the natural decomposition that occurs in nature?

11. Classify water pollutants into as few major groups as you can. Relate these groups to the topics presented in this chapter.

12. In your judgment, what are the most serious water pollution problems? Be ready to defend your points in class discussion.

13. What is natural osmosis? Explain the significance of the word "reverse" in reverse osmosis.

14. From your experience, see if you can add one additional example for each of the classes of pollutants listed in Table 18–1.

15. If electrical energy costs 1¢ per kilocalorie, what will be the cost to distill 4 liters (approximately 1 gal) of water from sea water if no attempt is made to utilize the heat of condensation of the water vapor?

16. What will be the BOD of a stream that contains the equivalent of 0.0001 per cent C as oxidizable organic matter?

17. If the biological half-life of DDT is 8 years, how long will it take to reduce the amount of DDT in a human from 100 mg/kg to 3.1 mg/kg?

SUGGESTIONS FOR FURTHER READING

Behrman, A. S., "Water is Everybody's Business," Doubleday & Co., Garden City, N.Y., 1968.
Hills, E. S. (ed.), "Arid Lands: A Geographical Appraisal," Methuen & Co. (New York), 1969.

Howells, G. P., and Kneipe, T. J., "Water Quality in Industrial Areas: Profile of a River," *Environmental Science and Technology*, Vol. 4, p. 26 (1970).

Keller, E., "Fish Kills," *Chemistry*, Vol. 41, No. 9, p. 8 (1968).

Keller, E., "Nuclear-Powered Desalting in the Middle East," *Chemistry*, Vol. 42, No. 2, p. 7 (1969).

Keller, E., "The DDT Story," *Chemistry*, Vol. 43, No. 2, p. 8 (1970).

Leopold, L. B., and Langheim, W. B., "A Primer on Water," U.S. Government Printing Office, Washington, D.C., 1960.

Newman, Frank, "A Water Pollution Study," *Chemistry*, Vol. 42, No. 1, p. 28 (1969).

Overman, M., "Water," Doubleday & Co., Garden City, N.Y., 1968.

Slabaugh, W. H., "Clay Colloids," *Chemistry*, Vol. 43, No. 4, p. 8 (1970).

19

AIR POLLUTION

You are probably breathing polluted air this very minute. Over 43 million Americans live in 300 cities described by the United States Public Health Service as suffering from "major" air pollution. The PHS also reports that every community of over 50,000 population has an air pollution problem of some sort. If you just now had a breath of clean, fresh air, you are in a rapidly shrinking minority.

Increasing and widespread urbanization, industrialization, and use of the automobile have spread air pollution across our country. It is increasingly becoming a cause of concern to the general public. Isolated incidences of devastating air pollution, however, have been known for centuries in several communities of the world.

Air pollution was so bad in 17th century London that John Evelyn published a treatise called *The Smoake of London* (Figure P–4). The title page of this book bears a quotation from Lucretius, a Roman poet, translated, "How easily the heavy potency of carbons and odors sneaks into the brain!", indicating air pollution was a problem in the 1st century B.C., too. Shakespeare was perhaps impressed by the pestilence of early 17th century London air when he wrote,

> ". . . this most excellent canopy, the air,
> look you, this excellent o'erhanging
> firmament, this majestical roof fretted
> with golden fire, why, it appears
> no other thing to me but a foul
> and pestilent congregation of vapors."
>
> —*Hamlet*, Act II, Scene II

London experienced its worst air pollution in 1952, and lesser crises in 1956, 1957, and 1962. The mixing of the nearby warm Gulf Stream and the frigid Arctic currents creates frequent fog in the London area. The fog of December 5 to 8, 1952, however, was unusual. It was especially thick and cold. To counteract the cold, Londoners fired their coal heaters hotter and hotter. The smoke spread out near the ground instead of vanishing into the upper atmosphere. The fog grew thicker; the percentages of sulfur dioxide, particulate matter, and other pollutants climbed to 10 times the average readings.

FIGURE 19-1   Smog over New York City. A heavy haze hangs over Manhattan Island, viewed from the roof of the RCA Building. The Empire State Building is barely visible in the background. (Wide World Photos, Inc.)

For five days 12 million people filled their lungs with this poison. A state of frenzy resulted as deaths from bronchitis jumped to nine times the usual rate and deaths from pneumonia increased by fourfold. Over 4000 deaths were attributed to the air pollution. Autopsies on approximately 1000 Londoners who died in their homes or on the streets showed, almost without exception, that the victims were already suffering from serious heart or lung ailments. Those fatally affected were mainly adults aged 45 and older, but mortality of infants under one year also rose. Many young and apparently healthy persons were made violently ill.

A government investigation revealed that coal smoke reacting with the atmosphere generated the deadly pollutants. Sulfur oxides and tarry particles in contact with the lungs caused severe irritation, a starvation of oxygen, and finally heart failure. Although the report offered suggestions for preventing subsequent disasters, another London smog in January, 1956, killed 500 Londoners, and a third in December, 1956, caused an estimated 400 deaths. In December, 1962, a $3\frac{1}{2}$ day smog accounted for 133 more deaths.

Neither has the United States been spared its share of air pollution disasters. In late October, 1948, Donora, Pennsylvania, had a five day siege of extreme air pollution. Before rain cleansed the air, over 800 domestic animals died and 43 per cent of the population became ill; 5910 people became sick; 20 died, where the normal rate was 2 every 5 days. The spasmodic air pollution crises in New York City and the normal smoggy conditions in Los Angeles attract appreciable attention.

Air pollution damage near Copperhill, Tennessee, is as dramatic in scope

FIGURE 19-2   Copperhill Basin (Ducktown), Tennessee, as photographed in 1943, shows what can happen to the ecology of an area from poisonous smelter fumes. This plant converted the poisonous fumes to useful acids in 1917, but such damage had already been done that a hundred square miles or more continued to erode. In 1943, the vegetation had started to come back very slowly (notice the cactus in the foreground). (U.S. Forest Service photograph.)

as a natural disaster such as an earthquake or a tidal wave (Figure 19–2). Copper ore has been mined and smelted in this area since 1847. In the early years, large quantities of sulfur dioxide, a by-product, were discharged directly into the atmosphere:

$$Cu_2S + O_2 \xrightarrow{\text{Heat}} 2Cu + SO_2$$

Copper ore,   Air
containing
copper (I)
sulfide

Before the situation was corrected, the SO_2 killed all of the vegetation for miles around the smelter. Today, the sulfur is reclaimed in the exhaust stacks to make sulfuric acid, but the denuded soil remains a monument to the misuse of the atmosphere.

The widespread interest in air pollution today stems not from the momentary isolated tragedies but from the fairly recent realization that our whole atmosphere is becoming polluted. The long-range effects of these primarily invisible pollutants on materials and the health of plants, animals, and human beings is just beginning to be understood. The lung cancer rate in large metropolitan areas is twice as great as the rate in rural areas, even after full allowance is made for differences in cigarette smoking habits. The serious pulmonary disease emphysema shot up *eightfold* during the decade of the sixties. Esthetic, psychological, and weather effects of air pollution are being evaluated. How depressing it is to be enshrouded by smog day after day. How repulsive it is for a new building to become clothed in dirt, filth, and decay. However, only after we comprehend the role and extent of air pollution can we evaluate wisely its relative importance.

SOURCES OF POLLUTANTS IN THE AIR

Automobiles, industry, and electric power plants are the main sources of air pollutants from man-controlled processes. Volcanic action, forest fires, and dust storms are natural sources of air pollutants, but these contribute very little compared to the man-made sources. A summary of the principal sources of emissions in the United States in 1968 is shown in Figure 19–3.

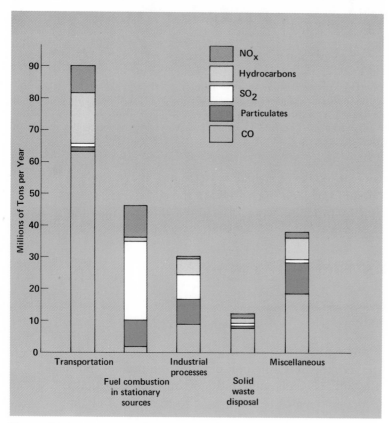

FIGURE 19–3   Nationwide Emissions of Pollutants, 1968.

POLLUTANT PARTICLE SIZE

Pollutant particles may be grouped in three sizes: fundamental (such as single molecules, ions, or atoms), aerosols (a thousand up to about a million atoms, ions, or small molecules per particle), and particulates (suspended particles

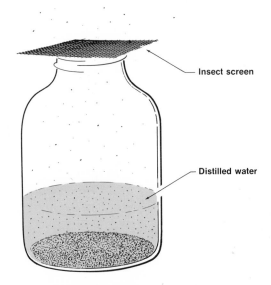

Insect screen

Distilled water

FIGURE 19-4   A dust sample collector for use in the backyard. The open jar containing water is exposed to the air for a known interval. The water is then evaporated and the residue weighed. Fifteen days is sufficient exposure time. Balances sensitive to 0.001 g should be used; the area (cm^3) of the mouth of the jar must be known.

composed of more than a million atoms, ions, or molecules). Aerosols range in size from 0.001 to 1 micron in diameter (1 micron = 10^{-4} centimeter or 10^{-6} meter); molecules are smaller; particulates are larger. Pollutants dispersed as molecules, ions, or atoms are invisible. Aerosols and particulates serve as carriers and collectors of the chemically active SO_2, nitrogen oxides, ozone, hydrocarbons, and other pollutant molecules.

AEROSOLS

Aerosols, which can be either liquid or solid particles, are small enough to remain suspended in the atmosphere for long periods of time. Aerosol-size particles cannot be seen individually by even the most powerful optical microscopes, but in large numbers they appear as smoke, dust, clouds, fog, mist and sprays. Aerosols have much greater surface areas than a single lump of the material. Consider a cube of coal 1 centimeter on an edge. Such a cube would have a surface area of 6 square centimeters. If this cube is subdivided into a sextillion (1,000,000,000,000,000,000,000) smaller cubes, each cube would be 1 micron on an edge, about the size of an aerosol particle. The surface area is now 60,000,000 square centimeters—about 1.5 acres. Because of this large surface area, aerosols have enormous capacities to absorb and concentrate gases on their surfaces. Many times aerosols absorb toxic gases, provide the water medium for a reaction to occur, and cause devastating results when breathed. Specific examples and the important role that aerosols play in smog formation will be cited later.

PARTICULATES

Particulates are the sum total of air pollution to most people because these particles are large enough to be seen. They range in size from 1 to 10 microns in diameter. Over 130 million tons of soot, dust, and smoke particulates were

361

FIGURE 19-5   The General Electric CF6-6D engines that power this McDonnell Douglas DC-10 were designed to eliminate exhaust smoke and reduce by one-half the noise level at takeoff. American Airlines advertises that it spent in a ten-year-period an amount equal to 43 per cent of its profits on noise and air pollution control systems. (Courtesy of American Airlines.)

deposited into the atmosphere of the United States in 1968 by automobiles and industry. Trash burning, forest fires, and jet aircraft deposited another 30 million tons. Average suspended particulate concentrations in the United States range from 0.00001 gram per cubic meter of air (g per m^3) in remote, rural areas to about six times that value in urban locations. In heavily polluted areas, concentrations up to 0.002 g per m^3, or 200 times the usual values, have been measured.

Particulates may cause physical damage to certain materials. Particles whipped by the wind grind exposed materials by abrasive action. Particles settling in electronic equipment can break down the resistance and foul contacts and switches. Dust settling in paint detracts from its beauty and hastens its deterioration by allowing water to reach the surface underneath. Particles may interfere physically with one or more of the clearance mechanisms in the respiratory tract of man and animals (inhibiting the ciliary transport of mucus, for example). People who have asthma or emphysema know that heavy concentrations of particles in the air increase discomfort. In extremely polluted regions, these diseases often lead to death.

Particulates may also injure humans or animals because they are intrinsically toxic. The effects of lead were described in Chapter 10. Lead compounds are emitted in automobile exhaust. Particles containing fluorides, commonly emitted from aluminum-producing and fertilizer factories have caused weakening of bones and loss of mobility in animals which have eaten plants covered with the dust. Leaves of bean plants dusted with cement kiln particles (0.00047 g per cm^2 per day) have been known to wilt significantly.

Particulates absorb and scatter light; consequently, they can reduce the

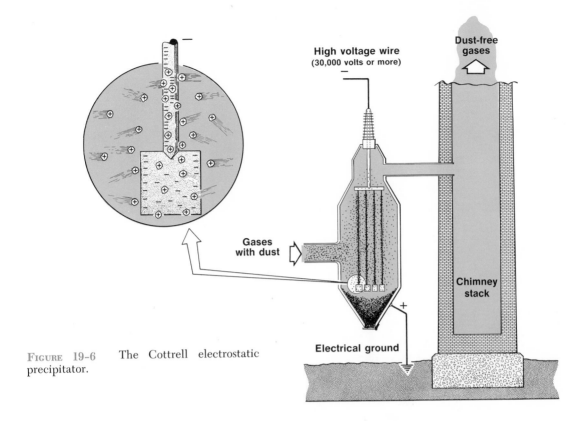

High voltage wire
(30,000 volts or more)

Dust-free gases

Gases with dust

Chimney stack

Electrical ground

FIGURE 19-6   The Cottrell electrostatic precipitator.

amount of light reaching the earth to warm it. The amount of particulate matter in the air has increased since 1964, and this may explain why the mean global temperature has been decreasing over the past few years.

Particles are removed from the atmosphere by gravitational settling and by rain and snow. They can be prevented from entering the atmosphere by treating industrial emissions by one or more of a variety of physical methods such as filtration, centrifugal separation, spraying, ultrasonic vibration, and electrostatic precipitation.

One of the most efficient ways to remove both particulates and aerosols from the air is electrostatic precipitation. Electrostatic precipitators can remove aerosols and dust particles smaller than 1 micron from plant exhaust gases. A Cottrell electrostatic precipitator is illustrated in Figure 19–6. The central wire is connected to a source of direct current and high voltage (about 50,000 volts). As dust or aerosols pass through the strong electrical field, the particles attract ions which have been formed in the field, become strongly charged, and are attracted to the walls of the tubes. The precipitated solid falls to the bottom where it is collected.

SMOG—INFAMOUS AIR POLLUTION

The poisonous mixture of smoke, fog, air, and other chemicals was first called *smog* in 1911 by Dr. Harold de Voeux in his report on a London air pollution

FIGURE 19-7 Photochemical smog (a brown haze) enveloping the city of Los Angeles. (Los Angeles Times photo.)

disaster that caused the deaths of 1150 people. Through the years, smog has been a technological plague in many communities and industrial regions.

What general conditions are necessary to produce smog? Although the chemical ingredients of smogs often vary, depending upon the unique sources of the pollutants, certain geographical and meteorological conditions exist in nearly every instance of smog.

There must be a period of windlessness so that pollutants can collect without being dispersed vertically or horizontally. This lack of movement in the ground air can occur when a layer of warm air rests on top of a layer of cooler air. This sets the conditions for a *thermal inversion,* which is an abnormal temperature arrangement for air masses. If the warmer air is on the bottom nearer the warm earth, which is usual, the warmer, less dense air rises and transports most of the pollutants to the upper atmosphere where they are dispersed. When the warmer air is on top, as in thermal inversion, the cooler, more dense air retains its position nearer the earth and vertical movement is stagnated. If the land is bowl-shaped (surrounded by mountains, cliffs, or the like), horizontal movement of the air mass is also hindered.

When these natural conditions exist, man supplies the killing ingredients by combustion and evaporation in automobiles, electrical power plants, space

364

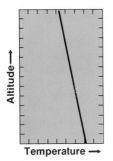
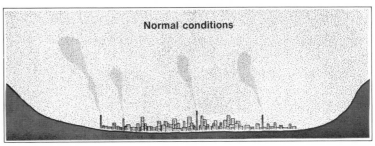

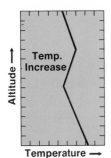
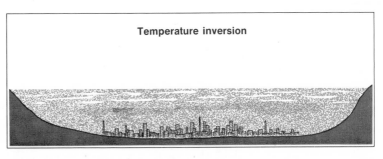

FIGURE 19-8   A diagram of a temperature inversion layer over a city. Warm air over a polluted air mass effectively acts as a lid, holding the polluted air over the city until the atmospheric conditions change. The line on the left of the diagram indicates the relative air temperature.

heating, and industrial plants. The chief pollutants are sulfur dioxide (from burning coal and some oils), and nitrogen oxides, carbon monoxide, and hydrocarbons (chiefly from the automobile). Add to these ingredients the radiation from the sun, and a massive case of smog is in the offing.

Two general kinds of smog have been identified. One is the chemically reducing type that is derived largely from the combustion of coal and oil, which contains sulfur dioxide mixed with soot, fly ash, smoke, and partially oxidized organic compounds. This is the *London type*, which is diminishing in intensity and frequency as less coal is burned and more controls are installed. A second type of smog is the chemically oxidizing type, typical of Los Angeles; it is called *photochemical* smog because light—in this instance sunlight—is important in initiating the photochemical process. This smog is practically free of sulfur dioxide but contains substantial amounts of nitrogen oxides, ozone, ozonated olefins, and organic peroxide compounds, together with hydrocarbons of varying complexities.

The exact reaction scheme by which primary pollutants are converted into the secondary pollutants found in photochemical smog is still not completely understood, but the reactions shown in Figure 19-9 account for the major secondary pollutants. The process is thought to begin with the absorption of a quantum of ultraviolet light by nitrogen dioxide, which causes its breakdown into nitric oxide and atomic oxygen. The very reactive atomic oxygen reacts with molecular oxygen to form ozone (O_3), which is then consumed by reacting with nitric oxide to form the original reactants—nitrogen dioxide and molecular

365

$$NO_2 + light \longrightarrow NO + O \cdot$$
$$O \cdot + O_2 + M \longrightarrow O_3 + M$$
$$O_3 + NO \longrightarrow NO_2 + O_2$$
$$O \cdot + Hc \longrightarrow HcO \cdot$$
$$HcO \cdot + O_2 \longrightarrow HcO_3 \cdot$$

$$HcO_3 \cdot + Hc \longrightarrow RCHO \text{ or } R\overset{\overset{\displaystyle O}{\|}}{C}R$$
$$HcO_3 \cdot + NO \longrightarrow HcO_2 \cdot + NO_2$$
$$HcO_3 \cdot + O_2 \longrightarrow O_3 + HcO_2 \cdot$$

$$HcO_3 \cdot + NO_2 \longrightarrow R-\overset{\overset{\displaystyle O}{\|}}{C}-O-O-N\overset{\nearrow O}{\searrow O} + \text{other products}$$
$$\text{(PAN)}$$

FIGURE 19-9 Simplified reaction scheme for photo-chemical smog. Hc is hydrocarbon (olefin or aromatic); M is a third body to absorb the energy released from forming the ozone; among many possibilities, M could be a N_2 molecule, O_2 molecule, or solid particle. A species with a dot, as HcO, is a chemical radical—a very reactive chemical species.

oxygen. Atomic oxygen, however, also reacts with reactive hydrocarbons—olefins and aromatics—to form chemical radicals (species with unpaired valence electrons). Chemical radicals, in turn, react to form other radicals and secondary pollutants such as aldehydes (formaldehyde, acetaldehyde, acrolein), ketones, and peroxyacyl nitrates (PAN).

It is evident that further research is needed to clear up several obscure features of the overall scheme. While it is known that a few tenths of a part per million of nitrogen oxides and 1 ppm of reactive hydrocarbons suffice to initiate the process, the sources of all of the hydrocarbons that enter the smog-forming process cannot be accounted for. Furthermore, the exact nature and behavior of the radicals is not well understood at all. The ultimate fate of the substances involved is also not known.

The type of smog formed in London and around some industrial and power plants is thought to be caused by sulfur dioxide. Laboratory experiments have shown consistently that sulfur dioxide increases aerosol formation, particularly in the presence of irradiated hydrocarbon-nitrogen oxide-air mixtures. For example, irradiated mixtures of 3 ppm olefin, 1 ppm NO_2, and 0.5 ppm SO_2 at 50 per cent relative humidity form aerosols which have sulfuric acid as a major product. Even with 10 to 20 per cent relative humidity, sulfuric acid is a major product. Without the SO_2, olefins and nitrogen oxides do not cause appreciable aerosols.

How is the sulfuric acid formed? Most likely SO_2 is oxidized to SO_3. The SO_3 has less energy per mole than SO_2 (about 24 kcal per mole; 94.45 kcal are lost in forming one mole of SO_3 and only 70.96 kcal for a mole of SO_2). The oxidation of SO_2 to SO_3 is spontaneous and should go readily if oxygen is present and the energy of activation is available. Air supplies the oxygen and sunlight supplies the energy, but the mechanism is unknown. It has been confirmed experimentally that SO_2 radiated by sunlight forms a sulfuric acid aerosol. The speed of this reaction is increased by the presence of metals in the aerosols, hydrocarbons, and nitrogen oxides. Ferric oxide, Fe_2O_3, is known to catalyze the conversion of SO_2 to SO_3, and there is a considerable amount of rust available.

Once the SO_3 is formed in the presence of water, it is readily converted to sulfuric acid:

$$SO_3 + H_2O \longrightarrow H_2SO_4$$

The sulfuric acid formed in this kind of smog is very harmful to people suffering from bronchial diseases, asthma, or emphysema. At a concentration of 5 ppm for one hour, SO_2 can cause constriction of bronchial tubes. A level of 10 ppm for one hour can cause severe distress. In the 1962 London smog, readings as high as 1.98 ppm of SO_2 were recorded. The sulfur dioxide and sulfuric acid are thought to be the primary causes of deaths in the London smogs.

SULFUR DIOXIDE

Sulfur dioxide is produced when sulfur or sulfur-containing substances are burned in oxygen:

$$S + O_2 \longrightarrow SO_2(g)$$

Most of the sulfur dioxide in the atmosphere comes from sulfur-containing fuels burned in power plants, from smelting plants treating sulfide ores, and from sulfuric acid plants. Coal contains sulfur in several forms: as elemental sulfur (S), iron pyrites (FeS_2), and as organic compounds, such as mercaptans (compounds containing—SH groups). According to the National Air Pollution Control Association (NAPCA), the burning of fuels by the nation's power plants accounted for 46 per cent (23 million tons) of the SO_2 emitted into the atmosphere in 1966. One study found a concentration of 2200 ppm SO_2 in the stack of a coal-burning power plant. Concentrations as high as 2.9 ppm have been recorded up to one-half mile from power plants. If coal and petroleum containing up to 5 per cent sulfur are burned, a 1000-megawatt electric power plant could emit 600 tons of SO_2 per day. Much of the SO_2 pollution in this country is isolated in seven industrialized states (New York, Pennsylvania, Michigan, Illinois, Indiana, Ohio, and Kentucky), which account for almost 50 per cent of the nation's total SO_2 output.

In spite of the large output of SO_2 in the United States per year, its concentration rarely exceeds a few parts per million. This is because SO_2 has a relatively short atmospheric life. The SO_2 and sulfuric acid formed from it are readily dissolved in rivers, lakes, and streams and can increase the acidity considerably.

$$H_2O + SO_2 \rightleftharpoons \underset{\substack{\text{Weak} \\ \text{acid}}}{H_2SO_3} \rightleftharpoons H^+ + HSO_3^-$$

$$H_2SO_4 \longrightarrow H^+ + HSO_4^-$$

Often streams contain ions that help to offset the acidity increase. Sulfite is one such ion.

$$\underset{\textit{Sulfite}}{SO_3^{2-}} + H^+ \longrightarrow \underset{\textit{Bisulfite}}{HSO_3^-}$$

FIGURE 19–10   Rose leaves in Independence, Missouri, show marginal and inter-veinal necrotic injury from sulfur dioxide. (From publication no. AP-71, National Air Pollution Control Administration, HEW Public Health Service, 1970.)

If the pH of streams varies too much, aquatic life suffers. Salmon, for example, cannot survive if the pH is as low as 5.5. The lower limit of tolerance for most organisms is a pH of 4.0. In the late 1950's and early 1960's, certain sections of the Netherlands had precipitation with a pH less than 4.

The corrosion of steel is promoted by sulfur dioxide. In a study involving 100-gram panels of steel exposed at several sites in Chicago in 1964, the amount of weight loss by corrosion was very definitely related to the SO_2 concentration in the air. Particulate matter, high humidity, and temperature are also factors which determine the extent of corrosive action.

The effects of SO_2 on vegetation show up as bleached spots, suppression of growth, leaf drop, and reduction in yield. Plant damage has been noticed 52 miles downwind from a smelting operation which emitted large quantities of SO_2. The SO_2, and SO_3 formed from it, enter the stomata of the leaves, hydrolyze, and dehydrate the tissue as sulfuric acid; sulfates are then formed, causing yellowing and leaf drop.

Chemical means are efficient in removing SO_2. One method involves heating limestone to produce lime (calcium oxide) and reacting this with the SO_2 to produce calcium sulfite, which can be removed from the stack by an electrostatic precipitator.

$$CaCO_3 \xrightarrow{\text{Heat}} CaO + CO_2$$
Limestone   Lime

$$CaO + SO_2 \longrightarrow CaSO_3 \text{ (solid)}$$
Calcium sulfite

FIGURE 19-11   (A) When air pollution is minimal, tall stacks help to keep the pollution away from people. This photograph of the stacks of Consolidated Edison in New York City makes it all too clear that tall stacks are no real solution when a temperature inversion exists. Note the United Nations Secretariat Building in the right-hand corner. (Wide World Photos, Inc.) (B) TVA's Bull Run Steam Plant, five miles east of Oak Ridge in east Tennessee, has one of the largest generating units in the world, with a capacity of 950,000 kilowatts. At full operation the plant burns 7584 tons of coal (a large trainload) every day. Nearly 600 million gallons of water a day flow through the plant from Melton Hill Lake to condense spent steam from the turbines. Bull Run has an 800-foot chimney. The plant went into regular operation in 1967.

Much of the petroleum used in the Eastern United States, where SO_2 pollution is a problem, is Caribbean residual fuel oil, which has an average sulfur content of 2.6 per cent. While technology has developed refining techniques for reducing the sulfur content to 0.5 per cent, this procedure would increase the cost of the oil by about 35 per cent. The process involves the formation of hydrogen sulfide by bubbling hydrogen through the oil in the presence of metallic

369

catalysts (platinum-palladium, for example). Residual fuel oil produces less than 10 per cent of the total utility power. In 1968, New York's Consolidated Edison switched to low-sulfur oil from Liberia and Nigeria at a cost of $3 million and an increase of $7.5 million in a $63 million fuel bill.

Despite all of these methods, and more under development, the equilibrium concentration of SO_2 in the air is increasing nationwide on an average of about 6 to 7 per cent per year. Some cities—Washington, D.C., for instance—have reduced SO_2 emissions, while in others the emissions are on the increase. The SO_2 concentration depends, of course, upon the amount of sulfur-containing coal burned and the output of SO_2 by chemical industries.

NITROGEN OXIDES

There are eight known oxides of nitrogen, three of which are recognized as important components of the atmosphere. These are: nitrous oxide (N_2O), nitric oxide (NO), and nitrogen dioxide (NO_2).

Most of the nitrogen oxides emitted are in the form of NO, a colorless reactive gas. In a typical combustion process involving air as the oxidant, some of the atmospheric nitrogen reacts with oxygen to produce NO:

$$N_2 + O_2 + \text{Heat} \longrightarrow 2NO$$

Since the formation of nitric oxide is endothermic (21,600 cal per mole of NO), it follows that a higher combustion temperature would produce relatively more NO. This will be an important point to consider in Chapter 21 in our discussion of the automobile and its nitrogen oxide emissions, since one way to achieve greater burning efficiency of fuels in automobile engines is to operate at high temperatures.

In the atmosphere NO reacts rapidly with atmospheric oxygen to produce NO_2:

$$2NO + O_2 \longrightarrow 2NO_2$$
Nitrogen dioxide

If NO_2 does not react photochemically, it can react with water vapor in the air to form nitric and nitrous acids:

$$2NO_2 + H_2O \longrightarrow HNO_3 + HNO_2$$
Nitric acid   Nitrous acid

In addition, nitrogen dioxide and oxygen yield nitric acid:

$$4NO_2 + 2H_2O + O_2 \longrightarrow 4HNO_3$$

These acids in turn can react with ammonia or metallic particles in the atmosphere to produce nitrate or nitrite salts. For example,

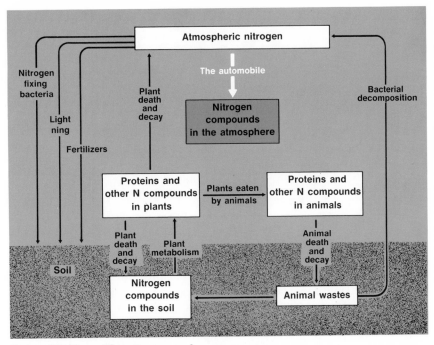

FIGURE 19-12   The nitrogen cycle.

$$NH_3 \quad + HNO_3 \longrightarrow \quad NH_4NO_3$$

Ammonia   *Nitric*   *Ammonium*

acid   *nitrate*

The acids or the salts, or both, ultimately form aerosols, which eventually settle from the air or dissolve in raindrops. Nitrogen dioxide, then, is a primary cause of haze in urban or industrial atmospheres because of its participation in the process of aerosol formation. Normally nitrogen dioxide has a lifetime of about three days in the atmosphere.

In laboratory studies, nitrogen dioxide in concentrations of 25 to 250 ppm inhibits plant growth and causes defoliation. The growth of tomato and bean seedlings is inhibited by 0.3 to 0.5 ppm NO_2 applied continuously for about 10 to 20 days.

In a concentration of 3 ppm for 1 hour, nitrogen dioxide causes bronchio-constriction in man, and short exposures at high levels (150 to 220 ppm) produce fibrotic changes in the lungs that produce fatal results.

CARBON MONOXIDE

Carbon monoxide (CO) is the most abundant and widely distributed air pollutant found in the atmosphere. It appears to be almost exclusively a man-made pollutant because more than any other air pollutant, carbon monoxide is closely related to the total population of an area.

It is produced in combustion processes when carbon or some carbon-containing compound is burned in an insufficient amount of oxygen:

$$2C + \quad O_2 \longrightarrow 2CO$$

Limited

supply

371

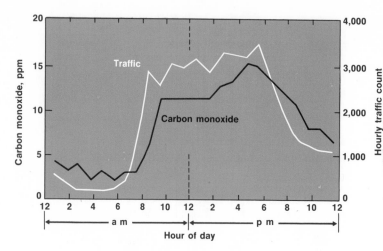

FIGURE 19–13   Hourly average carbon monoxide concentration and traffic count in midtown Manhattan. (From Johnson, K. L., Dworetzky, L. H., and Heller, A. N.: "Carbon Monoxide and Air Pollution from Automobile Emissions in New York City." Science, Vol. 160, No. 3823, p. 67 [1968].)

In the United States combustion sources of all types dump about 2.1×10^{14} grams of carbon monoxide per year into the atmosphere (about 5×10^{21} grams). In terms of local emission, each person accounts for about 1400 lbs per year in Los Angeles and 600 lbs per year in New York City. For every 1000 gallons of gasoline burned, 2300 lbs of CO are emitted.

The huge emissions of CO are a mystery. There is enough CO emitted to raise the concentration 0.03 ppm yearly, but apparently it is not increasing at all. Although data on background concentrations of CO are still very limited, best estimates of the maximum global level are of the order of 0.1 ppm. The background concentration in cities is higher; in heavy traffic sustained levels of 50 or more ppm are common; and instantaneous concentrations of 150 ppm and higher have been found. NAPCA data collected from 1964 through 1966 in five major cities show an average of 7.3 ppm background for offstreet sites. The variation of CO levels over a 24-hour period is shown in Figure 19–13. However, in spite of increased emission of CO, the global level does not seem to be rising.

It is not known for certain where all the carbon monoxide goes once it gets into the atmosphere. There are a number of possible explanations. One of these is that carbon monoxide is slowly oxidized by molecular oxygen (O_2) in the lower atmosphere:

$$CO + O_2 \longrightarrow CO_2 + \underset{\substack{\textit{Oxygen} \\ \textit{atom}}}{O}$$

The oxygen atom then reacts with other species. This reaction has a high energy of activation (51,000 cal per mole), however, and only occurs readily at temperatures above 500°C or with suitable catalysts.

The danger of carbon monoxide to man is the strong bond it forms with the iron in hemoglobin. (The structure of heme is a part of Figure 9–13, Chapter 9.)

$$CO + HbFeO_2 \longrightarrow HbFeCO + O_2$$

By combining with hemoglobin, CO starves the body of oxygen. This produces listlessness and can lead to death. About 30 ppm for 8 hours is sufficient to cause headache and nausea.

TABLE 19-1 Concentration of CO in Atmosphere
Versus Percentage of Hemoglobin (Hb) Saturated°

CO concentration in air	0.01%	0.02%	0.10%	1.0%
Percentage of hemoglobin molecules saturated with CO†	17	20	60	90

° A few hours of breathing time is assumed.
† Normal human blood contains up to 5% carboxyhemoglobin (COHb).

The percentage of hemoglobin rendered useless by a specific concentration of CO in the air is given in Table 19–1. Since the reaction of CO with hemoglobin is reversible, the administration of pure oxygen can replace the CO with O_2 on hemoglobin.

$$O_2 + HbFeCO \longrightarrow HbFeO_2 + CO$$

Oxygen tanks have been installed at busy intersections of some cities for use of traffic policemen who must stand for hours in the swirl of auto exhaust fumes. Every half hour the men have to take an "oxygen break."

HYDROCARBONS

As we have seen in Chapter 7, hydrocarbons come in all shapes and sizes from methane, CH_4, to octane, C_8H_{18}, and beyond to molecules containing many more carbon atoms. Some have all single bonds, some have double bonds, and a few have triple bonds. Aromatic hydrocarbons have bonds that can be described as intermediate between a single and a double bond. Literally hundreds of these hydrocarbons and their oxygen, sulfur, nitrogen, and halogen derivatives find their way into the atmosphere.

Trees and plants silently breathe turpentine, pine oil, and thousands of other fragrances into the air. These are hydrocarbons or hydrocarbon derivatives. Bacterial decomposition of organic matter emits very large amounts of marsh gas, principally methane. Man contributes his share of 15 per cent (of the total global emissions; a greater quantity in urban areas) through incomplete incineration, leakage of industrial solvents, unburned fuel from the automobile, incomplete combustion of coal and wood, and through petroleum processing, transfer, and use.

Benzo (α) pyrene (BaP) and some other polynuclear hydrocarbons are commonly emitted in coal smoke as well as in cigarette smoke. The carcinogenic effects of these compounds will be discussed in the next chapter.

Although BaP and its polynuclear counterparts have received considerable publicity, other hydrocarbons and hydrocarbon derivatives play equally important roles in air pollution. The chlorinated hydrocarbons, widely used as pesticides, at least partially counteract their beneficial aspects by killing birds and fish, and by generally polluting our streams. Some of these, such as DDT, are found in the atmosphere.

In the section on smog we discussed in detail the role of olefins and aromatic hydrocarbons in photochemical smog formation. In a study made in Los Angeles in 1970, an average of 0.106 ppm (maximum of 0.33 ppm) aromatics was found in that city's atmosphere. (About 38 per cent of the total was toluene and 40 per cent the more reactive dialkyl- and trialkylbenzenes.) These compounds are about as reactive as propylene and higher-molecular-weight olefins in causing smog formation. The automobile is responsible for emitting most of these hydrocarbons to the atmosphere. In fact, the automobile emits more than 200 different hydrocarbons and hydrocarbon derivatives. Most of these come from unburned gasoline and from reactions among hydrocarbons, oxygen, and nitrogen during the combustion process. Aromatic compounds constitute an average of 15 to 30 per cent of the components of regular and premium grade gasoline sold in the United States. Several nonleaded gasolines contain as much as 40 per cent aromatics; their exhaust products have about the same percentages. About 85 per cent of these have been shown to be photochemically reactive.

WHAT DOES THE FUTURE HOLD?

There will undoubtedly be an abatement of air pollution in the future; the sheer pressures of population increase will demand it. But life also will undoubtedly have to be different. Perhaps the first major change will be the disappearance of the automobile from the city, followed by a gradual modification of the power plant of the automobile until it is relatively nonpolluting.

The costs of abating pollution are already high, and they are likely to increase much more. It has been estimated that during the first half of the 1970's, all levels of government will spend about $1.6 billion on programs aimed at cleaning the air. Added to this will be $2.75 billion to upgrade fuels for autos, $0.5 billion for chemical industry to lower its emissions, and $2.64 billion for motorists to control the emission from their own automobiles. All this amounts to $7.49 billion, and most of this cost will be paid for directly by the public.

Most pollution exists because we demand the benefits of a technology which, for the most part, has given little consideration to the long range effects of its products. When industry, automobile manufacturers, or power plant operators add equipment to stop noxious waste products from getting into the air, the costs are added to the already considerable manufacturing expense without adding one cent to the market value of the product being made. The cost to the consumer, however, will go up and will be reflected in the increased price of consumer goods, automobiles, and electrical rates. This is a high price, of course; but it is not too high if we are to have clean air.

New and more stringent laws can be expected. The Federal Clean Air Act of 1963 was designed to provide financial assistance to state and local agencies for developing, establishing, or improving air pollution control programs. The Air Quality Act of 1967 set up 57 air quality regions within the United States for the purpose of monitoring the amounts of various pollutants in the air, designating an air quality standard for each pollutant, and developing a plan by which the standard will be implemented. New laws will be more specific, regulating the amounts of pollutants that can be emitted by various industries. Special courts will be required to handle pollution cases.

In the final analysis, we all pollute the atmosphere, and much of the pollution is due to the misapplication of chemical techniques. Although we are becoming aware of the problems and although it is within the capabilities of chemical technology to eradicate most forms of air pollution, the process will be very slow. We have time . . . perhaps.

QUESTIONS

1. The formation of photochemical smog involves very reactive free radicals. What structural feature makes free radicals very reactive?

2. Write a balanced chemical equation for the burning of iron pyrite (FeS_2) in coal to sulfur dioxide and Fe_2O_3, iron(III) oxide.

3. What conditions are necessary for thermal inversion?

4. What are the basic chemical differences between New York (or London) smog and Los Angeles smog?

5. What are the major sources of the following compounds in the air?

   (a) carbon monoxide
   (b) sulfur dioxide
   (c) nitrogen oxides

6. What is a photochemical reaction? Give an example.

7. If air pollutants rise from earth into the atmosphere, why do they not continue on into space?

8. What effects does weather have on local air pollution problems? on regional air pollution problems?

9. Knowing the chemistry of photochemical smog formation, list some ways to prevent its occurrence.

10. What part do aerosols play in the formation of smogs?

11. Report on a recent article on air pollution.

12. Describe an air pollution problem in your community. How can this problem be solved?

13. Discuss the merits of abatement versus eradication of air pollution.

14. Of the air pollutants—particulates, sulfur dioxide, carbon monoxide, ozone, and nitrogen dioxide—

   (a) which is in greatest concentration in the air normally?
   (b) which is normally a secondary pollutant?
   (c) which is emitted almost exclusively by man-controlled sources?
   (d) which can be removed from emissions by centrifugal separators?

15. Define:

   (a) micron
   (b) ppm
   (c) PAN

16. With a population of 200 million in the United States, how much can we expect air pollution to cost us individually during the first five years of the 1970's?

17. Which is more effective in producing smog—paraffinic (all single bonds) or olefinic (some double bonds) hydrocarbons? Why?

18. Why is natural gas a good substitute for oil and coal as far as air pollution is concerned? What problem is related to its substitution?

19. What is the approximate concentration of air pollutants in the atmosphere during smoggy conditions?

20. What do you think the future holds for our atmosphere?

SUGGESTIONS FOR FURTHER READING

Paperbacks

Cook, L. M. (ed.), "Cleaning Our Environment—A Chemical Basis For Action," A. C. S. Committee on Chemistry and Public Affairs, American Chemical Society, Washington, D.C., 1970.

Giddings, J. C., and Monroe, M. B., Editors, "Our Chemical Environment," Harper and Row, New York (1972).

Stoker, H. S., and Seager, S. L., "Environmental Chemistry—Air and Water Pollution," Scott, Foresman and Company, Glenview, Ill. (1972).

Turk, A., Turk, J., and Wittes, J. T., "Ecology, Pollution, Environment," W. B. Saunders Co., Philadelphia (1972).

Winn, I. J., Editor, "Basic Issues in Environment," Charles E. Merrill Publishing Co., Columbus, Ohio (1972).

Articles

Beard, R. R., and Wertheim, G. A., "Behavioral Impairment Associated with Small Doses of Carbon Monoxide," *American Journal of Public Health*, 57:2012 (1967).

Hall, H. J., and Bartok, W., "NO_x Control From Stationary Sources," *Environmental Science and Technology*, Vol. 5, No. 4 (1971), p. 320.

Newell, R. E., "The Global Circulation of Atmospheric Pollutants," *Scientific American*, Vol. 224, No. 1 (1971), p. 32.

"Rivers of Dust," *Chemistry*, Vol. 42, No. 11 (1969), p. 24.

Wolf, P. C., "Carbon Monoxide-Measurement and Monitoring in Urban Air," *Environmental Science and Technology*, Vol. 5, No. 3 (1971), p. 212.

20

FOUR "GENS" AND A "GIN"

Every day of our lives we pollute our bodies with foreign substances which were not intended to be there. These chemical compounds may come from air and water pollution; the additives we find in our foods; the chemicals we come in contact with as we work; or some act of habit, enjoyment, or escapism we indulge in from time to time.

In this chapter we will look at the chemistry of some substances that generally make life more miserable. They produce such diverse effects as malignancies, mutations, allergic reactions, hallucinations, and alcoholic stupor.

CARCINOGENS—CANCER

Cancer is any abnormal cell growth which does not respond to normal growth-control mechanisms. Cancer is found in all forms of plant and animal life. Although the various forms of this disease are many, several facts are apparent. Individuals differ in their susceptibility, and cancer appears to have many causes.

A cancer manifests itself in three ways. (1) The rate of cell growth (that is, the rate of cellular multiplication) in cancerous tissue differs from the rate in normal tissue. (2) Cancerous cells spread to other tissues; they know no bounds. Normal liver cells divide and remain a part of the liver. Cancerous liver cells may leave the liver and be found, for example, in the lung. (3) Most cancer cells show partial or complete loss of specialized functions. Although located in the liver, cancer cells no longer perform the functions of the liver.

Cancer of epithelial tissue is called a *carcinoma*, whereas cancer in the connective tissues is classified as a *sarcoma*. Cancer from one tissue can spread to other tissues as *metastases*, which, in man, are found in many locations such as in the lung, liver, stomach, bone, brain, and kidney. An abnormal growth is classified as cancerous, or malignant, when microscopic examination shows that it is invading neighboring tissues or when metastases are found. An abnormal growth is benign if it is localized at the original site.

There is a direct link between certain viruses and cancers. Cancer can also be induced by repeated exposure to radiation, such as sunlight and x-rays which produce skin cancer. In addition, certain cancers can be produced by chemicals.

When a certain chemical is found to be the cause of a cancer, that chemical is said to be *carcinogenic*. Much of our early knowledge of the causes of cancer came from studies in which the disease was linked to a person's occupation, which in turn was linked to some chemical he came in contact with daily.

It was first noticed in 1775 that persons employed as chimney sweeps in England had a higher rate of skin cancer than the general population. It was not until 1933, however, that benzo(α)pyrene, $C_{20}H_{12}$, (a 5-ringed aromatic hydrocarbon) was isolated from coal dust and shown to be carcinogenic.

Benzo(α)pyrene

In 1895, a German physician, Rehn, noted three cases of bladder cancer in employees of a factory which manufactured dye intermediates in the Rhine Valley. Rehn attributed these cancers to his patients' occupation. These and other cases confirmed that at times as many as 30 years passed between the time of the initial employment and the occurrence of bladder cancer. The principal product of these factories was aniline. While aniline was first thought to be the carcinogenic agent, it was later shown to be noncarcinogenic. It was not until 1937 that continuous, long-term treatment with 2-naphthylamine, one of the suspected dye intermediates, in dosages of up to 0.5 g per day produced bladder cancer in dogs. Since then other dye intermediates have been shown to be carcinogenic.

Aniline   *2-Naphthylamine*

TABLE 20-1   Latent Periods of Some Occupational Cancers

Type of Cancer	Average Latency in Years
Skin	
Tar and soot	20–24
Solar radiation	20–30
X-radiation	7
Lung	
Asbestos	18
Nickel	22
Tar fumes	16
Bladder	
Aromatic amines	11–15

TABLE 20-2   A Sample of Carcinogenic Compounds[°]

Compound	Use or Source	Formula of a Representative Compound
Carbon tetrachloride	Dry cleaning agent, solvent	CCl_4
Dioxane	Solvent in chemical industry, cosmetics, glues, deodorants	
Cyclamates	Artificial sweeteners	
Aromatic amines		
2-Naphthylamine	Optical bleaching agent	
Benzidine	Polymer formation, manufacture of dyes, stain for tissues, rubber compounding	
N-2-Fluorenylacetamide	Herbicide	
Azo dyes		
N, N-Dimethyl-p-phenyl-azoaniline	Yellow dye (once used as a food color)	
Nitrosoamines		
N-Ethyl-N-nitroso-n-butylamine	Insecticide, gasoline and lubricant additive	
Polynuclear compounds		
Benzo(α)pyrene	Cigarette and coal smoke	
Aflatoxin	Mold on peanuts and other plants (*Aspergillus flavus*)	

[°] Most of the evidence is based on studies with mice.

A vast amount of research effort has verified the carcinogenic behavior of a large number of diverse chemicals. Some of these are listed in Table 20–2. This research has led to the formulation of a few generalizations concerning the relationship between chemicals and cancer.

For example, carcinogenic effects on lower animals are extrapolated to man. The mouse has come to be the classic animal for studies of carcinogenicity. Strains of inbred mice and rats have been developed which are uniform and show a standard response.

Some carcinogens are relatively nontoxic in a single, large dose, but may be quite toxic, often increasingly so, when administered continuously. Thus, much patience, time, and money must be expended in carcinogenic studies. The development of a sarcoma in humans, from the activation of the first cell to the clinical manifestation of the cancer, takes from 20 to 30 years. With life expectancy of an average person in the United States now set at about 70 years, it is not surprising that the number of deaths due to cancer is increasing.

Cancer does not occur with the same frequency in all parts of the world. Breast cancer occurs with lower frequency in Japan than in the United States or Europe. Cancer of the stomach, especially in males, is more common in Japan than in the United States. Cancer of the liver is not widespread in the Western Hemisphere but accounts for a high proportion of the cancers in the Bantu in Africa and in certain populations in the Far East. The widely publicized incidence of lung cancer is higher in the industrialized world and is increasing at an appreciable rate.

To illustrate the importance of location in determining the intake of a potentially carcinogenic compound, consider the data in Table 20–3. Since polynuclear hydrocarbons like benzo(α)pyrene (BaP) are products of incomplete combustion of organic matter, they occur in the smoke of burning leaves, in tobacco smoke, and in smoked foods as well as in the air of heavily industrialized cities. In 1967, the estimated annual emissions of BaP in the United States were 422 tons from burning coal, oil, and gas, 20 tons from refuse burning, 19 tons from industries (petroleum catalytic cracking, asphalt road mix, and the like), and 21 tons from motor vehicles.

Some compounds cause cancer at the point of contact: benzo(α)pyrene, for example, which has already been mentioned. Other compounds cause cancer in an area remote from the point of contact. The liver is particularly susceptible to such compounds. Since the original compound does not cause cancer on contact, some other compound made from it must be the cause of cancer. For example, it appears that the substitution of a NOH group for a N—H group

TABLE 20–3	ESTIMATED INHALED BENZO(α)PYRENE	
	By a Male Non-smoker in	Micrograms per Year (+60, if Man Smokes 20 Cigarettes per Day)
	Missouri State Forest	0.1
	San Francisco	14.0
	Detroit	110.0
	London	320.0

in an aromatic amine derivative (such as an amide, $-\overset{\overset{\text{H}}{|}}{\text{N}}-\overset{\overset{\text{O}}{\|}}{\text{C}}-$) produces at least one of the active intermediates for carcinogenic amines. If R denotes a two or three ring aromatic system, then the process can be represented as follows:

$$\underset{\substack{\textit{Inactive} \\ \textit{on contact}}}{\overset{\overset{\text{H}}{|}}{\text{RNCOCH}_3}} \longrightarrow \underset{\substack{\textit{Active on} \\ \textit{contact}}}{\overset{\overset{\text{OH}}{|}}{\text{RNCOCH}_3}} \longrightarrow \underset{\substack{\textit{Other unknown} \\ \textit{intermediates}}}{\underline{\text{RX?} \rightarrow \text{RY?}}} + \text{tissue} \longrightarrow \text{Tumor cell}$$

As indicated by the variety of chemicals in Table 20–2 many molecular structures produce cancer. This has hindered the isolation of many carcinogenic substances. Even closely related structures can produce totally different effects. The 2-naphthylamine mentioned earlier is carcinogenic, but repeated testing gives negative results for 1-naphthylamine.

1-Naphthylamine
(Noncarcinogenic)

2-Naphthylamine
(Carcinogenic)

For some types of cancer there are discrete and distinct periods which ultimately result in cancer. These may be identified as the initiation period, the development period, and the progression period. A single, minute dose of a carcinogenic polynuclear aromatic hydrocarbon, such as benzo(α)pyrene, applied to the skin of mice produces the permanent change of a normal cell to a tumor cell. This is the initiation step. No noticeable reaction occurs unless further treatment is made. If the area is painted repeatedly with noncarcinogenic croton oil, even up to one year later, carcinomas appear (Figure 20–1). This is the development period. Additional fundamental alterations in the nature of the cells

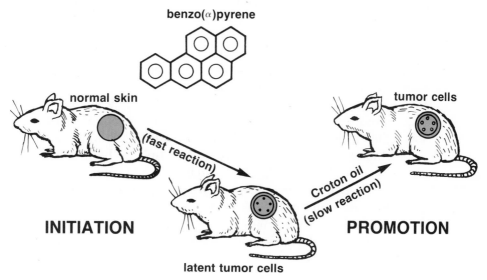

FIGURE 20-1   A second chemical can promote tumor growth in mice after an initiation period. Treatment of mice with croton oil produces no tumor nor does treatment with small quantities of benzo (α) pyrene alone.

381

occur during the progression period. If there is no initiator, there are no tumors. If there is initiator but no promoter, there are no tumors. If the initiator is followed by repeated doses of promoter, tumors appear. This seems to indicate that cancer cannot be contracted from chemicals unless repeated doses are administered or applied.

Just how do these toxic substances work? Cancer might be caused if the carcinogen combines with growth control proteins, rendering them inactive. During the normal growth process the cells divide and the organism grows to a point and stops. Cancer is abnormal in that cells continue to divide and portions of the organism continue to grow. One or more proteins are thought to be present in each cell with the specific duty of preventing replication of DNA and cell division. Virtually all of the carcinogens bind firmly to proteins, but so do some similar compounds which are noncarcinogenic. The specific growth proteins involved are not yet known for any of the carcinogens, despite considerable efforts to find them.

Another theory suggests that carcinogens react with and alter nucleic acids so the proteins ultimately formed on the messenger RNA are sufficiently different to alter the cell's function and growth rate. The carcinogen may be included in the DNA or RNA strands by covalent bonding or it may be entangled in the helix by a clathrate-type complex.° The carcinogenic compounds nitro-sodimethylamine and mustard gas have been shown to react with nucleic acids.

While researchers collect data in their laboratories and speculate on the theoretical structural causes of cancer, we can studiously avoid compounds known to cause cancer in man.

REPRODUCTION VIA CHEMICAL MUTAGENS

During reproduction, DNA (Chapter 12) normally is duplicated exactly; so exactly that when the transmission of a given DNA's code is given via messenger RNA molecules, protein synthesis in the ribosomes of the daughter cell is exactly like that in the original cell.

DNA is not always completely accurate in transmitting its code. Mistakes in the code do occur, and these are called *mutations*. Although mutations can occur spontaneously, there are other causes. In 1927, Herman Muller recognized that ionizing radiation produced mutations in some animals. Today we know that mutations can be induced by other forms of radiation, like x-rays, and by some chemicals as well. These chemicals are called *mutagens*.

Although many chemicals are under suspicion because of their mutagenic effects on laboratory animals, it should be emphasized that no one has yet shown conclusively that any chemical induces mutations in the germinal cells of man. Part of the difficulty of determining the effects of mutagenic chemicals in man is the extreme rarity of mutation. A specific genetic disorder may occur as infrequently as only once in 10,000 to 100,000 births. Therefore, to obtain

---

°The term is derived from the Latin, *clathratus*, meaning enclosed by bars. The helical structure of some molecules can physically entrap certain smaller molecules.

meaningful, statistical data, a carefully controlled study of the entire population of the United States would be required. In addition, the very long time between generations presents great difficulties, and there is also the problem of tracing a medical disorder to a single, specific chemical out of the tens of thousands of chemicals with which man comes in contact.

If there is no direct evidence for specific chemical mutagenic effects in man, why, then, the interest in the subject? The possibility of a deranged, deformed human race is frightening; the chance for an improved human body is hopeful; and the evidence for chemical mutation in plants and lower animals is established. A wide variety of chemicals is known to alter chromosomes and to produce mutations in rats, worms, bacteria, fruit flies, and other plants and animals. Some of these are listed in Table 20–4.

The horrible Thalidomide disaster in 1961 focused worldwide attention on chemically induced birth defects. Thalidomide, a tranquilizer and sleeping pill, caused gross deformities (flipper-like arms, shortened arms, no arms or legs, and other defects) in children whose mothers used this drug during the first two months of pregnancy. The use of this drug resulted in more than 4000 surviving

TABLE 20–4   MUTAGENIC SUBSTANCES AS INDICATED BY
EXPERIMENTAL STUDIES ON PLANTS AND ANIMALS

SUBSTANCE	EXPERIMENTAL RESULTS
Aflatoxin (from mold, *Aspergillus flavus*)	Mutations in bacteria, viruses, fungi, parasitic wasps, human cell cultures, mice
Benzo(α)pyrene (from cigarette and coal smoke)	Mutations in mice
Caffeine	Chromosome changes in bacteria, fungi, onion root tips, fruit flies, human tissue cultures
Captan (a fungicide)	Mutagenic in bacteria and molds; chromosome breaks in rats and human tissue cultures
Dimethyl sulfate (used extensively in chemical industry to methylate amines, phenols, and other compounds)	Methylates DNA base guanine; potent mutagent in bacteria, viruses, fungi, higher plants, fruit flies
LSD (lysergic acid diethylamide)	Chromosome breaks in somatic cells of rats, mice, hamsters, white blood cells of humans and monkeys
Maleic hydrazide (plant growth inhibitor; Trade names Slo-Gro, MH-30)	Chromosome breaks in many plants and in cultured mouse cells
Mustard gas (dichlorodiethyl sulfide)	Mutations in fruit flies
Nitrous acid (HNO_2)	Mutations in bacteria, viruses, fungi
Ozone (O_3)	Chromosome breaks in root cells of broadleaf plants
Solvents in glue (glue sniffing) (toluene, acetone, hexane, cyclohexane, ethyl acetate)	4% more human white blood cells showed breaks and abnormalities (6% versus 2% normal)
TEM (triethylenemelamine) (anticancer drug, insect chemosterilants)	Mutagenic in fruit flies, mice

383

malformed babies in West Germany, more than 1000 in Great Britain, and about 20 in the United States. With shattering impact, this incident demonstrated that a compound can appear to be remarkably safe on the basis of animal studies (so safe, in fact, that Thalidomide was sold in West Germany without prescription) and yet cause catastrophic effects in humans. While the tragedy focused attention on chemical mutagens, Thalidomide presumably does not cause genetic damage in the germinal cells, and is really not mutagenic. Rather, Thalidomide, when taken by a woman during early pregnancy, causes direct injury to the developing embryo. Even though Thalidomide is generally believed not to be mutagenic, the disaster spotlighted the serious potential hazards of supposedly nontoxic chemicals in our environment.

Experimental work on the chemical basis of the mutagenic effects of nitrous acid (HNO_2) has been very revealing. Repeated studies have shown that nitrous acid is a potent mutagen in bacteria, viruses, molds, and other organisms. In 1953, at Columbia University, Dr. Stephen Zamenhof demonstrated experimentally that nitrous acid attacks DNA. Specifically, nitrous acid reacts with the adenine, guanine, and cytosine bases of DNA by removing the amino group of each of these compounds. The eliminated group is replaced by an oxygen atom (Figure 20-2). The changing of the bases may garble a part of DNA's genetic message, and in the next replication of DNA the new base may not form a base pair with the proper nucleotide base.

For example, adenine (A) typically forms a base pair with thymine (T) (Figure 20-3). However, when adenine is changed to hypoxanthine, the new

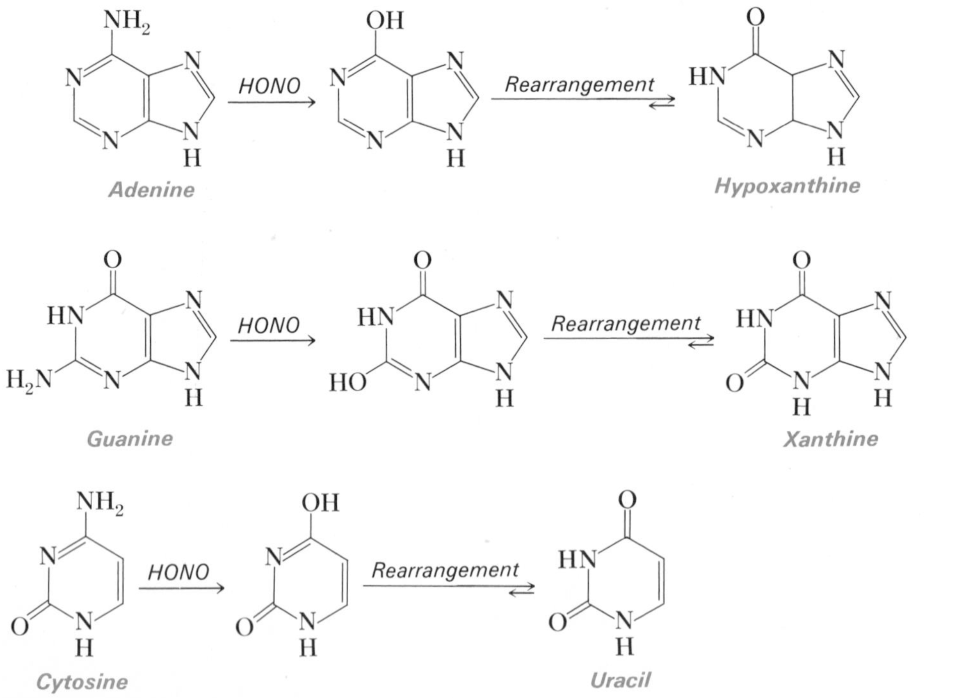

FIGURE 20-2   Reaction of nitrous acid with nitrogenous bases of DNA. Nitrogen gas (N_2) and water are also products of each reaction.

384

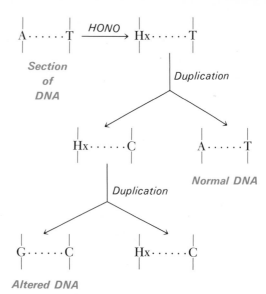

FIGURE 20–3   Alteration of DNA genetic code by base pairing nitrous acid-converted nitrogenous bases. The bases are adenine (A), cytosine (C), guanine (G), hypoxanthine (Hx), and thymine (T).

compound forms a base pair with cytosine (C). In the second replication, the cytosine forms its usual base pair with guanine (G). Thus, where an adenine-thymine (A-T) base pair existed originally, a guanine-cytosine (G-C) pair now exists. The result is an alteration in the DNA's genetic coding, so that a different protein could be formed later.

Do all of these findings mean that nitrous acid is mutagenic in humans? Not necessarily. We do know that sodium nitrite has been widely used as a preservative, color enhancer, or color fixative in meat and fish products for at least the past 30 years. It is currently used in such foods as frankfurters, bacon, smoked ham, deviled ham, bologna, Vienna sausage, smoked salmon, and smoked shad. The sodium nitrite is converted to nitrous acid by hydrochloric acid in the human stomach:

$$NaNO_2 + HCl \longrightarrow HNO_2 + NaCl$$

Some scientists theorize that the mutagenic effects of nitrous acid in lower organisms are sufficiently ominous to suggest strongly that the use of sodium nitrite in foods be severely curtailed, if not completely banned. A number of European countries already restrict the use of sodium nitrite in foods. The concern is that this compound, after being converted in the body to nitrous acid, may cause mutation in somatic cells (and possibly in germinal cells) and thus could possibly produce cancer in the human stomach. Other scientists doubt that nitrous acid is present in germinal cells and, therefore, seriously question whether this compound could be a cause of genetically produced birth defects in man. The uncertainty of extrapolating results obtained in animal studies to human beings hovers over the mutagenic substances.

Thus far, research has concentrated on the action of chemicals in causing mutations in bacteria, viruses, molds, fruit flies, mice, rats, human white blood cells, and so on. Perhaps in the next 10 to 20 years it will be demonstrated that these chemicals can produce transmissible alternation of chromosomes in human

385

germinal cells. Meanwhile, many scientists are pressing for a more vigorous research effort to expand our knowledge of chemically induced mutations and of their potentially harmful effects in man. One intriguing theory that will surely invoke experimental examination is the belief that some compounds cause cancer because they are first and foremost mutagenic. The supporting evidence at present is still inconclusive.

ALLERGENS AND ANTIHISTAMINES

Some people are allergic to pollen, mold, food, cosmetics, penicillin, and even cold, heat, and ultraviolet light. In the United States about 5000 people die yearly from bronchial asthma, at least 30 from the stings of bees, wasps, hornets, and other insects, and about 300 people die from ordinary doses of penicillin. The reason: *allergy*. About one in 10 suffers from some form of allergy; more than 16 million Americans suffer from hay fever. This means that many of you already are familiar with the symptoms of headaches, inflamed eyes, congested sinuses, sneezing, and the raw, endlessly running nose that accompanies hay fever.

An allergy is an adverse response to a foreign substance or to a physical condition that produces no obvious ill effects in *most* other organisms, including man. An *allergen* (the substance that initiates the allergic reaction) is, in most cases, a highly complex substance—usually a protein. Some are polysaccharides or complexes of a protein and a polysaccharide. Usually allergens have a molecular weight of 10,000 or more.

What is really the chemical cause of an allergy such as, say, hay fever? The details are at best sketchy now. But the overall process and a few of the details have been worked out reasonably well.

For a patient with pollen allergy, a pollen grain enters his nose and clings to the mucous membrane. The nasal secretions, acting on the pollen grain, release the grain's allergens and other soluble components, which penetrate the outer layer of the mucous membrane. The principal allergen of ragweed pollen, a major allergy-producer, has been isolated and is named ragweed antigen E. It is a protein with a molecular weight of about 38,000; it represents only about 0.5 per cent of the solids in ragweed pollen, but contributes about 90 per cent of the pollen's allergenic activity. A mere 1×10^{-12} gram of antigen E injected into an allergic person is enough to induce a response. The reason why antigen E, of all the ragweed pollen proteins, is so unusually reactive is as yet not understood.

The allergens come in contact with special cells in the nose and breathing passages to which are already attached a particular type of antibody present in unusually high concentration in allergic persons. We have a number of antibodies in our cells and bloodstream to react with and protect us from harmful parasites, bacteria, and the like. Before 1966, three antibodies, designated IgG, IgA, and IgM (Ig for immunoglobulin), were known. These are proteins with molecular weights of about 150,000, 180,000, and 950,000, respectively. In 1966, scientists isolated a previously unknown antibody, IgE. This globulin protein comprises only 0.001 per cent to 0.01 per cent of the globulin proteins in blood

FIGURE 20-4   In this Abbott Laboratories map, size of each circle represents amount of all late-summer and fall pollens found in the air in each city. Dark portions show amount of ragweed pollen. Shaded areas are regions of low pollen count.

serum, and it escaped discovery for a long time by hiding as an impurity in IgA antibodies. The molecular weight of 196,000 for IgE is very close to the molecular weight of 160,000 to 200,000 for IgA. The antibody IgE is formed in the nose, bronchial tubes, and gastrointestinal tract, and binds firmly to specific cells, called mast cells, in these regions. An allergic person has 6 to 14 times as much IgE in his blood serum as a nonallergic person. In nasal secretions of allergics, the concentration of IgE is 100 or more times greater than in the serum of the same person.

Antigen E from ragweed reacts with the IgE antibody attached to the mast cells, forming antigen-antibody complexes. Some think that the ragweed antigen E forms a bridge between two or more IgE molecules attached to a mediator-containing cell. This bridge, in turn, alters the configuration of parts of the attached IgE molecules.

By a series of events that are only crudely understood, the formation of these antigen-antibody complexes leads to the release of so-called "allergy mediators" from special granules in the mast cells. Some believe that the changed configuration of the cell caused by the antigen-antibody complex activates enzymes that ultimately cause the cell to release the allergy mediators. One of these mediators, and the most potent one found so far, is *histamine*, which is formed by the breakdown of the amino acid, histidine. Although it is widely distributed in the body, it is especially concentrated in the 250 to 300 granules

$$H_2NCH_2CH_2 \underset{\underset{H}{N}}{\overset{N}{\bigsqcup}}$$

Histamine

387

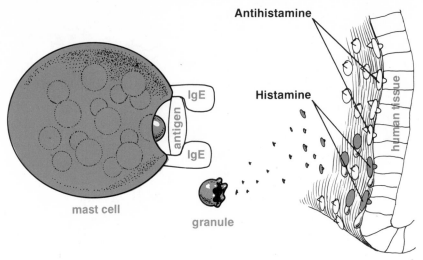

FIGURE 20-5   A postulated mechanism for the cause of and relief from hay fever.
The details are described in the text.

of the mast cells. Histamine accounts for many, if not most, of the symptoms of hay fever, bronchial asthma, and other allergies. This compound causes dilation of blood capillaries. In addition, it makes the capillaries more permeable to blood fluids. Thus, these fluids can readily leak out of the capillaries and cause swelling of the tissues. The compound causes contraction and spasm of smooth muscles (a serious difficulty in the bronchial tubes in asthma). It can produce skin swellings (hives) and stimulate the glands that secrete watery nasal fluids, mucus, tears, saliva, and so on.

The chemical mediators such as histamine must be released from the cell to cause the symptoms of allergy. The release mechanism is an energy-requiring process in which the granules may move to the outer edge of the living cell and, without leaving the cell, discharge their contents of histamine through a temporary gap in the cell membrane. This sends histamine on its way to produce the toxic effects of hay fever.

Although histamine, an allergenic compound, has been isolated and some understanding of the mechanism that leads to allergy is now known, many questions are still unanswered. Since IgE is a class of several proteins, what structural relationships cause an IgE to react, for example, with the allergen in ragweed pollen and another IgE to react with the allergen in lobster? Does heredity play a part in allergy? Are people allergic because their membranes are inherently thinner or weaker than those of nonallergics? If antigen E causes 90 per cent of the ragweed allergic activity, what causes the other 10 per cent? What chemical reactions does histamine undergo in order to produce its symptoms? These are but a few questions, and there are many others. Although hay fever, bronchial asthma, and other allergies may not be conquered, they are better understood and better treated today.

Treatment consists of three procedures: avoidance, desensitization, and drug therapy. The idea of avoidance is simple enough. If you are allergic to strawberries, do not eat strawberries; if you are allergic to pollen, move, say, from

Decatur, Illinois (ragweed pollen index: 114), to Seattle, Washington (index: 0.02); see Figure 20–4. Sometimes, however, avoidance is impractical.

Desensitization therapy is costly and inconvenient, since 20 or more injections are required to get what usually is a partial cure. One chemical idea of desensitization is to inject a blocking antibody that preferentially reacts with the allergen so that it cannot react with the IgE allergy-sensitizing antibody. This breaks the chain of events leading to the release of histamine or other allergy-producing mediators. Many small injections, spaced in time, are required to build up a sufficient level of the blocking antibody.

Epinephrine (Adrenalin), steroids, and antihistamines are effective drugs in treating allergies. The first two are particularly effective in treating bronchial asthma, while the antihistamines, introduced commercially in the United States in 1945, are the most widely used drugs for treating allergies. More than 50 antihistamines are offered commercially in the United States. Many of these contain, as does histamine, an ethylamine group, ($-CH_2CH_2N<$):

Pyribenzamine
(an important antihistamine)

These drugs act competitively by occupying the receptor sites normally occupied by histamine on cells. This, in effect, blocks the action of histamine. A new drug, disodium cromoglycate, acts not by blocking the action of histamine (as do the antihistamines), but by blocking the release of histamine and other mediators from the granules. Efforts are being made to find drugs with less troublesome side-effects than those often associated with antihistamines. The most troublesome side reactions are drowsiness, mental confusion, dizziness, headache, nervousness, rapid pulse, nausea, depression, blurred vision, and dryness of the mouth. No proved drug is available now that will relieve the allergy and completely avoid the side-effects.

HALLUCINOGENS

Hallucinogens produce temporary changes in perception, thought, and mood. These substances include mescaline, D-lysergic acid diethylamide (LSD), cannabidiol (the active component of marijuana), and a broadening field of more than 50 other substances. They are included in this chapter as a separate section because they can be toxic in comparatively small doses and because most of them are known to function in the body in more than one way. Also, they deserve a special section because of the current interest in them.

Several characteristics of some of the more famous hallucinogens are given in Table 20–5. Each of these substances is capable of disturbing the mind and producing bizarre and even colored interpretations of visual and other external

TABLE 20-5   SOME HALLUCINOGENIC CHEMICALS

CHEMICAL	ALIASES	NOTES	ACTIVE COMPOUND	FORMULA
LSD	Acid Crackers The chief The hawk	Very powerful hallucinogen	D-Lysergic acid diethylamide	
Mescaline		Extracted from mescal buttons from peyote cactus in South and Central America	3,4,5-Trimethoxy-phenethylamine	
Marijuana	Grass Pot Reefers Locoweed Hash Mary Jane	Leaves of the 5-leaf-per-frond marijuana plant (*Cannabis sativa*)	3,4-*trans*-tetra-hydrocannabinol	
Amphetamines	Pep pills Speed Bennies Ups	Eighteen times more active than mescaline	2,4,5-Trimeth-oxyamphetamine (for example)	

stimuli. Mescaline is one of the oldest known hallucinogens, having been isolated from the peyote plant in 1896 by Heffter. Indeed, as early as 1560 the Mexican Indians who ate or drank the peyote experienced "terrible or ludicrous visions; the inebriation lasting for two or three days and then disappearing." Today there are many substances that produce these experiences; the most powerful one known is LSD. Our brief discussion will be restricted to this compound.

Literally hundreds of scientists are doing research on the effects of hallucinogens, including LSD. They are probing the effects on chromosome alteration, nerve and brain function, tissue damage, psychological changes, and a host of other physiological and psychological effects. Scientists are not near a consensus on the toxic effects that LSD can cause. Collecting valid data is very difficult; many users of LSD overestimate the quantity of the drug they have used, the purity is often in question, and some take other drugs in addition. Even if the purity of the LSD is known, the results are difficult to interpret. For example, a study in which mice were given 0.05 to 1.0 microgram of LSD on the seventh day of pregnancy showed a 5 per cent incidence of badly deformed mouse embryos. Another study, probably equally valid and meticulous, showed no unusual fetal damage to rat embryos when 1.5 to 300 micrograms of LSD was administered during the fourth or fifth day of pregnancy. Babies with depressed skulls were aborted from two mothers who had taken LSD during the early weeks of pregnancy. Two other women, also LSD users, delivered full term, apparently normal babies. The debate over LSD's dangers is continuing and is not likely to be resolved soon.

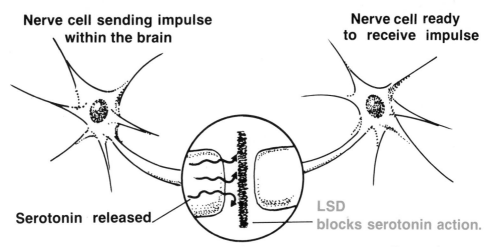

Nerve cell sending impulse within the brain

Nerve cell ready to receive impulse

Serotonin released

LSD blocks serotonin action.

FIGURE 20-6   LSD blocks the flow of serotonin from one brain nerve cell to another.

There are, however, some dangers that are well documented. These drugs destroy one's sense of judgment. Such things as height, heat, or even a moving truck may seem to hold no danger for the person under the influence of a hallucinogen. Excessive or prolonged use of LSD can cause a person to "freak out" and possibly sustain permanent brain damage. After one "trip," a user can experience another "trip" some time later, unexpectedly, without taking any more drug.

Out of the darkness of LSD-induced suicides and brain damage has come some new understanding of how the brain works. It is an interesting coincidence that soon after LSD was found localized in areas of the brain responsible for eliciting man's deep-seated emotional reactions, the compounds that are probably responsible for transmitting the impulse across synapses in the brain were discovered. Serotonin and norepinephrine are thought to act in a manner similar to that of acetylcholine, which was discussed earlier. These compounds carry the message from the end of one neuron to the end of another across the synapse. Somewhat later Dilworth Wooley discovered that LSD blocks serotonin action (Figure 20-6). This has led to an interesting theory to explain the LSD trip.

The theory of the hallucinogenic mechanism is, roughly, as follows: the LSD releases in some chemical way those emotional experiences that are generally hidden away, or chemically stored in the lower midbrain and the brain stem. Serotonin or norepinephrine, or both, inhibit the escape of these experiences into consciousness, as they turn off certain excitations. LSD interacts with serotonin or norepinephrine at the synapse and nullifies their blocking effect. Thus, these stored, previously rejected thoughts are allowed to enter the conscious part of the brain where the imaginary trip occurs.

This theory is built on the experimental fact that LSD blocks the action of serotonin; but there are other theories as well as some unexplained phenomena involved. Psychiatry is fraught with theories of why LSD causes an uplifting experience in some people and a frightening one in others. There are theories about dosage, purity of the compound, psychiatric state of the tripper, and

391

conditions of administration. If there is anything close to a consensus, it is that the more controlled and relaxed the surroundings, the "better" the trip will be; even this, however, is debatable. Many trips, even under medical supervision, have not been satisfactory.

GIN

Gin is a very simple alcoholic liquor, being nothing more than pure ethyl alcohol (ethanol, grain alcohol), pure water, and flavoring. The name comes from *genievre* (juniper); it was corrupted to "geneva," by which name it was formerly known in England, and later to "gin." The flavor is derived by allowing the alcoholic vapors to flow over the flavoring ingredients, usually juniper berries (although orange peel, caraway seeds and other flavoring materials are used to suit the individual manufacturer). The principal effects of gin are produced by the alcohol, and it, like the other "gen's," has a toxic effect on the human body. The concentration of ethyl alcohol in beverages is expressed in *proof*, which is twice the percentage of the alcohol.

Most people realize that ethyl alcohol is harmful, yet it has been around in beverage drinks since before recorded history and occupies a special place in our society. This is due in part to the fact that it can be so easily made; see Chapter 7.

Of the alcohols, only ethyl alcohol (ethanol, grain alcohol) has toxic effects mild enough to render it "safe" enough for consumption. Much of the alcohol a person consumes is absorbed by the stomach mucosa. The alcohol molecule, being small and unchanged, easily passes through the stomach lining and is immediately picked up by the bloodstream. About 58 per cent of the alcohol is absorbed in 30 minutes, 88 per cent in 1 hour, and 93 per cent in 90 minutes.

Several factors affect the absorption of alcohol by the stomach. For example, carbon dioxide hastens the absorption rate, so that champagne or whiskey and soda are more quickly absorbed than, say, straight whiskey. Certain foods such as milk slow down the absorption rate.

Over 90 per cent of the ethyl alcohol absorbed into the tissues is slowly oxidized to acetaldehyde, then to acetic acid, and finally to carbon dioxide and water, mainly in the liver although almost every cell in the body can accomplish the oxidation. The rate of oxidation is such that a 160 pound man could sip about $\frac{3}{4}$ ounce of whiskey each hour for 24 hours and not significantly increase the alcohol content of his blood. If the same man were to consume 2 or 3 ounces of whiskey or three bottles of beer in a short time, the alcohol concentration could arise to 0.05 per cent in his bloodstream. This amount of alcohol depresses the centers of the brain which control judgment and restraint. If the 160 pound man were to drink about 10 ounces of whiskey in a short time, the concentration of alcohol in his blood would rise to 0.20 per cent. This is sufficient alcohol to cause erratic emotional behavior and affect manual dexterity. A pint of whiskey can raise the alcohol concentration to about 0.30 per cent. Concentrations above this amount are extremely dangerous, since they lead to coma and possibly heart arrest.

In a government report released in 1972, alcohol abuse was called the

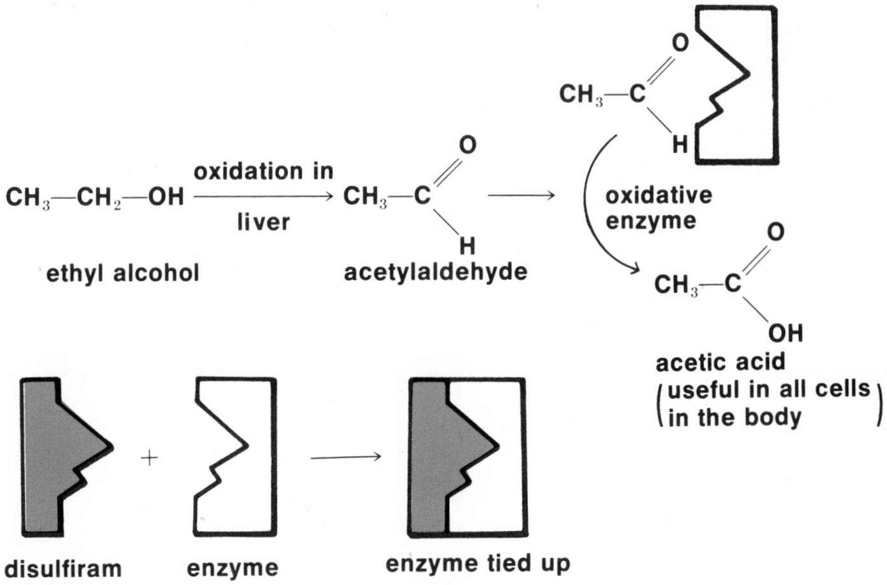

FIGURE 20-7   The blocking of ethyl alcohol oxidation by disulfiram (Antabuse). When the normal oxidative process is stopped, acetylaldehyde builds up in the body.

nation's greatest drug problem, warping nine million lives, causing 28,000 deaths on the nation's roadways annually, and costing $15 billion a year in lost jobs, welfare, impaired health and property damage. Of the estimated 95 million drinkers in the United States, most seem to use alcohol without damaging effects. However, the estimated 5 per cent of adult Americans who do have serious drinking problems shorten their life span by about 10 to 12 years and are the source of much personal tragedy.

The intoxicated person's staggering gait, stupor, and nausea are caused by the presence of acetaldehyde, but the chemical reactions involved are not fully understood. The compound disulfiram (*Antabuse*) is sometimes given as a treatment for chronic alcoholism because it blocks the oxidative steps beyond acetaldehyde. The accumulation of acetaldehyde causes nausea, vomiting, blurred vision, and confusion. This is supposed to encourage the partaker to avoid this severe sickness by avoiding alcohol. Interestingly, this drug was discovered by two researchers who took a dose for another purpose and got violently ill that evening at a cocktail party.

Individuals having a physiological dependence on alcohol will often drink liquids adulterated with alcohols other than ethyl alcohol. Methyl alcohol is the most commonly used.

Methyl alcohol (methanol or wood alcohol) is highly poisonous, and, unlike the other simple alcohols, it is, in effect, a *cumulative* poison in the human body. That is, it is only very slowly removed from the body and continues to do damage as long as it is in the body. Methyl alcohol has a specific toxic effect on the optic nerve, causing blindness when taken in large doses. After its rapid uptake by the body, there occurs an oxidation in which the alcohol is first converted into formaldehyde and then to formic acid, which is eliminated in the urine.

393

$$CH_3OH \xrightarrow{\text{Oxidation}} H-\overset{\overset{\displaystyle O}{\|}}{C}-H \xrightarrow{\text{Oxidation}} H\overset{\overset{\displaystyle O}{\|}}{C}OH$$

Methyl alcohol   Formaldehyde   Formic acid

This is a slow process and, for this reason, daily exposure to methyl alcohol can cause an extremely dangerous buildup of the alcohol in the body. The toxic effect on the optic nerve is thought to be caused by the oxidative products, formaldehyde and formic acid.

QUESTIONS

1. Should any laws and regulations be placed on the use of any of the following? Explain your answers.

   (a) LSD   (d) ethyl alcohol
   (b) marijuana   (e) aromatic amine dyes
   (c) cyclamates

2. Discuss some of the pros and cons of testing potentially toxic substances on animals.

3. What questions do you think need to be answered before the action of ethyl alcohol is understood?

4. Discuss the importance of individual differences in evaluating the carcinogenic or mutagenic nature of chemicals.

5. A new compound comes to your attention. It has the structural formula

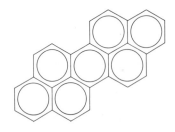

Would you be suspicious of this compound? Why?

6. Explain why a latent period in the carcinogenic effect of most substances makes evaluation of experimental data difficult.

7. In 1959, a certain insecticide widely used on the Oregon cranberry crop was discovered to be carcinogenic to mice. What was this chemical?

8. List as many ways as you can in which we can come in contact with benzo(α)pyrene.

9. What do the following compounds have in common?

   (a) saffrole   (c) cyclamates
   (b) butter yellow

10. Outline how nitrous acid can alter the genetic code transmitted by a DNA molecule.

11. How many molecules of antigen E are needed to induce a response in an allergic person?

12. In an allergic reaction, does the foreign particle, like a pollen grain, carry the chemicals which actually cause the symptoms of allergy? Explain.

13. List some of the problems associated with using LSD.

14. What are the oxidation products of ethyl alcohol? Which is responsible for most of the toxic effects of ethyl alcohol?

15. How does the drug Antabuse work?

SUGGESTIONS FOR FURTHER READING

Adams, E., "Poisons," *Scientific American*, Vol. 201, p. 76 (1959).

Greenberg, L. A., "Alcohol in the Body," *Scientific American*, Vol. 189, No. 6, p. 86 (1953).

Hoffer, A., and Osmond, H., "The Hallucinogens," Academic Press, New York, 1967.

Hueper, W. C., Jr., and Conway, W. D., "Chemical Carcinogenesis and Cancers," Charles C Thomas, Publisher, Springfield, Illinois, 1964.

Loomis, T. A., "Essentials of Toxicology," Lea and Febiger, Philadelphia, 1968.

McGrady, P., "The Savage Cell: A Report on Cancer and Cancer Research," Basic Books, New York, 1964.

Maxwell, K. E., "Chemicals and Life," Dickenson Publishing Co., Inc., Belmont, Calif., 1970.

Robinson, T., "Alkaloids," *Scientific American*, Vol. 201, No. 1, p. 113 (1959).

Sanders, H. J., "Chemical Mutagens," *Chemical and Engineering News*, Vol. 47, No. 21, p. 50 (1969), and Vol. 47, No. 23, p. 54 (1969).

Sanders, H. J., "Allergy—A Protective Mechanism Out of Control," *Chemical and Engineering News*, Vol. 48, No. 20, p. 84 (1970).

"The Allergic Mechanism," *Chemistry*, Vol. 44, No. 5, p. 23 (1971).

Weisburger, J. H., and Weisburger, E. K., "Chemicals as Causes of Cancer," *Chemical and Engineering News*, Vol. 44, No. 7, p. 124 (1966).

21

CHEMISTRY FOR THE FAMILY CAR

A significant amount of chemistry is involved in the production and operation of an automobile. Before an automobile can roll off the assembly line, metals must be taken from their ores; plastics must be synthesized and fabricated; paints must be formulated; sulfuric acid for the battery must be made; rubber must be synthesized, vulcanized, and formed into shape; and glass must be mixed, fired, molded and cut. Numerous other chemical reactions have to be carried out to produce the fuel the car will need.

After the car hits the road, a host of different reactions are involved in transferring the energy stored in the fuel to the drive wheels. In its wake the automobile leaves its exhaust, which undergoes a variety of chemical reactions such as smog formation.

In this chapter we shall look at some of this chemistry. The logical starting point is with the fuel.

GASOLINE

From the time petroleum was first discovered in the United States in 1859 until 1900 when the automobile became popular, most oil was refined to yield kerosene, a mixture of hydrocarbons in the C_8 to C_{13} range. This liquid was used principally as a fuel for lamps.

The internal combustion engines used in early automobiles were designed to burn a more volatile mixture of hydrocarbons, the C_6 to C_{10} fraction, which became known as *gasoline*. These lower molecular weight hydrocarbons evapo-

TABLE 21-1 Division of a Barrel of Crude Oil

	1920 %	1967 %
Gasoline	26.1	44.8
Kerosene	12.7	2.8
Jet Fuel	—	7.6
Heavy Distillates	48.6	22.2
Other	12.6	22.6
Total	100.0	100.0

rate easily, so that they mix readily with air in simple carburetors and burn fairly completely.

With the increasing popularity of the automobile, petroleum refiners had to shift the output of a barrel of crude oil from mostly kerosene to mostly gasoline (Table 21–1).

This dramatic increase in the amount of gasoline from a barrel of crude oil was accomplished by the discovery of chemical processes which convert non-gasoline molecules into ones which burn well in an automobile engine.

The first conversion discovered was *thermal cracking,* a process by which long chain hydrocarbon molecules were heated under pressure. This heating strains the bonds of the molecule so that it may break or "crack," thus forming two smaller molecules, some of which will be in the gasoline range.

$$C_{16}H_{34} \xrightarrow[\text{Heat}]{\text{Pressure}} C_8H_{18} + C_8H_{16}$$

An alkane   *An alkane*   *An alkene*
in the gasoline range

Later, cracking was carried out in the presence of certain catalysts which allowed the processes to proceed at lower pressures and to produce even higher yields of gasoline. Today, these catalysts include aluminum oxide and platinum, as well as certain acids and specially processed clays. Each refiner of petroleum has his own special methods which offer different advantages in cost and type of crude oil handled. The hydrocarbon molecules produced are much the same regardless of the methods used.

Cracking of petroleum fractions brought about not only an increase in the quantity of gasoline available from a barrel of crude oil, but also an increase in *quality*. That is, gasoline from a cracking process can be used at higher efficiency (in a high-compression engine) than can "straight run" gasoline, because the molecular structures of the hydrocarbons in the cracked gasoline allow them to oxidize more smoothly at high pressure. When the burning of the gasoline-air mixture is too rapid or irregular, preignition occurs in the combustion chamber, resulting in a small detonation which is heard as a "knock" in the engine. This knocking can spell trouble to the owner of an automobile, because it will eventually lead to the breakdown of the internal parts of the automobile's engine.

An arbitrary scale for rating the relative knocking properties of gasolines has been developed. Normal heptane, typical of straight run gasoline, knocks considerably and is assigned an octane rating of 0,

397

$$CH_3CH_2CH_2CH_2CH_2CH_2CH_3 \text{ octane rating} = 0$$

while isooctane (2,2,4-trimethylpentane) is far superior in this respect and is assigned an octane rating of 100.

$$
\begin{array}{ccc}
& CH_3 & CH_3 \\
& | & | \\
CH_3-C-CH_2-C-CH_3 & & \text{octane rating} = 100 \\
& | & | \\
& CH_3 & H
\end{array}
$$

To determine the octane rating of a gasoline, it is used in a standard engine and its knocking properties are recorded. This is compared to the behavior of mixtures of n-heptane and isooctane, and the percentage of isooctane in the mixture with identical knocking properties is called the octane rating of the gasoline. Thus, if a gasoline has the same knocking characteristics as a mixture of 9 per cent n-heptane and 91 per cent isooctane, it is assigned an octane rating of 91. This corresponds to a regular grade of gasoline. Since the octane rating scale was established, fuels have been developed which are superior to isooctane, so the scale has been extended well above 100.

A great deal of research has been directed toward reducing the knocking caused by gasoline, and it has been discovered that this can be avoided by two procedures. The first is to use only hydrocarbons with structures known to burn satisfactorily in a high-compression engine. Branched-chain structures are particularly desirable in this respect, and gasolines with appreciable amounts of aromatic hydrocarbons (20 to 40 per cent) also are less prone to knock. The second is to add "antiknock" agents to a more motley mixture of hydrocarbons to achieve the same effect. The best-known antiknock additive is tetraethyllead, $(C_2H_5)_4Pb$. Tetraethyllead is extremely toxic (see later section) and gasolines containing it ordinarily contain a dye to warn of its presence. Gasoline containing this compound is known as "ethyl" or "high octane" gasoline.

Antiknock additives in general act to slow down free radical reactions that take place in the combustion chamber of an engine. By slowing these reactions down, the gasoline-air mixture can burn more smoothly, producing more power with a reduced likelihood of damaging the engine. When tetraethyllead is used in gasolines, it also is necessary to add a halogen compound such as 1,2-dibromoethane, $BrCH_2-CH_2Br$, to assist in the removal of the lead from the engine. The burning of a mixture of these compounds produces lead bromide, $PbBr_2$, which is eliminated as a vapor in the engine exhaust and released into the atmosphere.

A typical antiknock mixture for gasolines consists of about 62 per cent tetraethyllead, 18 per cent ethylene dibromide, 18 per cent ethylene dichloride, and 2 per cent other ingredients, such as dye, petroleum solvent, and stability improver. Ethyl gasoline normally contains only a few tenths of a percent of such a mixture, corresponding to about 2 to 4 g of lead per gallon. Nevertheless, the total consumption of lead (and discharge into the atmosphere) amounts to millions of pounds per year.

WHAT MAKES BRAND X BETTER?

The ideal gasoline would be one that burns completely to carbon dioxide and water at the right time, and leaves no deposits in the engine. It would have sufficient vapor pressure (from lower boiling fractions) to ignite when the spark plug sparks, but not enough vapor pressure to cause a vapor lock in the fuel lines or fuel pump and stop the flow of gasoline. Moreover, it would have low gum and sulfur content. Gum formation is the result of oxidation and polymerization of unsaturated fuel components, particularly hydrocarbons with two or more double bonds per molecule. Most motor gasolines contain less than 3 mg of gum per 100 ml. Thiols, or mercaptans (—SH groups), present in crude oil sometimes find their way into the finished gasoline to provide the sulfur.

Other chemicals are added to gasoline to prevent the ignition of new fuel by glowing particles from a previous ignition. These additives are called *deposit modifiers*. Phosphorus compounds, such as tricresyl phosphate (TCP) and, more recently, boron compounds, have been used for this purpose. These alter the composition of the deposits and make them less likely to glow. The phosphorus compounds also prevent spark plug deposits from becoming so electrically conductive that the charge leaks away instead of firing the plug. Boron compounds raise the octane number of the gasoline by preventing sulfur compounds from rendering tetraethyllead ineffective.

Antioxidants such as phenylenediamine, aminophenols, dibutyl-p-cresol, and orthoalkylated phenols, are added to prevent the formation of peroxides that lead to knock and to gum formation. About 2 or 3 pounds of these additives are added to every 1000 barrels of gasoline.

The mechanism of antioxidation of automotive oils is very similar to the mechanism described for BHA and BHT in food additives (Chapter 13). Because copper ions catalyze gum formation, metal ion scavengers such as ethylenediamine are added to chelate the trace amounts of copper ions and render them ineffective. The copper gets into the gasoline from the copper tubing used for fuel lines and from brass parts of the engine (brass is an alloy of copper and zinc.)

Copper   Ethylene-   Chelated Cu^{2+} ion
ion   diamine

To inhibit water from corroding and rusting storage tanks, pipelines, tankers, and fuel systems of engines, *antirust agents* are added. Some of the most effective are sodium and calcium sulfonates, which are detergent molecules. These molecules coat metal surfaces with a very thin protective film that keeps water from contacting the surfaces, as shown in Figure 21–1. This also helps prevent gummy deposits in the carburetor, and combats carburetor icing during cold weather.

Anti-icing agents coat metal surfaces, as do antirust agents, and thus prevent

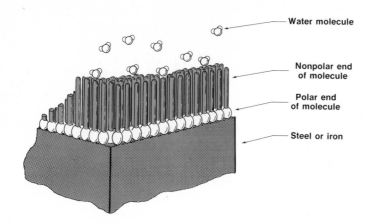

Water molecule

Nonpolar end
of molecule

Polar end
of molecule

Steel or iron

FIGURE 21-1   The action of surface active agents, such as rust in-
hibitors, mild antiwear agents, and
some de-icing agents.

ice particles from accumulating on surfaces, and/or depress the freezing point.
The freezing point depressants, which include alcohols and glycols, act in the
same manner as the antifreeze in the engine's cooling system. The small ice
particles pass harmlessly through the carburetor and into the engine where the
heat converts them to water vapor that eventually exits with the exhaust gases.

Detergents, which include alkylammonium dialkyl phosphates, are added
to prevent the accumulation of high-boiling components on the walls of the
carburetor. These deposits interfere with the air flow into the carburetor and
cause rough idling, frequent stalls, poor performance, and increased fuel consumption. The effectiveness of these detergents stems from their surface active
properties. The film of detergent provides a thin, nonpolar coating on the metal
surfaces which prevents high molecular weight, nonpolar gums from forming
thick deposits on the surfaces.

Upper cylinder lubricants are sometimes blended into gasoline. These are
usually light mineral oils or low viscosity naphthenic distillates (such as cyclo-
pentane) that dissolve away deposits in the intake system from the carburetor,
cylinders, top piston rings, and valves. (Nonpolar deposits dissolve in nonpolar
solvents.)

At one time *dyes* were added only to distinguish antiknock "ethyl" gasolines
from unleaded gasolines. Always present in very small amounts, they are now
used to distinguish grades and brands.

A common misconception about gasolines is the relationship between *energy
content* and *octane number.* All grades of gasoline, whether regular or high test,
have almost identical energy contents, about 29,000 kilocalories per gallon if
burned completely. Power losses are caused by incomplete burning and mechanical knocking due to preignition. The difference lies not in their energy
content but in their octane rating and completeness of burning, which are
determined by the amount of low-molecular-weight hydrocarbons as well as by
the kind and amount of additives present. Regular gasoline is about 91 to 94
octane rating and high test is about 97 to 101.

POLLUTION BY THE AUTO—A SPECIAL CASE

When we think about cleaning up the atmosphere, our first concern should
be the automobile. More than 100 million automobiles jam our roadways, each

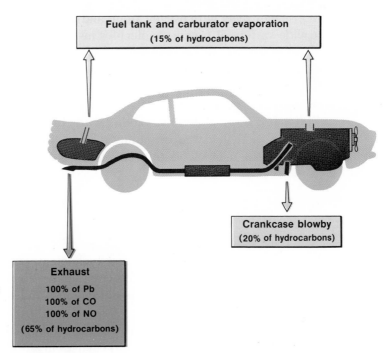

Fuel tank and carburator evaporation
(15% of hydrocarbons)

Crankcase blowby
(20% of hydrocarbons)

Exhaust
100% of Pb
100% of CO
100% of NO
(65% of hydrocarbons)

FIGURE 21-2   Sources of automobile pollutant emissions.

adding its share of pollutants to the atmosphere. It has been calculated that the automobile accounts for half of the hydrocarbons in the atmosphere, two thirds of the carbon monoxide, and two fifths of the nitrogen oxides (see Figure 19–3).

Automobile air pollutants enter the atmosphere in three major ways: from the exhaust, from the crankcase blowby (gases that escape around the piston rings), and by evaporation from the fuel tank and carburetor.

When gasoline is mixed with air, compressed in the combustion chamber, and ignited, many of the hydrocarbon molecules combine with oxygen to form CO, CO_2, and water vapor. Some of the hydrocarbon molecules fail to react,

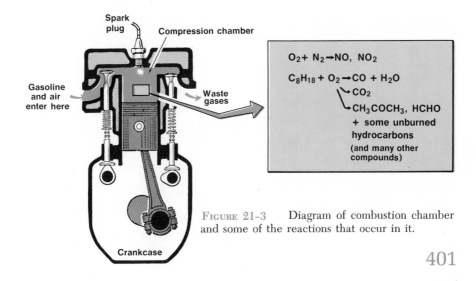

Spark plug

Compression chamber

Gasoline and air enter here

Waste gases

Crankcase

$$O_2 + N_2 \rightarrow NO, NO_2$$
$$C_8H_{18} + O_2 \rightarrow CO + H_2O$$
$$\hookrightarrow CO_2$$
$$\hookrightarrow CH_3COCH_3, HCHO$$
+ some unburned hydrocarbons
(and many other compounds)

FIGURE 21-3   Diagram of combustion chamber and some of the reactions that occur in it.

or else they react with oxygen atoms and other hydrocarbon fragments to produce a wide variety of organic molecules instead of burning (Figure 21–3).

Researchers at the Deepwater, New Jersey, Du Pont plant have found more than 200 different hydrocarbons in automobile exhaust.

Nitrogen combines with oxygen in the automobile engine to form a variety of nitrogen oxides, principally NO and a little NO_2. The tetraethyllead, $Pb(C_2H_5)_4$, forms $PbBr_2$ by reacting with ethylene dibromide. These combustion products, added to the evaporation losses from the carburetor and gas tank, comprise the noxious waste products of the automobile.

WAYS TO CONTROL EMISSION

What can be done to abate these emissions? Before discussing the specific methods, we should be aware that technology over the past two decades has developed many ways to solve this complicated air pollution problem. But every method increases the cost of the automobile, and some improvements increase the costs substantially.

Two approaches are presently being developed to meet recently established air pollution standards: mechanical changes in automobiles, and changes in fuel. Mechanical changes are made at relatively low cost, but fuel modifications offer advantages over mechanical devices. The most obvious advantage is that fuel changes would apply to all cars, old and new.

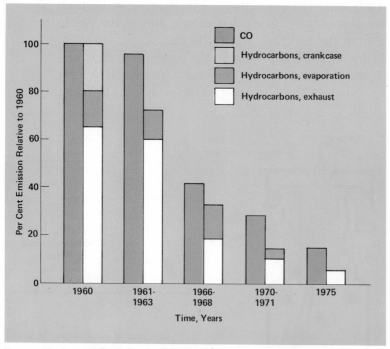

FIGURE 21–4   Effect of control devices on carbon monoxide and hydrocarbon emissions from the automobile, 1960–1971. The predictions for 1975 are based on current legislation.

Methods for reduction of hydrocarbon and carbon monoxide emissions in the exhaust are being developed along three lines: (1) the adjustment of fuel-air ratio, spark timing, and other variables; (2) the injection of air into the hot exhaust gases; and (3) the use of a catalytic converter. The first two methods have physical limits that will prevent complete conversion of hydrocarbons and carbon monoxide to carbon dioxide and water. Ideally, if the right amount of oxygen is present, complete conversion to CO_2 is expected. For example, the combustion of a mole of octane requires 12.5 moles of oxygen:

$$C_8H_{18} + 12.5O_2 \longrightarrow 8CO_2 + 9H_2O + heat$$

But the time for reaction in an automobile engine is so rapid that the octane and oxygen molecules cannot react completely. In the chaotic split second of ignition and reaction, many undesirable side reactions occur. These reactions produce the 100 or so hydrocarbons found in an automobile's exhaust which were not originally in the gasoline.

An important factor to consider is that less nitrogen is oxidized to oxides of nitrogen if the temperature within the combustion chamber is reduced to the lowest practical limit. One way to do this is to recycle part of the exhaust gas back into the engine. This reduces the peak combustion temperature and the amount of nitric oxide that forms. Prototypes of such recycling systems have reduced nitrogen oxide emissions by 80 per cent without causing the exhaust to exceed 275 ppm hydrocarbons and 1.5 per cent CO. Coincident with the recirculation of about 15 per cent of the exhaust and the reduction of nitrogen oxides by 88 per cent is a cut in power output by 16 per cent and a decrease in fuel economy by 15 per cent. Changes in timing can increase power and economy, but the amount of nitrogen oxides produced also increases. Fuel injected specifically at the spark plug gap will produce low hydrocarbon, CO, and NO_x emissions, but then particulate emissions become a problem.

Considerable research is being done to develop catalysts for converting CO, hydrocarbons, and NO_x to less harmful products, such as N_2, H_2O, and CO_2. A major problem is finding a catalyst that remains active in the presence of the lead normally found in gasoline.

Perhaps the best method of removing lead from the exhaust is to remove it from the gasoline. One company has marketed a lead-free gasoline for many years. Its gasoline contains about 30 per cent aromatics—very reactive, smog-producing compounds. Another factor which is slowing the conversion to no-lead gasoline is the high cost of changing production methods. At an estimated cost of $4.25 billion—roughly 40 per cent of the present gross investment in refining equipment in the United States—refining equipment could be updated to a point where lead would no longer be necessary in gasoline.

Yet the cost may not be too high. The lead absorbed by the lungs of a typical urban child in a city of 2.5 million population (the air there contains about 3 micrograms of lead per cubic meter) will be about 20 to 25 micrograms per day. This represents about 8 per cent of the maximum capacity of the body to excrete lead. Other sources of lead include food, beverages and water.

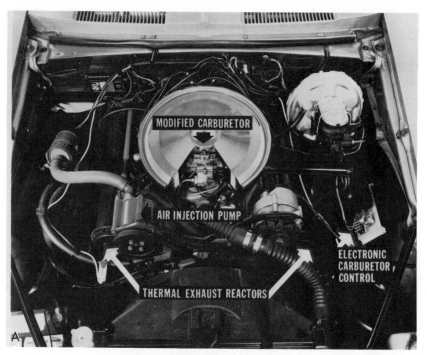

1975 EMISSION SYSTEM

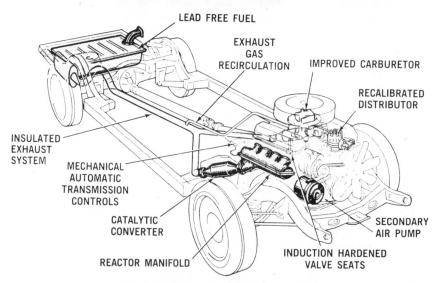

LEAD FREE FUEL

EXHAUST GAS RECIRCULATION

IMPROVED CARBURETOR

RECALIBRATED DISTRIBUTOR

INSULATED EXHAUST SYSTEM

MECHANICAL AUTOMATIC TRANSMISSION CONTROLS

CATALYTIC CONVERTER

REACTOR MANIFOLD

INDUCTION HARDENED VALVE SEATS

SECONDARY AIR PUMP

NOTE: THE 1976 (76-1) EMISSION SYSTEM WILL USE 1975 SYSTEM COMPONENTS BUT WILL BE TOTALLY RECALIBRATED FOR NOx REDUCTIONS.

B

FIGURE 21–5   Automotive hardware for pollution control.

NEW KINDS OF CARS?

Changes in the types of automobiles on the streets of the United States are inevitable, and some are taking place now. For example, several thousand automobiles in this country are now burning natural gas instead of gasoline. Natural gas, which is mostly methane, CH_4, burns cleanly to CO_2 and water. A necessary device, positioned over the carburetor, costs about $350. The problem here is that our surplus of natural gas in this country is rapidly diminishing, a potential long range problem.

One of the very few ways to eliminate combustion processes is to use electricity in some way to power our transportation devices. Early attempts to develop electric cars in the United States were made in Boston in 1888, 28 years after the development of the first storage battery in 1860. By 1912, about 6000 electric passenger and 4000 commercial vehicles were being manufactured annually in this country. The electric car lost out in competition with the internal combustion engine during the 1920's. The problems of short range (about 20 miles), low speeds (20 miles per hour maximum), 8 to 12 hours recharge time for the batteries, and relatively high price combined to eliminate the electric car from the race for leader in transportation.

At present, more than 100,000 rider-type, battery powered, materials-handling vehicles are operating in U.S. plants and warehouses where it is vital to prevent air pollution from internal combustion engines. Despite this positive start, attempts to employ electric vehicles for street and highway use generally have been commercial failures. The energy storage capacities of conventional batteries (lead-acid, nickel-iron, and silver-zinc) are too limited or expensive to provide an acceptable energy source for electric passenger cars.

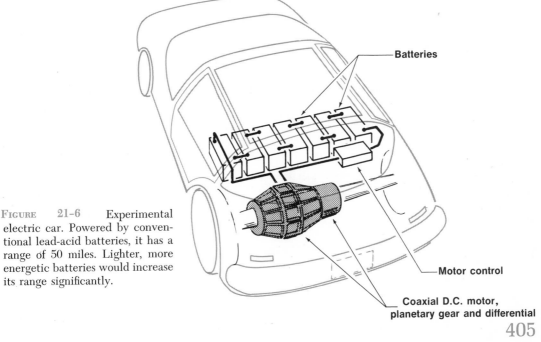

Batteries

FIGURE 21-6 Experimental electric car. Powered by conventional lead-acid batteries, it has a range of 50 miles. Lighter, more energetic batteries would increase its range significantly.

Motor control

Coaxial D.C. motor, planetary gear and differential

Many new systems are in the development stage. For example, zinc-air, lithium-chlorine, sodium-sulfur, sodium-air, lithium-carbon fluoride, lithium-nickel halide, and various kinds of fuel cells show some promise of being perhaps 10 or 15 times more powerful than the battery systems now in use.

Perhaps after 1985, the electrically powered car may be most attractive, but sufficiently powerful and durable batteries and fuel cells for this purpose do not now exist.

We should not rule out the possibility that, by the turn of the century, cheap, fusion-generated electricity may make hydrogen available from the electrolysis of water. Nothing could be cleaner than a hydrogen-powered car!

$$2H_2(g) + O_2(g) \longrightarrow 2H_2O(g) + heat$$

QUESTIONS

1. What are the essential differences between high and low octane gasolines with respect to energy content, additives, and completeness of burning?

2. Write a balanced equation for the complete combustion of 2,2,4-trimethylpentane.

3. What is meant by high detergency (HD) gasoline? Specifically, what do the detergents do for a gasoline engine?

4. What is the purpose of each of these substances in gasoline?

   (a) ethylenediamine   (c) ethylene glycol
   (b) TCP   (d) dyes

5. Why is it so difficult to avoid the formation of either carbon monoxide or nitrogen oxides in the combustion process in an automobile engine?

6. Describe the antipollution devices on your car.

7. What compounds give "ethyl" gasoline its name?

8. Describe an effective way of eliminating carbon monoxide from the exhaust of an automobile.

SUGGESTIONS FOR FURTHER READING

Medeiros, R. W., "Lead From Automobile Exhausts," *Chemistry*, Nov. 1971, p. 7.
Wolf, P. C., "Carbon Monoxide Measurement and Monitoring in Urban Air," *Environmental Science and Technology*, Vol. 5, No. 3 (1971), p. 212.

Part 7

Chemistry Out Beyond

22

SPACE CHEMISTRY

Our knowledge of the chemistry of space (astrochemistry) has just entered upon an era during which it can be expected to undergo enormous growth. Although we have not yet been able to venture very far into space, we impatiently gaze at its wonders, receive its radio, light, and meteor messages, and painstakingly study them along with lunar samples to expand our knowledge of this vast universe.

There is one hypothesis which is used in practically all studies of the chemistry of space: the basic physical and chemical principles governing the behavior of matter are the same in all parts of the universe and are independent of both time and place. The relationships developed on earth are assumed to apply to all matter everywhere under similar conditions. This hypothesis is supported by an enormous amount of evidence. Observations of the motions of the planets were explained by the physical theories developed by Sir Isaac Newton (1642–1727), which applied equally well to phenomena on earth. In 1859, Kirchoff and Bunsen invented the spectograph and subsequently showed that each kind of matter has a characteristic and typical set of wavelengths of light which it can emit or absorb. Soon afterwards this instrument was used in combination with telescopes to analyze the light given off by distant stars. Such studies revealed that stars were made up of the same atomic and molecular constituents found on the earth.

METEORITES

Before the dramatic sampling of lunar material in 1969, the analysis of meteorites offered man his only physical contact with extraterrestrial matter. It is estimated that about 8000 tons of extraterrestrial material accumulate on the earth each day. Evidence indicates that this process has been going on at the same rate for at least 400 million years. Most of this matter is inaccessible for study because it either strikes the earth as micrometeorites or enters the

TABLE 22-1   TYPICAL CHEMICAL COMPOSITION FOR
METEORITES (PERCENT BY WEIGHT)

IRON		STONE	
Iron	90%	Oxygen	43%
Nickel	7%	Silicon	21%
Cobalt	0.3%	Magnesium	17%
Other trace	2.7%	Iron	13%
elements		Calcium	2%
		Aluminum	2%
		Other trace	2%
		elements	

atmosphere as larger particles that disintegrate before they hit the earth. It has been estimated that approximately 2000 macrometeorites large enough to be identified strike the earth each year; however, only a few—10 to 20 a year—are actually recovered.

Meteorites can be divided into three classes based on the ratio of metal to silicate (minerals containing ionic structures such as SiO_4^{4-}, $Si_2O_7^{6-}$, $Si_3O_9^{6-}$, $Si_6O_{18}^{12-}$ anions). These are *irons, stones,* and *stony irons.* Typical chemical compositions are given for iron and stony meteorites in Table 22-1; the stony iron meteorites have intermediate values. Stony meteorites comprise 92 per cent of the known falls, irons comprise 6 per cent, and stony irons 2 per cent. Figure 22-1 shows typical examples of these groups.

Radioactive dating techniques have consistently shown meteorites to be

FIGURE 22-1   Various types of meteorites. (A) iron; (B) stony iron; (C) stone; (D) carbonaceous. (Courtesy of Floyd Getsinger Slides, Phoenix, Arizona.)

approximately 4.5 billion years old. They are at least one billion years older than the oldest rocks found on the surface of the earth and equal to the assumed age of earth itself. Thus it is reasonable to believe that meteorites contain a cosmological record which, if deciphered by man, could tell him how the solar system was formed.

One theoretical picture of the origin of meteorites begins with the gaseous nebula from which our sun was made; it suggests that there was a cooling period in which the less volatile, gaseous materials were condensed into stony meteorites. The more volatile gaseous material, if trapped in the condensed material, subsequently escaped into the background of interstellar matter.

What do meteorites tell about the possibility of life elsewhere in the universe? Some freshly fallen stony meteorites have been found to contain carbonaceous materials, including complicated organic compounds (Figure 22–1). There is little doubt that these organic compounds are extraterrestrial in origin. However, it is as possible that they were synthesized from atoms and smaller molecules in the presence of high energy radiation in space as it is that they are truly fragments of another living world. British chemists claim to have found sporopollenin in meteorites. Sporopollenin is a covering material for spores or pollen grains. If this material did indeed ride to the earth on the meteorite, and is not an example of earthly contamination, it represents very ancient life forms.

THE ELEMENTAL COMPOSITION OF THE SUN

Our sun is one star in a group of almost 150 billion stars known as the Milky Way Galaxy (Figure 22–2). The only star close enough for us to see other than as a point of light is our sun. The enormous size of the sun is rather hard to visualize. It is about 800,000 miles in diameter, compared to earth's 8000 mile diameter and Jupiter's 88,000 miles. Over 1,000,000 volumes the size of earth could be placed inside the sun! Giant flames of glowing gas, called prominences, leap out from the surface to distances of 100,000 miles (Figure 22–3). The solar corona, a very thin atmosphere of glowing gas, reaching a diameter of over 2,000,000 miles, is too dim to see ordinarily (Figure 22–4). Beyond this the rarefied atmosphere fades into the interstellar medium.

Of what is the sun composed chemically? Spectrographic analysis of the sun, begun over 100 years ago, indicates qualitatively what elements or compounds are present, and the intensity of the lines provides a quantitative measure of their relative amounts.

The surface temperature of the sun is 10,000°F and its interior temperature is several million degrees. The current assumption about the sun's interior, based on fluid motions within the sun, is that our analysis of the surface is representative of the sun as a whole. A recent and important advance in the chemical assay of the sun is man's ability to sample actual particles of matter shot off from the sun which have been transported towards earth by the solar wind. Satellites, solar wind experiments from Apollo missions, and lunar samples themselves contain quantitative samplings of solar material.

409

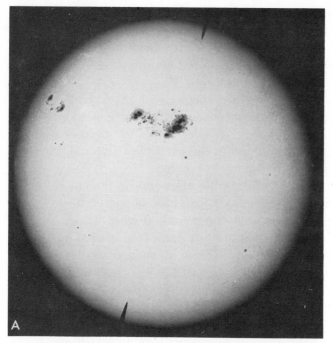

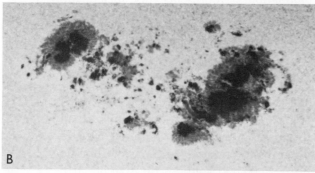

FIGURE 22-2   Photograph of the sun. Dark spots, called sun spots, are regions of solar storms and are cooler than the surrounding surface. This picture was taken April 7, 1947. (Copyright by the California Institute of Technology and Carnegie Institution of Washington. Photograph from the Hale Observatories.)

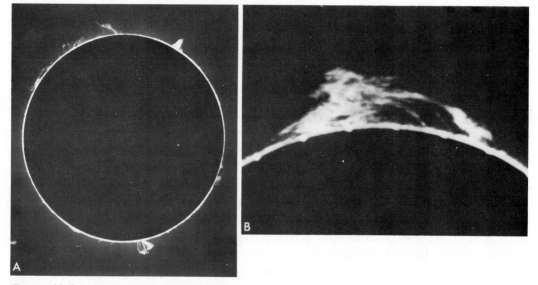

FIGURE 22-3   (A) Prominences at the edge of the sun. Photographed December 9, 1929. (B) Closeup of prominences. Photographed August 18, 1947. (Copyright by the California Institute of Technology and Carnegie Institution of Washington. Photographs from the Hale Observatories.)

FIGURE 22-4 Solar corona, photographed June 8, 1918, at Green River, Wyoming, during a total solar eclipse. The corona, a vacuum by earth standards, consists of glowing gases originating from the sun and activated by the sun. (Copyright by the California Institute of Technology and Carnegie Institution of Washington. Photograph from the Hale Observatories.)

TABLE 22-2   RELATIVE ELEMENTAL ABUNDANCES IN THE ATMOSPHERE OF THE SUN°

ELEMENT	RELATIVE NUMBER OF ATOMS (H = 1×10^{12})
Hydrogen	1×10^{12} (1,000,000,000,000)
Helium	9×10^{10}
Oxygen	5×10^{8}
Carbon	3.5×10^{8}
Nitrogen	8.5×10^{7}
Silicon	3.5×10^{7}
Iron	3.2×10^{6}
Aluminum	2.5×10^{6}
Sodium	1.5×10^{6}
Chromium	3.0×10^{5}
Phosphorus	2.7×10^{5}
Nickel	1.2×10^{5}
Potassium	1.1×10^{5}
Manganese	7.6×10^{4}
Titanium	3.2×10^{4}
Vanadium	8.3×10^{3}
Cobalt	3.4×10^{3}
Scandium	1.1×10^{3}

° Lambert and Warner, Oxford, 1968

411

TABLE 22-3   RELATIVE NUMBER OF ATOMS IN THE UNIVERSE
(BASED ON SILICON = 1000)°

Hydrogen	25,000,000	Calcium	49
Helium	3,000,000	Sodium	44
Oxygen	25,000	Nickel	27
Neon	14,000	Phosphorus	10
Carbon	9,300	Chromium	7.8
Nitrogen	2,400	Manganese	6.8
Silicon	1,000	Potassium	3.2
Magnesium	910	Chlorine	2.6
Sulfur	380	Cobalt	1.8
Argon	150	Titanium	1.7
Iron	150	Fluorine	1.6
Aluminum	95		

° From spectral data and meteorite analysis.

Studies of the evidence at hand have led to estimates of the relative abundances in the sun's atmosphere of some of the major elements (Table 22–2). The sun is roughly 90 per cent hydrogen and 9 per cent helium with several remaining elements constituting about 1 per cent. These figures indicate the relative numbers of atoms, not their relative contributions to the mass of the sun.

Because the sun is a star of a sort that is quite common in the known universe, it is assumed that the composition of the sun is, to a first approximation at least, roughly the same as the composition of the universe as a whole. Table 22–3 gives an estimate of the relative number of atoms of the more common elements in the universe.

MATTER IN INTERSTELLAR SPACE

A very important aspect of astrochemistry is that, taking an overall view, matter is scattered very thinly in space. For example, the solar system is a better vacuum than can be achieved in earthbound laboratories. The earth-moon system is 99.99953 per cent empty. The solar system as a whole is even more empty. Actually a straight line randomly drawn through space would be highly unlikely to intersect any matter at all.

At the beginning of the twentieth century, scientists had no knowledge of the kinds of interstellar material except for what had been observed within our own solar system. Since then great clouds of gaseous material have been located within our galaxy (Figure 22–5). Even though clouds are a concentration of matter, they exist in the high vacuum of space. How can scientists make definite statements about gas and dust particles that are thousands of light years away? (A light year is the distance light, traveling 186,000 miles per second, will travel in one year; 5.85×10^{12} miles.) The answer lies in the light transmitted by starlight and in the electromagnetic energy emitted from interstellar particles.

The story begins at about the turn of the twentieth century with J. Hartmann of Potsdam, Germany, who was studying the orbital motion of the double star, Delta Orionis. The star gave a typical continuous emission spectrum with certain wavelengths (lines) missing owing to the absorption by low energy atoms in the

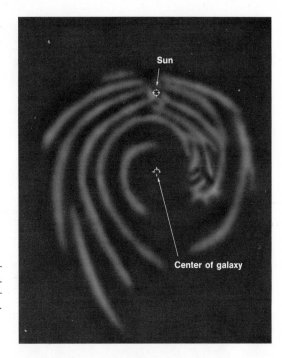

Sun

Center of galaxy

FIGURE 22-5   Darkened areas represent the concentration of hydrogen clouds located in our galaxy.
The hydrogen is "observed" by characteristic electromagnetic radiation. Note the position of the sun.

outer atmosphere of the star. As the star rotated in its orbit about its dark
companion, these spectral lines shifted a bit on the spectral photograph. As the
star moved away from us the line shifted toward the longer wavelengths, and
as it moved toward us it shifted toward the shorter. This is termed the Doppler
shift, after the scientist who first described the related phenomenon in sound.
Recall that a horn sounds as though it has a higher pitch as it approaches you
than when it is receding.

Hartmann found one line, the calcium line at 393.4 nanometers, that did
not display the Doppler shift. This could only be explained if the calcium atoms
were in the space between Delta Orionis and earth (Figure 22–6). Since this
concentration of calcium cannot be found in the earth's atmosphere nor in the
regions of the solar system, it must necessarily follow that it is in interstellar
space. Further study revealed additional lines due to calcium.

Soon, other elements, including sodium, potassium, titanium, and iron, were
discovered in interstellar space with the same technique.

Only recently has radioastronomy opened up a new and powerful approach
to interstellar studies. Hydrogen atoms at very low temperatures, close to
$-273°C$, can absorb and emit a very small quantum of energy relative to visible
light. Remember that the quantum energy emitted grows smaller as the wavelength increases. The wavelength emitted by or absorbed by elemental hydrogen
at these extremely low temperatures is 21 cm (about 8 inches) and is in the
radio region. The large, efficient radio antennas and receivers, products of space
age technology, enable us to "see" interstellar hydrogen. There are two ways
to do this. First, we may note the energy emission and absorption characteristics
of hydrogen atoms. The radio receiver can be directed to regions containing
the hydrogen gas to observe the intensity of the 21-cm energy emission. Secondly,
there are large and powerful natural radio generators in space that give a spread

413

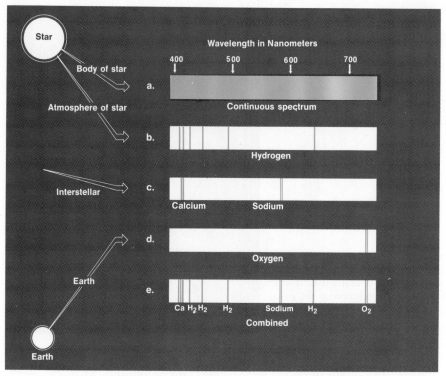

FIGURE 22-6   Spectral lines from interstellar space. A star produces a continuous spectrum (a) at its surface. Some of the light does not escape its atmosphere (b). Some of the light is absorbed between the star and earth (c) and in the atmosphere of the earth (d). Lines from interstellar space can be sorted out from the combined spectrum (e) owing to the Doppler effect.

of electromagnetic energy in this region of the spectrum. If the radio receiver is directed toward such a radio source, and if there is an intervening cloud of hydrogen atoms, they will appear as an absorption "line" in the radio spectrum at 21 cm.

By the middle of 1971, radio astronomers had discovered 13 different kinds of molecules in interstellar space. They are:

CO	Carbon monoxide
NH_3	Ammonia
OH	Hydroxyl radical
CN	Cyanide radical
CS	Thiocarbonyl radical
H_2O	Water
HCN	Hydrogen cyanide
HCHO	Formaldehyde
HCOOH	Formic acid
CH_3OH	Methyl alcohol
$HC{\equiv}CCN$	Cyanoacetylene
CH_3NO	Formamide
?	Exogen

414   Exogen is simply a name given to a molecular species yet to be identified. The

existence of these species, and perhaps a multitude of others, indicates that there are many chemical reactions going on in the vacuum of space.

GALAXIES

Chemical differentiation exists between galaxies as well as within them. Some galaxies which are elliptical in shape contain less than 0.01 per cent of their total mass in the form of interstellar matter. Spiral galaxies contain more; our galaxy, for example, contains about 2 per cent interstellar matter. Some galaxies, termed irregular galaxies, will have as much as one-half of their total mass in the form of interstellar matter (see Figure 22–7).

As one contemplates the chemical composition of galaxies, the question arises as to whether there are totally different kinds of matter than that which we are accustomed to in our solar system. White dwarf stars have densities greater than 10^8 g per cubic cm (100,000,000 g per cubic cm). It is impossible for atomic structures, as pictured by modern atomic theory, to exist under such densities. The electron clouds within the atoms would be crushed, and the material would be an assembly of nuclear particles rather than an assembly of atoms or molecules. In 1963, cosmic radio emissions (quasars) were described in the British journal *Nature*. Although there is much to be learned about these cosmic energy sources, it appears that they emit energy over a broad range of the electromagnetic spectrum. Some of the output from these energy sources

FIGURE 22–7   A, Sombrero Galaxy (NGC 4594) in Virgo at 41 million light years (photographed through the Palomar 200-inch telescope); B, spiral galaxy M 81 in Ursa Major at 8.6 light years (photographed through the Palomar 200-inch telescope). These galaxies, similar to our own, generally contain Population I stars in their spiral arms and Population II stars in their central globular clusters. Even though the concentration of stars appears dense in these photographs, it has been calculated that there would be only a chance of *one* star collision (between two individual stars) if these galaxies moved through each other. (Copyright by the California Institute of Technology and Carnegie Institution of Washington. Photographs from the Hale Observatories.)

415

(pulsars) fluctuate in a pulsating fashion, with pulse rates of a few milliseconds up to several seconds. Some have been observed to gradually change in intensity over a period of months. Although the theory is certainly tentative at this time, the observed properties can be explained if some galactic centers have far greater densities and violent changes within them than anything thus far experienced by man.

THE ATMOSPHERES OF COLD HEAVENLY BODIES

The ability of a planet or a planetary satellite to hold a gaseous atmosphere around its surface depends on the mass of that body and the average velocity of the atmospheric particles. The average velocity of the particles, in turn, is a function of temperature. The minimum velocity corresponding to the energy necessary for the gas molecule to escape from the planet is the *escape velocity*. The escape velocity for objects leaving the earth is very nearly the same for an atom, a molecule, an ion, or even a multistage rocket. Escape velocity from the earth's moon (5400 miles per hour) can be compared with the escape velocities from the planets in Table 22–4.

A second factor controlling the ability of a heavenly body to maintain an atmosphere is its surface temperature. If surface temperatures are above the boiling point of atmospheric gases and particle velocities are considerably less than the required escape velocity, there could be no pools of the liquid gases on the surface; but the gases would be maintained in the atmosphere with only a few of the most energetic particles escaping. Oxygen, nitrogen, and other gases in the earth's atmosphere fit this description. It has been estimated that if the escape velocity is five times the average velocity of the molecules, it would require 25 billion years for the earth to lose all such molecules from its atmosphere. At 0°C the average velocity of nitrogen molecules is about 1100 miles

TABLE 22–4   COMPARISON OF TERRESTRIAL AND OUTER PLANETS OF THE SOLAR SYSTEM

	TERRESTRIAL				OUTER				
	Mercury	Venus	Earth	Mars	Jupiter	Saturn	Uranus	Neptune	Pluto
Average Solar Distance (millions of km)	57.8	108	105	228	780	1,430	2,880	4,550	11,700
Planet Diameter (km)	4,950	12,400	12,700	6,870	140,000	115,000	51,000	50,000	6,440
Period of Solar Orbit (earth years)	0.24	1.60	1.0	1.88	12	30	84	165	?
Period of Axis Rotation (earth days)	58	243	1	1	0.42	0.42	0.45	0.66	27.3
Density (g/cc)	5.5	4.8	5.5	4.0	1.3	0.7	1.1	1.6	7.7
Escape velocity (miles/hour)	8,100	23,000	25,000	11,000	130,000	81,000	47,000	51,000	9,000

per hour compared to the escape velocity from the earth of 25,000 miles per hour.

THE PLANETS

The small planets—Mercury, Venus, Earth, Mars, and Pluto—have higher densities, and presumably very different chemical compositions, than the large planets—Jupiter, Saturn, Uranus and Neptune—even though the chemical compositions of planetary crusts other than the earth's are almost unknown. While space exploration programs may well bring samples back to the earth in the not too distant future, we are presently dependent upon earthly observatories and space probes to gather spectrographic data.

Venus

Venus has a size and density that are quite close to those of the earth. While the two planets may have similar gross compositions, Venus has a dense atmosphere, high surface temperature, and a uniform cloud cover. We are quite certain that the atmosphere of Venus is very dense. Although the first Soviet space probes crushed upon landing, the later successful probe measured the pressure at the surface of Venus at about 90 earth atmospheres.

The atmosphere of Venus, opaque to visible light, appears as a nearly uniform white cloud cover; it is 57 km (35 miles) deep. The atmosphere accounts for a greenhouse effect on the planet. Visible and ultraviolet light from the sun transmit large amounts of energy to the surface of the planet where the radiant energy is transformed into heat. The resulting thermal energy is not easily lost through the dense atmosphere, and, as a result, the temperature of the surface is much higher than it would be without such an atmosphere. The atmosphere is believed to be responsible for a nearly constant temperature of about $500°C$ over the entire surface of the planet.

The major atmospheric gas determined thus far is carbon dioxide; it is certainly in excess of 30 per cent and may well be in excess of 80 per cent. Other gases found in the Venusian atmosphere are listed in Table 22–5 together with the gaseous constituents of the atmospheres of the other planets.

TABLE 22-5   GASES IN PLANETARY ATMOSPHERES

PLANET	GASES IDENTIFIED
Mercury	CO_2
Venus	CO_2, CO, HCl, HF, H_2O, O_2
Earth	N_2, O_2, H_2O, Ar, CO_2, Ne, He, CH_4, Kr, N_2O, H_2, O, O_3, Xe
Mars	CO_2, CO, H_2O
Jupiter	H_2, CH_4, NH_3
Saturn	H_2, CH_4, NH_3
Uranus	H_2, CH_4
Neptune	H_2, CH_4
Pluto	none

417

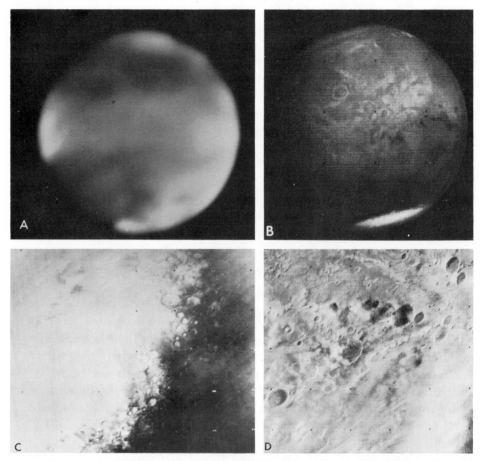

FIGURE 22-8   A, The planet Mars, photographed in blue light through the Palomar 200-inch telescope. (Copyright by the California Institute of Technology and Carnegie Institution of Washington. Photograph from the Hale Observatories.) B, Photograph of Mars taken by Mariner 7, August 4, 1969, at a distance of 293,200 miles from the planet's surface. Prominent in the picture is the south polar ice cap. C, Photograph of the edge of the south polar ice cap of Mars made by Mariner 7. The deposit is thought to be either a layer of frozen carbon dioxide a few feet thick or a layer of water ice less than an inch deep. Craters on the edge show covered south-facing slopes and exposed north-facing slopes. D, Mariner 7 photograph of the south polar ice cap of Mars, showing a wide variety of crater sizes and forms. A pair of large crates resembling a footprint can be seen clearly in the right-center portion of the picture; the sunset shadow line (evening terminator) appears to the east (right) of the picture. (B, C, and D courtesy of the Jet Propulsion Laboratory, California Institute of Technology, National Aeronautics and Space Administration, Pasadena, California.)

Essentially nothing is known of the surface chemistry of Venus because the planet cannot be observed directly, and only limited information has been obtained from the space probes because of the great surface pressure. Our only information about the surface has been obtained from radar mapping with wavelengths of 12.5 cm and 3.8 cm. These results can be explained by the presence of mountains, craters, or fields of boulders.

Mars

Mars, the fourth planet from the sun, is reddish orange with bluish green regions and white polar caps (Figure 22–8). This planet has a density of 4.0 g per cc, probably has a metallic core, a solid mineral mantle, and a crust made of lighter materials like the earth. Comparison of its density to that of earth (5.52 per cc) is somewhat misleading since there is less compression of matter in the smaller gravitational field of Mars. Its diameter is approximately one-half that of earth. The surface of the crust of the planet is similar to that of the earth. There is no direct evidence at all for any liquid on the surface of this planet. Mars has a cratered surface which is obviously weathered away in places. This, along with erosion effects, can be explained in terms of a very thin atmosphere. Some of the erosion effects are evidently the result of prevailing winds, as indicated by the uneven destruction of crater walls. Three kinds of terrain are observed: (1) the cratered surface, indicating extreme age, (2) a chaotic surface accounted for by recent changes from either weather or movements within the planet, and (3) a featureless surface which has been smoothed by recent changes. That the changes in color from place to place may indicate gross surface chemical composition is uncertain. One theory explains the light areas in terms of limonite, a hydrated oxide of iron (Fe_2O_3), and the darker regions in terms of larger particles of the same compound. Dust clouds are often observed, and these are sometimes so intense that they obscure large sections of the planetary features.

The atmosphere of Mars is very thin. A typical pressure at the surface of the planet would be six millibars (a bar is nearly one earth atmosphere of pressure—0.987 atm—so a millibar would be about 1/1000 of an earth atmosphere) or about 0.006 of that at the surface of the earth. The martian atmosphere is composed mostly of carbon dioxide with a smaller amount of carbon monoxide. Water appears to constitute no more than 0.01 per cent of the atmosphere. The air glow observed in ultraviolet light reveals atomic carbon, hydrogen, and oxygen high in the atmosphere. These atoms result from the photodecomposition of CO_2, CO, and H_2O by the high energy photons from the sun.

The temperature on the surface of Mars varies over a wide range, depending on the latitude, longitude, season, time of day, and weather conditions. In the light-colored equatorial region a range of $-70°$ to $20°C$ ($-112°$ to $68°F$) has been observed. The darker regions are always slightly warmer ($-27°C$) than the adjacent lighter ones. The polar cap has a temperature of about $-123°C$; this is in good agreement with the frost temperature (the point at which a solid condenses from a gas) of *carbon dioxide* at 6.5 millibars of pressure. Conditions

419

are not present for the existence of liquid water, unless it is in the form of a very concentrated solution.

The polar caps on Mars are very thin by earth standards, matching the thin atmosphere. If the "snow" is pure carbon dioxide, spectral evidence indicates a depth of from several tens of grams per square centimeter to a thin film. As the sunlight melts the solid, a spotty coverage results, as is often observed after a snow on earth is partially melted. The view of the cratered surface is enhanced under such circumstances. It is likely that the polar caps contain some water; the hydrate $CO_2 \cdot 6H_2O$ is known to be stable under the conditions apparently existing in these frozen areas.

Jupiter

Jupiter, which is the largest planet, having a mass more than 300 times that of earth, is thought to be 80 per cent hydrogen. This conclusion is based on its mass, planetary radius, and rotational data.

It is reasonable to conclude that such a planet would probably contain the other elements in cosmic abundances. We conclude then, even though helium has not been observed directly, that Jupiter is approximately 20 per cent helium, with the remaining elements present in relatively small amounts. A reasonable model of the internal structure of Jupiter would consist of a solid core of hydrogen and helium at a temperature somewhere between 1000° and 10,000°C. One may question how these light gases could be solids at such high temperatures. The answer lies in the tremendous pressures that exist owing to gravitational forces in the interior of such a large mass.

Electromagnetic radiations, both emitted and reflected, indicate that Jupiter has an atmosphere of hydrogen, helium, methane, neon, and ammonia.

A reasonable model of the atmosphere of Jupiter suggests that its visible clouds are made of ammonia crystals, which condense at the temperature and density of the cloud top. There is a slow increase in temperature downward to the surface of the planet. The observed clouds of NH_3 may be 50 km thick, but toward the surface it would become too hot for solid ammonia to exist. Below, there would be another deck of clouds composed of water. Between these layers, ammonia rain storms would occur. The surface of the planet would be several hundred kilometers below the upper visible clouds at a temperature of about 700°C and a pressure of several thousand earth atmospheres.

Saturn

Saturn is distinguished by its rings, (Figure 22–9). It now appears that the rings are composed of crystals of ammonia. Eight absorption bands in the spectrum taken from Saturn's rings have been identified as identical with those obtained from laboratory samples of solid ammonia.

The atmosphere of Saturn displays similar banding patterns to those of Jupiter. Clouds appear and are used to measure the planet's rotation. The planet, like Jupiter, rotates in approximately 10 hours. The evidence available indicates that Saturn has an atmosphere composed of the lighter elements in approximately

FIGURE 22-9   Jupiter (A) and Saturn (B).

the solar abundances. It is interesting to note that the methane spectral bands are stronger and the ammonia bands weaker in Saturn, Uranus, and Neptune than in Jupiter. As with Jupiter, the interior of Saturn is assumed to be essentially solid hydrogen.

OUR MOON

The lunar samples from the Apollo space missions are unique; they are different in composition from any other known material. This, along with the moon's density (3.4 g per cc) reinforces the idea that mammoth chemical separations took place in the formation and aging of the heavenly bodies. The moon may be an especially fruitful place to inquire into the history of the solar system. While the oldest rocks on earth can be dated back 3.5 billion years, some of the lunar rocks, Figure 22–10, appear to be 4.5 billion years old, the generally accepted age of the solar system. Based on the same dating techniques, other

421

FIGURE 22-10   The lunar rocks are low density vesicular material, being as full of empty spaces as volcanic pumice. This specimen is on display at the Smithsonian Institution in Washington, D.C. (Courtesy of Fred H. Zerkel, *Chemical and Engineering News.*)

TABLE 22-6   ELEMENTAL ABUNDANCES IN LUNAR SAMPLES FROM THE APOLLO MISSION (PER CENT BY WEIGHT)

	VESICULAR (FINE-GRAINED)	CRYSTALLINE (COARSE-GRAINED)	BRECCIA	FINE MATERIAL; SOIL	FOR COMPARISON: EARTH CRUST	A STONE METEORITE
Si	21.5	18.4	19.8	20.2	27.72	21
Fe	15.7	15.2	14.8	12.5	5.00	13
Ca	10.1	11.0	11.0	9.6	3.63	2
Ti	6.5	5.8	5.6	4.1	0.44	–
Al	4.0	5.4	5.7	7.3	8.13	2
Mg	3.7	3.4	2.8	4.6	2.09	17
Na	0.35	0.32	0.32	0.33	2.83	–
K	0.22	0.09	0.09	0.11	2.59	–

NOTES:

Vesicular rocks—a rock containing many open spaces, formed by gas in molten material, gas such as volcanic pumice. The open spaces in bread are analogous.

Breccia—a rock containing small sharp material that is held together by a cementlike material. Oxygen was not determined directly and is the other major element present. The balance of the material is essentially oxygen. Analyses of 54 other elements were obtained, ranging from chlorine at 0.2 per cent (0.01 in earth crust) to antimony at 0.005 parts per million (0.2 ppm in earth crust).

lunar rocks are 3.5 billion years old, indicating that a cataclysmic event may have occurred at that time in the earth-moon region. Constant change due to weathering and movements within the earth contrast sharply to the relatively static situation on the moon. The lunar surface is subjected only to meteorite impact, solar wind, and thermal changes. The nonuniform interior of the moon apparently changes relatively little. The ancient record is at least partially there, and one can well understand the excitement of the scientists in their efforts to read that record.

Table 22–6 gives the elemental abundances of the major elements on the surface of the moon. There is general agreement that the lunar rocks were formed by crystallization from molten silicate melts. Under such conditions fractional crystallization could be expected to produce large-scale chemical separations in the solid materials produced. The volatile elements and compounds would escape into space, because of the small lunar gravity. If the present sampling is representative, the moon is a "dehyrated cinder." High melting elements such as titanium are enriched over earth-crust standards and low melting elements such as sodium are depleted.

The lunar soil and dust that gave rise to the well publicized astronaut tracks is evidently produced by the constant rain of meteorites and solar wind. Upon impact of a meteorite, its kinetic energy is converted into thermal energy. If concentrated enough, melting occurs and some molten material may be thrown out. A free liquid in space assumes a spherical shape and if solidification occurs

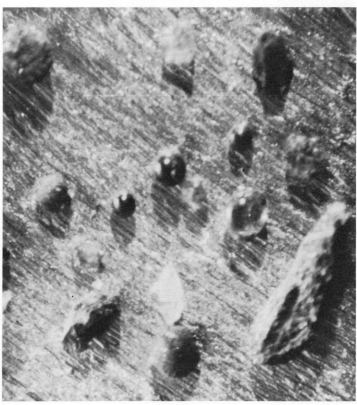

FIGURE 22-11 Lunar glass spherule, the largest of which is 0.4 mm in diameter. This sample was among those collected by Astronauts Neil A. Armstrong and Edwin E. Aldrin, Jr., on the Apollo 11 lunar landing mission. (Courtesy of the National Aeronautics and Space Administration.)

in flight, a spherule is produced. Such glassy spherules, (Figure 22–11) are very abundant in the grey lunar soil. In the same photograph, note that irregular particles also are plentiful. The texture of the lunar soil ranges from a fine-grained silt to coarse sand.

The origin of the heat necessary to produce the igneous lunar material has not yet been determined. The interior of the moon is not as thermally active as the earth. Uneven concentrations of mass beneath the lunar surface, called mascons, also occur. These occur in the basins of the large circular maria, almost as if massive bodies plunged into the moon at these points. Movements within the mantle and crust would, in time, erase such mascons in the earth, while they have been obviously preserved in the moon. Evidence has been found for the existence of mascons on Mars and Mercury also.

There is no evidence of any life forms on the moon. The total carbon concentration is less than 200 parts per million, well below that found in many meteorites. Organic carbon is present in considerably smaller proportions.

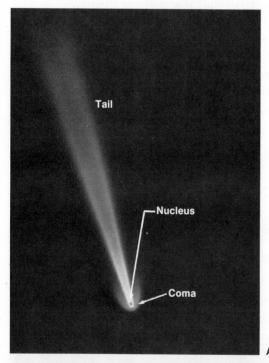

Tail

Nucleus

Coma

A

Figure 22–12   *A*, The structure of a comet. *B*, The comet Mkros, photographed between August 17 and 27, 1957, through a 48-inch telescope. Note the changing structure of the comet over this 19-day period. (Copyright by the California Institute of Technology and Carnegie Institution of Washington. Photographs from the Hale Observatories.)

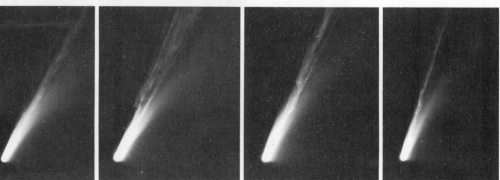

B

TABLE 22-7   POSSIBLE COMPOSITION OF COMET NUCLEUS

TYPE OF MATERIAL	EXAMPLES
Volatile inorganic	H_2, N_2, Ne, Ar, H_2O, NH_3, CO, H_2O_2, N_2H_4, H_2S
Volatile organic	CH_4, C_2H_2, C_2H_4, HCN, CH_3OH, C_6H_6 (also higher organic molecules)
Nonvolatile inorganic	Silicates, metallic oxides, metals, carbon grains
Radicals	OH, NH, C_7, C_3, CH_2, HO_2

COMETS

There are probably on the order of 100 billion comets in our solar system. From an orbital study of a few of them, it is known that comets are among the least massive objects in the solar systems, yet they are the largest things in the solar system except for the sun itself. The reason for this lies in the comet's structure (Figure 22-12).

The coma and tail are visible and are composed of atoms, ions, molecules, and small particles ejected from the nucleus. The coma of a large comet may be 100,000 miles in diameter with the tail stretching out for distances greater than 40,000,000 miles—greater than the distance between earth and Mercury! The small, relatively dense nucleus, the mass of which may be deduced from orbital mechanics, may be no more than a few miles in diameter.

Spectrographic analysis of light emitted from comets has identified a number of chemical species (Table 22-7). The comet is certainly an icy snowball of the lighter molecules with an unknown amount of heavier elements and their compounds.

IS THERE EXTRATERRESTRIAL LIFE?

Man has an imperfect understanding of the processes which lead from the atoms of the cosmos to organized living cells functioning in an organism. However, with the growth of biochemistry, it is possible for him to develop theories on the chemical conditions necessary for the origin, growth, and reproduction of life forms.

No plausible theory for the production of biochemicals from simple atoms and molecules has been advanced that could operate in an oxidizing atmosphere such as exists today on earth. But if the earth did condense from the solar nebula, it is likely that at first it had, like the giant planets have now, a reducing atmosphere composed of H_2, He, and NH_3. Laboratory experiments with such a reducing atmosphere demonstrate that an electrical discharge will induce the formation of some amino acids. The point to be made is that the earth is not peculiar in having the atoms from which to construct life or the conditions necessary for the production of simple biochemicals.

Does life exist elsewhere in our solar system? The harsh conditions that exist on our neighboring planets almost certainly preclude mammalian life as we know it. However, microorganisms in the laboratory have not only existed, but, indeed, multiplied under simulated Martian conditions. Even on Jupiter or on Jove, one

"Ammonia! Ammonia!"

FIGURE 22-13 Since the discovery of organic compounds, such as methyl alcohol and formic acid, in interstellar space, it is no longer necessary to postulate the beginnings of life on a terrestrial body. This may place the origins of life forms on a wide cosmological "geography" indeed. (Drawing by R. Grossman; © 1962 by *The New Yorker Magazine*, Inc.)

of its satellites, Jovian bacteria, but not Jovian cows, are reasonable possibilities. Thus far the moon appears sterile, but we cannot yet assume that all other bodies in the solar system are so.

The chances of finding other planets hospitable for life increase tremendously as we move outside of our solar system. There are approximately 150 billion stars in our galaxy, and the number of galaxies at present appears to be limitless. Based on the evolution of stars, it has been estimated that at least one billion of the 150 billion stars in our galaxy have habitable planets. Some have estimated that there must be $100,000,000,000,000,000,000$ (10^{20}) planetary systems in the *known* universe similar to ours. The chances appear to be overwhelming, from purely a statistical point of view, that the universe is rich in suitable habitats of life, even life as we know it.

It is also entirely possible that not all life forms require water as the fundamental fluid of the organized cell. It is possible that ammonia could serve the water role in a different type of living matter (Figure 23–13).

QUESTIONS

1. Why is the density of a planet related to its chemical composition?

2. Based upon escape velocities, which planet would be expected to have the greatest percentage of light molecules in its atmosphere?

3. Define the following terms: escape velocity, red shift, free radical, stony meteorite, solar wind.

4. Why is a spectrograph useful in astronomical studies?

5. Why are crystals of ammonia (m.p. $-77.8\,°C$) a reasonable constituent of the atmosphere of Jupiter but not of Venus?

6. What evidence do we have that the pressure at the surface of Venus is higher than at the surface of the Earth?

7. How are the polar caps of Mars different from those of the earth?

8. What type of evidence is used to support the claim that the most abundant element in the sun is hydrogen?

9. How many different kinds of evidence can you cite that should be useful in formulating theories concerning space chemistry?

10. Distinguish between two types of meteorites based on their chemical composition.

11. List the kinds of particles and objects, in order from small to large, that you can expect to find in the universe.

12. Explain the statement: "Earth chemistry is very much an example of 'special-case chemistry.'"

13. Suggest a reason why the most abundant element in the universe, hydrogen, is not the most abundant element on earth.

14. Recall a personal experience when you observed the Doppler effect in sound vibrations.

15. What is a way to observe interstellar hydrogen?

16. In terms of chemical composition, why is the interior of the sun less likely to be as different from its surface as the center of the earth is from its surface?

17. What assumptions must be made in order to give a list of the relative number of atoms for each element in the universe (Table 27–5)?

18. What is a quasar? A pulsar?

19. Terrestrial planets have oxidizing atmospheres, in contrast to the giant planets which have reducing atmospheres. Relate this to the definitions of oxidation and reduction and write simple chemical equations to illustrate typical reactions in each type of atmosphere.

SUGGESTIONS FOR FURTHER READING

Ashbrook, J., "Astronomical Scrapbook, The Gradual Recognition of Helium," *Sky and Telescope*, Vol. 36, No. 2, p. 87 (1968).

Cadle, R. D., and Roberts, W. O., "In Pursuit of Atmospheres," *Chemistry*, Vol. 42, No. 1, p. 6 (1969).

Cameron, A. G. W., "Interstellar Communications," W. A. Benjamin, Inc., New York, 1963.

Gamow, G., "A Planet Called Earth," Viking Press, New York, 1963.

Heide, F., "Meteorites," University of Chicago Press, Chicago, 1964.

Keller, E., "Origin of Life, Part III: On Other Planets?", *Chemistry*, Vol. 42, No. 4, p. 8 (1969).

Keller, E., "Meteorites and the Early Solar System," *Chemistry*, Vol. 42, No. 5, p. 20 (1969).

Merril, P. W., "Space Chemistry," University of Michigan Press, Ann Arbor, 1963.

Page, T. (ed.), "Stars and Galaxies," Prentice-Hall, Inc., Englewood Cliffs, N.J., 1962.

Shklovskii, I. S., and Sagan, C., "Intelligent Life in the Universe," Holden-Day, Inc., San Francisco, 1966.

Wood, J., "Meteorites and the Origin of Planets," McGraw-Hill Book Co., Inc., 1968.

Wood, J., "What the Moon is Made of," *Chemistry*, Vol. 43, No. 8, p. 22 (1970).

Appendix A

THE INTERNATIONAL SYSTEM OF UNITS (SI)

Since 1960, a coherent system of units known as the SI system, bearing the authority of the International Bureau of Weights and Measures, has been in effect and is gaining acceptance among scientists. It is an extension of the metric system which began in 1790, with each physical quantity assigned an SI unit uniquely. An essential feature of both the older metric system and now the newer SI is a series of prefixes that indicate a power of ten multiple or submultiple of the unit.

I. UNITS OF LENGTH.   The standard unit of length is the *meter*. It was originally meant to be one ten-millionth of the distance along a meridian from the north pole to the equator. However, the lack of precise geographical information necessitated a better definition. For a number of years the meter was defined as the distance between two etched lines on a platinum-iridium bar kept at $0°C$ ($32°F$) in the International Bureau of Weights and Measures, at Sèvres, France. The inability to measure this distance as accurately as desired prompted a recent redefinition of the meter as being a length equal to 1,650,763.73 times the wavelength of the orange-red spectrographic line of $^{86}_{36}Kr$.

The meter (39.37 inches) is a convenient unit with which to measure the height of a basketball goal (3.05 meters), but it is unwieldy for measuring the parts of a watch or the distance between continents. For this reason, prefixes are defined in such a way that, when placed before the meter, they define distances convenient for man's purposes. Some of the prefixes with their meanings are:

nano—1/1,000,000,000 or 0.000000001
micro—1/1,000,000 or 0.000001
milli—1/1000 or 0.001
centi—1/100 or 0.01
deci—1/10 or 0.1
deka—10
hecto—100
kilo—1000
mega—1,000,000

Should give abbreviations!

428   The corresponding units of length with their abbreviations are:

TABLE A–1   ENGLISH–SI EQUIVALENTS° (Conversion Factors)

Length:	1 inch (in)	=	2.54 centimeter (cm)
	1 yard (yd)	=	0.914 meter (m)
	1 mile (mi)	=	1.609 kilometers (km)
			(1,609 meters)
Volume:	1 ounce (oz)	=	29.57 milliliters (ml)
	1 quart (qt)	=	0.946 liter (l)
	1.06 quart (qt)	=	1 liter (l)
	1 gallon (gal)	=	3.78 liters
Mass (weight)°°:	1 ounce (oz)	=	28.35 grams (g)
	1 pound (lb)	=	453.4 grams (g)
	1 ton (tn)	=	907.2 kilograms (kg)

° Commonly used English units are used.

°° Mass is a measure of the amount of matter, whereas weight is a measure of the attraction of the earth for an object at the earth's surface. The mass of a sample of matter is constant, but its weight varies with position and velocity. For example, the space traveler, having lost no mass, becomes weightless in earth orbit. Although mass and weight are basically different in meaning, they are often used interchangeably in the environment of the earth's surface.

nanometer (nm)—0.000000001 meter
micrometer (μm)—0.000001 meter
millimeter (mm)—0.001 meter
centimeter (cm)—0.01 meter
decimeter (dm)—0.1 meter
meter (m)
dekameter (dkm)—10 meters
hectometer (hm)—100 meters
kilometer (km)—1000 meters
megameter (Mm)—1,000,000 meters

Since the prefixes are defined in terms of the decimal system the conversion from one metric length to another involves only shifting the decimal point. Mental calculations are quickly accomplished.

How many centimeters are in a meter? Think: Since a centimeter is the one-hundreth part of a meter, there would be 100 centimeters in a meter.

Conversion of measurements from one system to the other is a common problem. Some commonly used English—SI equivalents (conversion factors) are given in Table A–1.

II. UNITS OF MASS.
The primary unit of mass is the kilogram (1000 grams). This unit is the mass of a plantinum-iridium alloy sample deposited at the International Bureau of Weights and Measures. One pound contains a mass of 453.4 grams (a five-cent nickel coin contains about 5 grams).

Conveniently enough, the same prefixes defined in the discussion of length are used in units of mass, as well as in other units of measure.

III. UNITS OF VOLUME.
The SI unit of volume is the cubic meter (m^3). However, the volume capacity used most frequently in chemistry is the liter, which is defined as 1 cubic decimeter (1 dm^3). Since a decimeter is equal to 10 centimeters (cm), the cubic decimeter is equal to $(10 \text{ cm})^3$ or 1000 cubic centimeters (cc). One cc, then, is equal to one milliliter (the thousandth part of a liter). The ml (or cc) is a common unit that is often used in the measurement of medicinal and laboratory quantities.

429

Appendix B

CENTIGRADE OR CELSIUS TEMPERATURE SCALE

The system of measuring temperature which is used in scientific work is based on the centigrade or Celsius temperature scale. This temperature scale was defined by Anders Celsius, a Swedish astronomer, in 1742. The Celsius scale is based upon the expansion of a column of mercury which occurs when it is transferred from a cold standard temperature to a hot standard temperature. The cold standard temperature is the temperature of melting ice and is defined as 0°C. The hot standard temperature is the temperature at which water boils under standard conditions of pressure; it is defined as 100°C. The expansion of the mercury is assumed to be linear over this range, and the distance through which the mercury column expands between 0°C and 100°C is divided into 100 equal parts, each corresponding to one degree.

On the scale commonly used in the United States, the Fahrenheit scale (Daniel Fahrenheit), the freezing mark is 32° and the boiling mark is 212°. Obviously, the ice is no hotter when a Fahrenheit thermometer is stuck into the system than when a Celsius thermometer is used. The marks of the tubes are simply different names for the same thing.

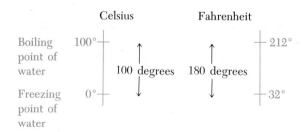

Because both systems are used in our environment, it is necessary at times to convert from one system to another.

Note that 100 Celsius degrees = 180 Fahrenheit degrees or 1 C degree = 1.8 or $\frac{9}{5}$ F degrees.

But if we wish to convert temperature from one scale to another, we must remember that $0\,^\circ$C is the same as $32\,^\circ$F. Therefore, 32 must be added to the calculated number of degrees Fahrenheit in order to revert back to the start of the counting in the Fahrenheit system.

Therefore,

$$^\circ\text{F} = \frac{\frac{9}{5}\,^\circ\text{F}}{1\,^\circ\text{C}}\,(^\circ\text{C}) + 32,$$

which can be rearranged to

$$^\circ\text{C} = \tfrac{5}{9}\,(^\circ\text{F} - 32)$$

when one wishes to make the conversion in the opposite direction.

Example Problem.   Convert $50\,^\circ$F to the corresponding Celsius temperature:

$$\begin{aligned}
^\circ\text{C} &= \tfrac{5}{9}(^\circ\text{F} - 32)\\
&= \tfrac{5}{9}\,(50 - 32)\\
&= \tfrac{5}{9}\,(18)\\
&= 10\,^\circ\text{C}
\end{aligned}$$

Appendix C
STOICHIOMETRIC CALCULATIONS

The bases for stoichiometric calculations were presented in Chapter 4. Problems of a more complex nature and a systematic approach to their solution are presented in the following examples. Finally, a list of exercise problems is given for further study.

Example 1
Balanced Equations Express Particle-Number Ratios

In the reaction of hydrogen with oxygen to form water, how many molecules of hydrogen are required to combine with 19 oxygen molecules?

Solution: A chemical equation can only be written for a reaction if the reactants and products are identified and the respective formulas determined. In this problem, the formulas are known and the equation is:

$$H_2 + O_2 \longrightarrow H_2O.$$

It is evident that one molecule of oxygen contains enough oxygen for two water molecules and the equation, as written, does not account for what happens to the second oxygen atom. As it is, the equation is in conflict with the conservation of atoms in chemical changes. This conflict is easily corrected by balancing the equation:

$$2H_2 + O_2 \longrightarrow 2H_2O$$

Now, all atoms are accounted for in the equation and it is obvious that two hydrogen molecules are required for each oxygen molecule. In other words, two hydrogen molecules are equivalent to one oxygen molecule in their usage. This can be expressed:

$$2 \text{ hydrogen molecules} \begin{array}{c} \text{are} \\ \text{equivalent} \\ \text{to} \end{array} 1 \text{ oxygen molecule;}$$

or,

$$2H_2 \text{ molecules} \sim O_2 \text{ molecule;}$$

or,

$$\frac{2H_2 \text{ molecules}}{O_2 \text{ molecule}},$$

which can be read as two hydrogen molecules per one oxygen molecule.

Using now the factor-label approach of multiplying by equivalent factors and cancelling units, the solution is readily achieved.

? H_2 molecules = $19O_2$ molecules

? H_2 molecules = $19O_2$ ~~molecules~~ $\times \dfrac{2H_2 \text{ molecules}}{O_2 \text{ ~~molecule~~}}$ = $38H_2$ molecules

Note 1: The reader is likely to say at this point that the method is cumbersome and that he can quickly see the answer to be $38H_2$ molecules without "the method." However, problems to follow are made much easier if a systematic method of approach is used.

Example 2
Laboratory Mole Ratios Identical With Particle-Number Ratios

How many moles of hydrogen molecules must be burned in oxygen (the reaction of example 1) to produce 15 moles of water molecules (about a glassful)?

Solution: The balanced equation,

$$2H_2 + O_2 \longrightarrow 2H_2O,$$

tells us that two molecules of hydrogen produce two molecules of water;

or,

$$2 \text{ molecules hydrogen} \sim 2 \text{ molecules water,}$$

and,

$$1 \text{ molecule hydrogen} \sim 1 \text{ molecule water.}$$

It is obvious then that the number of water molecules produced will be equal to the number of hydrogen molecules consumed regardless of the actual number involved. Therefore,

$$6.02 \times 10^{23} \text{ molecules of hydrogen} \sim 6.02 \times 10^{23} \text{ molecules of water.}$$

Since 6.02×10^{23} is a number called the mole, it follows that one mole of hydrogen molecules will produce one mole of water molecules. The general conclusion, then, is: the ratio of particles in the balanced equation is the same as the ratio of moles in the laboratory. The solution to the problem logically follows:

$$\left. \begin{array}{r} \text{? moles of hydrogen} \\ \text{molecules} \end{array} \right\} = \left\{ \begin{array}{c} 15 \text{ ~~moles~~} \\ \text{~~water molecules~~} \end{array} \right\} \times \dfrac{1 \text{ mole hydrogen molecules}}{1 \text{ ~~mole water molecules~~}}$$

$$= 15 \text{ moles hydrogen molecules}$$

Note 2: Again the solution to the problem looks simple enough without resorting to the factor-label method. However, in examples 3 and 4, the numbers become such that a quick mental solution is not readily achieved by most students.

Example 3
Mole Weights Yield Weight Relationships

How many grams of oxygen are necessary to react with hydrogen to produce 270 grams of water?

Solution: From the balanced equation,

$$2H_2 + O_2 \longrightarrow 2H_2O,$$

the mole ratio between oxygen and water is immediately evident and is one mole of oxygen molecules per two moles of water molecules, or

$$\frac{1 \text{ mole oxygen molecules}}{2 \text{ moles water molecules}}.$$

This mole ratio can be changed into a weight ratio since the mole weight can be easily calculated from the atomic weights involved. One molecule of oxygen (O_2) weighs 32 amu (16 amu + 16 amu). Therefore a mole of oxygen molecules weighs 32 grams. Similarly, two moles of water weigh 36 grams [2(16 + 2 × 1)]. Therefore the weight ratio is:

$$\frac{1 \text{ mole oxygen molecules} \times \dfrac{32 \text{ grams oxygen}}{\text{mole oxygen molecules}}}{2 \text{ moles water molecules} \times \dfrac{18 \text{ grams water}}{\text{mole water molecules}}}$$

or,

$$\frac{32 \text{ grams oxygen}}{36 \text{ grams water}}.$$

This weight relationship is exactly the conversion factor needed to answer the original question:

$$? \text{ grams oxygen} = 270 \text{ grams water} \times \frac{32 \text{ grams oxygen}}{36 \text{ grams water}}$$

$$= 240 \text{ grams oxygen}$$

Note 3: It should be observed that a weight relationship could be established between any two of the three pure substances involved in the reaction, regardless of whether they are reactants or products.

Example 4.   How many molecules of water are produced in the decomposition of 8 molecules of table sugar? The equation is:

$$C_{12}H_{22}O_{11} \longrightarrow C + H_2O$$

Solution: Balance the equation,

$$C_{12}H_{22}O_{11} \longrightarrow 12C + 11H_2O$$

$$? \text{ molecules of water} = 8 \text{ molecules sugar} \times \frac{11 \text{ molecules water}}{1 \text{ molecule sugar}}$$
$$= 88 \text{ molecules water}$$

Example 5.   How many grams of mercuric oxide are necessary to produce 50 grams of oxygen by the reaction:

$$2HgO \longrightarrow 2Hg + O_2$$

Solution:

(a) weight of two moles of HgO $= 2(201 + 16) = 2(217) = 434$ g

(b) weight of 1 mole of $O_2 = 2(16) = 32$ g

(c) ?g HgO $= 50$ g oxygen $\times \dfrac{434 \text{ g mercuric oxide}}{32 \text{ g oxygen}} = 678$ g mercuric oxide

Example 6.   How many pounds of mercuric oxide are necessary to produce 50 pounds of oxygen by the reaction:

$$2HgO \longrightarrow 2Hg + O_2$$

Solution: Note that the problem is the same as Example 5 except for the units of chemicals. Also note that the conversion factor of Exercise 5,

$$\frac{434 \text{ g mercuric oxide}}{32 \text{ g oxygen}}$$

can be converted to any other units desired,

$$\frac{434 \text{ g mercuric oxide} \times \dfrac{\text{pound}}{454 \text{ g}}}{32 \text{ g oxygen} \times \dfrac{\text{pound}}{454 \text{ g}}} = \frac{434 \text{ pounds mercuric oxide}}{32 \text{ pounds oxygen}}$$

It is evident that the ratio, $\dfrac{434}{32}$, expresses the ratio between weights of mercuric oxide and oxygen in this reaction regardless of the units employed.

$$? \text{ pounds mercuric oxide} = 50 \text{ pounds oxygen} \times \frac{434 \text{ pounds mercuric oxide}}{32 \text{ pounds oxygen}}$$
$$= 678 \text{ pounds of mercuric oxide}$$

EXERCISES

1. What weight of oxygen is necessary to burn 28 g of methane, CH_4? The equation is:

$$CH_4 + 2O_2 \longrightarrow CO_2 + 2H_2O$$

2. Potassium chlorate, $KClO_3$, releases oxygen when heated according to the equation:

$$2KClO_3 \longrightarrow 2KCl + 3O_2.$$

What weight of potassium chlorate is necessary to produce 1.43 grams of oxygen?

3. Fe_3O_4 is a magnetic oxide of iron. What weight of this oxide can be produced from 150 grams of iron?

435

4. Steam reacts with hot carbon to produce a fuel called water gas; it is a mixture of carbon monoxide and hydrogen. The equation is:

$$H_2O + C \longrightarrow CO + H_2$$

What weight of carbon is necessary to produce 10 grams of hydrogen by this reaction?

5. Iron oxide, Fe_2O_3, can be reduced to metallic iron by heating it with carbon.

$$2Fe_2O_3 + 3C \longrightarrow 4Fe + 3CO_2$$

How many tons of carbon would be necessary to reduce 5 tons of the iron oxide in this reaction?

6. How many grams of hydrogen are necessary to reduce 1 pound (454 grams) of lead oxide (PbO) by the reaction:

$$PbO + H_2 \longrightarrow Pb + H_2O$$

7. Hydrogen can also be produced by the reaction of iron with steam.

$$4H_2O + 3Fe \longrightarrow 4H_2 + Fe_3O_4$$

What weight of iron would be needed to produce one-half pound of hydrogen?

8. Tin ore, containing SnO_2, can be reduced to tin by heating with carbon.

$$SnO_2 + C \longrightarrow Sn + CO_2$$

How many tons of tin can be produced from 100 tons of SnO_2?

INDEX

Numbers in *italics* indicate a Figure; *t* following a page number indicates a Table.

PERIODIC CHART OF THE ELEMENTS

Groups: IA · IIA · IIIB · IVB · VB · VIB · VIIB · VIII · IB · IIB · IIIA · IVA · VA · VIA · VIIA · O

No.	Symbol	Atomic Weight	Uncertainty
1	H	1.00797	± 0.00001
2	He	4.0026	± 0.0005
3	Li	6.939	± 0.0005
4	Be	9.0122	± 0.0005
5	B	10.811	± 0.003
6	C	12.01115	± 0.00005
7	N	14.0067	± 0.00005
8	O	15.9994	± 0.0001
9	F	18.9984	± 0.0005
10	Ne	20.183	± 0.0005
11	Na	22.9898	± 0.00005
12	Mg	24.312	± 0.0005
13	Al	26.9815	± 0.00005
14	Si	28.086	± 0.001
15	P	30.9738	± 0.00005
16	S	32.064	± 0.003
17	Cl	35.453	± 0.001
18	Ar	39.948	± 0.0005
19	K	39.102	± 0.0005
20	Ca	40.08	± 0.005
21	Sc	44.956	± 0.0005
22	Ti	47.90	± 0.005
23	V	50.942	± 0.0005
24	Cr	51.996	± 0.001
25	Mn	54.9380	± 0.00005
26	Fe	55.847	± 0.003
27	Co	58.9332	± 0.00005
28	Ni	58.71	± 0.005
29	Cu	63.54	± 0.005
30	Zn	65.37	± 0.005
31	Ga	69.72	± 0.005
32	Ge	72.59	± 0.005
33	As	74.9216	± 0.00005
34	Se	78.96	± 0.005
35	Br	79.909	± 0.002
36	Kr	83.80	± 0.005
37	Rb	85.47	± 0.005
38	Sr	87.62	± 0.005
39	Y	88.905	± 0.0005
40	Zr	91.22	± 0.005
41	Nb	92.906	± 0.0005
42	Mo	95.94	± 0.005
43	Tc	(99)	
44	Ru	101.07	± 0.005
45	Rh	102.905	± 0.00005
46	Pd	106.4	± 0.05
47	Ag	107.870	± 0.003
48	Cd	112.40	± 0.005
49	In	114.82	± 0.005
50	Sn	118.69	± 0.005
51	Sb	121.75	± 0.005
52	Te	127.60	± 0.005
53	I	126.9044	± 0.00005
54	Xe	131.30	± 0.005
55	Cs	132.905	± 0.0005
56	Ba	137.34	± 0.005
57	La	138.91	± 0.005
72	Hf	178.49	± 0.005
73	Ta	180.948	± 0.0005
74	W	183.85	± 0.005
75	Re	186.2	± 0.05
76	Os	190.2	± 0.05
77	Ir	192.2	± 0.05
78	Pt	195.09	± 0.005
79	Au	196.967	± 0.0005
80	Hg	200.59	± 0.005
81	Tl	204.37	± 0.005
82	Pb	207.19	± 0.005
83	Bi	208.980	± 0.0005
84	Po	(210)	
85	At	(210)	
86	Rn	(222)	
87	Fr	(223)	
88	Ra	(226)	
89	†Ac	(227)	
104		(257)	

Lanthanum Series

No.	Symbol	Atomic Weight	Uncertainty
58	Ce	140.12	± 0.005
59	Pr	140.907	± 0.0005
60	Nd	144.24	± 0.005
61	Pm	(147)	
62	Sm	150.35	± 0.005
63	Eu	151.96	± 0.005
64	Gd	157.25	± 0.005
65	Tb	158.924	± 0.0005
66	Dy	162.50	± 0.005
67	Ho	164.930	± 0.0005
68	Er	167.26	± 0.005
69	Tm	168.934	± 0.0005
70	Yb	173.04	· 0.005
71	Lu	174.97	± 0.005

Actinium Series

No.	Symbol	Atomic Weight	Uncertainty
90	Th	232.038	± 0.0005
91	Pa	(231)	
92	U	238.03	± 0.005
93	Np	(237)	
94	Pu	(242)	
95	Am	(243)	
96	Cm	(247)	
97	Bk	(247)	
98	Cf	(249)	
99	Es	(254)	
100	Fm	(253)	
101	Md	(256)	
102	No	(253)	
103	Lw	(257)	

Atomic Weights are based on C^{12} —12.0000 and Conform to the 1961 Values

Printed in U.S.A.